T0360662

Diagnostics of Laboratory and Astrophysical Plasmas Using Spectral Lineshapes of One-, Two-, and Three-Electron Systems

Diagnostics of Laboratory and Astrophysical Plasmas Using Spectral Lineshapes of One-, Two-, and Three-Electron Systems

Eugene Oks
Auburn University, USA

NEW JERSEY • LONDON • SINGAPORE • BEIJING • SHANGHAI • HONG KONG • TAIPEI • CHENNAI • TOKYO

Published by

World Scientific Publishing Co. Pte. Ltd.
5 Toh Tuck Link, Singapore 596224
USA office: 27 Warren Street, Suite 401-402, Hackensack, NJ 07601
UK office: 57 Shelton Street, Covent Garden, London WC2H 9HE

Library of Congress Cataloging-in-Publication Data
Names: Oks, E. A. (Evgeniĭ Aleksandrovich), author.
Title: Diagnostics of laboratory and astrophysical plasmas using spectral lineshapes of one-, two-, and three-electron systems / Eugene Oks, Auburn University, USA.
Description: Singapore ; Hackensack, NJ : World Scientific, [2017] | Includes bibliographical references and index.
Identifiers: LCCN 2016054152| ISBN 9789814699075 (hardcover ; alk. paper) | ISBN 9814699071 (hardcover ; alk. paper)
Subjects: LCSH: Plasma spectroscopy. | Plasma diagnostics. | Plasma astrophysics. | Spectral line formation.
Classification: LCC QC718.5.S6 O365 2017 | DDC 530.4/4--dc23
LC record available at https://lccn.loc.gov/2016054152

British Library Cataloguing-in-Publication Data
A catalogue record for this book is available from the British Library.

Typeset by Stallion Press
Email: enquiries@stallionpress.com

Printed in Singapore

In memory of recently passed away

G.V. Sholin and V.P. Gavrilenko

with appreciation of their contribution to the area
of the Stark broadening of spectral lines in plasmas

Contents

Introduction

"Thaumas took as his wife Electra, daughter of Okeanos,
whose stream is deep, and she bore swift Iris..."

(Hesiod, *Theogony*, lines 265-6 — about *Iris, the goddess of rainbow*
and thus, the first spectroscopist)

Studies of plasmas have very broad practical applications across various areas of physics and technology. Examples are (but not limited to) magnetically-controlled fusion, laser-controlled fusion, X-ray lasers, powerful Z-pinches (used for producing X-ray and neutron radiation, ultra-high pulsed magnetic fields, and for X-ray lasing), low-temperature technological discharges for plasma chemistry (including plasma-surface processing for manufacturing microchips and for nano-technologies), interaction of powerful laser radiation or coherent microwave radiation with plasmas and gases, and astrophysics (including solar physics). An indispensable part of plasma research and its applications is *plasma diagnostics*, i.e., a compendium of methods for the experimental determination of various plasma parameters and of parameters of various electric and magnetic fields in plasmas.

In other areas of physics, the corresponding subarea is called *measurements methods*. The reason why in plasma physics it is called *diagnostics* is the following. The same observed signal from a plasma (such as a spectrum of the electromagnetic radiation) could correspond to many different sets of parameters of the plasma and/or of fields in it — like in the medicine, where a set of patient symptoms could correspond to different illnesses or disorders. Because of this analogy, i.e., because of the absence of one-to-one correspondence between the signals/symptoms and the underlying cause in both plasma physics and medicine, the medical term *diagnostics* is used in plasma physics.

Among various methods of plasma diagnostics, the methods employing lineshapes of spectral lines emitted from plasmas play a very important role for at least four reasons. First, these methods are *non-intrusive* (sometimes called *non-perturbing*) because they did not affect parameters of a plasma (and/or of fields in it) to be measured — in distinction to the overwhelming majority of other diagnostic methods. Second, these methods often are the *most informative.* Third, they do not depend on model assumptions about the plasma state (such as local thermodynamic equilibrium, or partial local thermodynamic equilibrium, or the corona model, or the collisional-radiative model, etc.) — in distinction to the diagnostic methods using various ratios of intensities (such as ratios of integrated intensities of several spectral lines, or line-to-continuum ratios, ratios of intensities of recombination continua, etc.). Fourth, sometimes there is *no other method* for the experimental determinations of a particular parameter of the plasma and/or a particular field in the plasma.

In these methods, the role of "probes" is played by radiating atoms or ions. Probes have to be well-calibrated for achieving the required accuracy in the experimental determination of a particular parameter. Among atoms and ions, the best-calibrated are those having the simplest electronic structure: atoms or ions having one or two or three electrons and thus the most accurately described by quantum mechanics both as isolated systems and as the systems immersed in various plasmas.

The absence of one-to-one correspondence between the observed lineshapes and the underlying parameters of a plasma and/or of fields in it poses a significant challenge for the interpretation of the experimental data and leads to a kind of a "gap" between theorists and experimentalists. Typically, theorists solve the so-called "*direct problem*": given a set of parameters of a plasma and/or of fields in it, they calculate, e.g., a Stark profile of a particular spectral line. They can repeat it for many other sets of parameters and produce a huge multi-parametric set of Stark profiles of a particular spectral line. Then theorists calculate a convolution of these profiles with other broadening mechanisms and thus dramatically increase both the size of the set of calculated profiles and the number of parameters, on which the set depends. However, interpreting an experimental profile of the spectral line is an "*inverse problem*": given the experimental profile, *to find a unique set* of the parameters of a plasma and/or fields in it corresponding to the observed profile.

Books on plasma spectroscopy, usually focus on presenting solutions of the direct problem, but pay zero or little attention to the solutions inverse problem — despite it is the latter that is most necessary in practice. An exception is Griem's book of 1974 (*Spectral Line Broadening by Plasmas*, Academic Press, New York), where Chapter 4 presented solutions of the inverse problem for certain parameters to be measured. However, first, the scope of parameters was limited to the electron density, the temperature and some parameters of some plasma waves. Second, but more importantly, over the subsequent four decades, very significant advances were made both in the theory of spectral line broadening by plasmas and in approaches to solving the corresponding inverse problem.

Hutchinson's book of 2005 (*Principles of Plasma Diagnostics*, Cambridge University Press, Cambridge) discussed the inverse problem in spectral lineshapes from plasmas only in a couple of sections of only one out of nine chapters (Chapter 6), and the discussion was cursory. Kunze's book of 2009 (*Introduction to Plasma Spectroscopy*, Springer, Berlin) discussed the inverse problem in spectral lineshapes from plasmas only in a few sections of only one out of 10 chapters (Chapter 10), and the discussion was relatively brief as well. Those are fine books — just they had a focus different from presenting solutions of the inverse problem in spectral lineshapes from plasmas.

As for other books having "plasma spectroscopy" in the title, such as Oks' book of 1995 (*Plasma Spectroscopy: The Influence of Microwave and Laser Fields*, Springer, Berlin), Griem's book of 1997 (*Principles of Plasma Spectroscopy*, Cambridge University Press, Cambridge), and Fujimoto's book of 2004 (*Plasma Spectroscopy*, Clarendon Press, Oxford), they presented lots of theoretical results, but paid zero or little attention to solutions of the inverse problem in spectral lineshapes from plasmas — because this was not the focus of those books.

The present monograph has the following three important features distinguishing it from the above-mentioned books. First, practically its *entire focus* is on presenting solutions for the inverse problem in spectral lineshapes from plasmas. Second, this monograph significantly *expands the scope of parameters* of plasmas and/or fields in it to be measured (such as effective charge of ions, parameters of low-frequency electrostatic turbulence in magnetized plasmas, parameters of Langmuir turbulence in magnetized plasmas, parameters of transverse laser-induced electromagnetic fields in plasmas). Third, for some parameters of plasmas and/or fields in it, this monograph presents *new, more advanced diagnostic methods* than the methods covered (though briefly) in the previous books.

In summary, the present book is an advanced tool for experimentalists using spectral lineshapes for diagnostics and for theorists helping the experimentalists in interpreting the experimental line profiles. This concerns both laboratory and astrophysical plasmas.

The book is divided into two parts. Part 1 is dedicated to plasmas that do not contain the electrostatic turbulence, i.e., turbulence represented by oscillatory electric fields. Part 2 is devoted to plasmas containing oscillatory electric fields. Both parts contain practical advice on how to interpret spectral line profiles observed in varieties of laboratory and astrophysical plasmas. They also contain numerous examples of the corresponding interpretations of various laboratory plasma experiments and astrophysical observations. The underlying theories are presented mostly in Appendices.

Part I

Non-Turbulent Plasmas

Chapter 1

Electron Density

1.1 Introductory Remarks

The electron density N_e is the primary plasma parameter. In laboratory and astrophysical plasmas, it varies over 20 orders of magnitude (in distinction to any other plasma parameter): from $N_e \sim (10^3\text{–}10^4)\,\text{cm}^{-3}$ in astrophysical objects called H II regions to $N_e \sim (10^{23}\text{–}10^{24})\,\text{cm}^{-3}$ in plasmas produced by very powerful lasers. (For comparison, the electron temperature T_e of these two extreme types of plasmas varies only by about three orders of magnitude: from $\sim$1 eV in H II regions to $\sim$1 keV in plasmas produced by very powerful lasers.) Obviously, there is no single universal method for measuring N_e that can be applied over 20 orders of magnitude. The overwhelming majority of method for measuring N_e are based on Stark broadening (SB) of various spectral lines by ion and electron microfields in plasmas.

SB of spectral lines in plasmas is controlled by the electron density N_e and to some extent by the electron (T_e) and ion (T_i) temperatures. Therefore, it was used for measuring the electron density in a very broad range by choosing appropriate spectral lines. The experimental determination of the electron density relies mostly on measuring the Stark width because it is typically by an order of magnitude greater than the Stark shift.

Hydrogenic radiators are the most appropriate for this purpose. The overwhelming majority of the states of hydrogenic radiators possess permanent dipole moments, thus, making these radiators more sensitive to ion and electron microfields in plasmas than non-hydrogenic radiators. The underlying theories are presented in Appendices A–E, G–I, and L.

In practice, there could be competing broadening mechanisms, such as, Doppler, instrumental and Zeeman broadenings, as well as a self-absorption. Therefore, the task is to find practical methods for extracting the Stark width despite the competing broadening effects, some of which could actually exceed the SB in certain situations. These methods are the focus of the subsequent sections.

1.2 Using the Intense (low-n) Lines of Hydrogenic Spectral Series

In hydrogen atoms and hydrogenlike ions (hereafter, hydrogenic atoms, for brevity), in spectral series, i.e., in radiative transitions between the upper states of various principal quantum number n and the fixed lower state of the principal quantum number n_0, the most intense are the lines, corresponding to $n = n_0 + 1$ (α-line), $n_0 + 2$ (β-line), n_0+3 (γ-line), n_0+4 (δ-line) — in order of diminishing intensity. So, on the first glance, it seems that one should use the most intense lines, like the α-line or the β-line.

However, there is a trade-off. The SB of the α-line is the smallest out of the spectral series. The SB increases along the spectral series: roughly speaking, as $\sim n^2$ for the ion broadening and $\sim n^4$ for the electron broadening. Therefore, while the absolute intensity (i.e., the wavelength-integrated intensity) of the spectral lines decreases as n grows, the sensitivity of the lines with respect to the electric microfields increases. Because of this situation, the usage of the low-n lines of the spectral series is appropriate only for a relatively dense plasmas ($N_e > 10^{14}$ cm^{-3}), where the ion and electron microfields are relatively large. Then the SB in its turn could be sufficiently large to be detected (despite the competing broadening mechanisms) and to allow deducing the electron density from the experimental line profiles.

If one has an experimental shape of only one spectral line, such as, e.g., the α-line or the β-line, it could be difficult to extract the contribution of the SB from the experimental shape containing contributions of other broadening mechanisms (except for plasmas of super-high densities where the SB would be by orders of magnitude greater than the competitors). Therefore, for an unambiguous extraction of the contribution of the SB, it would be best to have experimental profiles of at least two spectral lines — such as, e.g., both the α-line and the β-line, or both the β-line and the γ-line.

Also, it should be kept in mind that the β-line and the δ-line are relatively insensitive to the temperature, while the α-line and the γ-line are sensitive to the temperature. This is because the α-line and the γ-line have an intense central Stark component (i.e., the component that does not have a linear Stark shift under a static electric field). For this reason, the Stark width of the α-line and the γ-line has a significant contribution from the dynamical part of the electron and ion microfields, resulting in some temperature dependence of their widths. In distinction, the Stark width of the β-line and the δ-line is controlled primarily by the quasistatic SB (i.e., by the SB due to the quasistatic part of the ion microfield, which is the primary part of the ion microfield at $n^6 N_e > 10^{19}\,\mathrm{cm}^{-3}$).

The most rigorous analysis of experimental profiles would require calculating convolutions of tabulated Stark profiles with profiles due to other broadening mechanisms (moreover, in strongly magnetized plasmas, the convolutions would not be adequate and even more complicated calculations would be necessary — see Appendix E). This would entail creating huge multiparametric tables of theoretical profiles and a very difficult task of picking up the one (or several) providing the best fit to the experimental profiles — frequently there would be no unique solution. Before (or sometimes, instead of) this procedure, it might be a good idea to perform a simplified analysis described below — the analysis that would very significantly (by orders of magnitude) reduce the number of theoretical profiles to be subsequently calculated via rigorous convolutions or sometimes would be sufficient by itself.

The Full Width at Half Maximum (FWHM) of the total theoretical profile $\Delta\lambda_{1/2}(n, n_0)$ can be estimated from the relation

$$[\Delta\lambda_{1/2}(n, n_0, N_e)]^{3/2} = [\Delta\lambda_{1/2\mathrm{S}}(n, n_0, N_e)]^{3/2} + [\Delta\lambda_{1/2\mathrm{NS}}(n, n_0)]^{3/2}. \quad (1.1)$$

Here, $\Delta\lambda_{1/2\mathrm{S}}$ is the theoretical Stark FWHM and $\Delta\lambda_{1/2\mathrm{NS}}$ is the theoretical non-Stark FWHM (i.e., the theoretical FWHM due to other broadening mechanisms). Relation (1.1) works well at the range $n^6 N_e > 10^{19}\ \mathrm{cm}^{-3}$ and can also be used for estimates even at values of $n^6 N_e$ somewhat below $10^{19}\ \mathrm{cm}^{-3}$.

For a typical, practically important situation where $\Delta\lambda_{1/2\mathrm{NS}}$ (n, n_0) is dominated by the Doppler broadening and thus can be approximated as the Doppler FWHM $\Delta\lambda_{1/2\mathrm{D}}(n, n_0)$, the analysis further simplifies as follows. For any kind of FWHM $\Delta\lambda_{1/2}$, we introduce its scaled, dimensionless counterpart,

$$\delta\lambda_{1/2} = \Delta\lambda_{1/2\mathrm{D}}(n, n_0)/\lambda(n, n_0), \quad (1.2)$$

where $\lambda(n, n_0)$ is the unperturbed wavelength of the spectral line. For the Doppler broadening, the corresponding scaled FWHM $\delta\lambda_{1/2\mathrm{D}}$ does not depend on n and n_0. Let us divide Eq. (1.1) by $[\lambda(n, n_0)]^{3/2}$ and write the result for two spectral lines: one originating from the level of the principal quantum number n_1, another — from the level of the principal quantum number n_2:

$$[\delta\lambda_{1/2}(n_1, n_0, N_e)]^{3/2} = [\delta\lambda_{1/2\mathrm{S}}(n_1, n_0, N_e)]^{3/2} + [\delta\lambda_{1/2\mathrm{D}}]^{3/2}.$$
$$[\delta\lambda_{1/2}(n_2, n_0, N_e)]^{3/2} = [\delta\lambda_{1/2\mathrm{S}}(n_2, n_0, N_e)]^{3/2} + [\delta\lambda_{1/2\mathrm{D}}]^{3/2}. \quad (1.3)$$

By subtracting the second line of Eq. (1.3) from its first line, we get the relation,

$$[\delta\lambda_{1/2}(n_1, n_0, N_e)]^{3/2} - [\delta\lambda_{1/2}(n_2, n_0, N_e)]^{3/2} = [\delta\lambda_{1/2\mathrm{S}}(n_1, n_0, N_e)]^{3/2} - [\delta\lambda_{1/2\mathrm{S}}(n_2, n_0, N_e)]^{3/2}, \quad (1.4)$$

that does not contain the contribution of the Doppler broadening. Then by comparing the experimental value of the left side of Eq. (1.4) with the theoretical value of the right side of Eq. (1.4), calculated using formulas for the theoretical Stark widths presented below, one can easily determine the electron density N_e. Finally, substituting the obtained value of N_e either in the first or in the second line of Eq. (1.3), one find also the Doppler width and thus the temperature using the well-known relation:

$$\begin{aligned} \delta\lambda_{1/2\mathrm{D}} &= 2[2(\ln 2)T/(\mathrm{Mc}^2)]^{1/2} \\ &= 7.715 \times 10^{-5}[T(\mathrm{eV})/M(\mathrm{amu})]^{1/2}. \end{aligned} \tag{1.5}$$

Here, T and M are the temperature and the mass of the radiating atom/ion (hereafter, radiator), "amu" stands for "atomic mass unit".

It should be emphasized that the Doppler broadening can be caused not only by the thermal motion, but also by the non-thermal motion in plasmas, and by the combination of both thermal and non-thermal motions — especially in astrophysical plasmas. (Non-thermal motions are associated with a gas-dynamic turbulence, which differs in principle from a plasma turbulence: in the plasma turbulence particles oscillate around their equilibrium positions; besides, in astrophysical plasmas, characteristic spatial scales of the gas-dynamic turbulence, as a rule, exceed the Debye radius, which is a characteristic spatial scale of the plasma turbulence; therefore, both the gas-dynamic and plasma turbulences, generally speaking, can simultaneously exist in astrophysical plasmas independent of each other.)

In both laboratory and astrophysical plasmas, the low-n lines of spectral series could be optically thick — especially the α-lines. In frames of a frequently used model, which assumes a weak dependence of the source function S_λ on the wavelength λ (see, e.g., [1–3]), it is possible to take into account all three main mechanisms affecting the halfwidths $\Delta\lambda_{1/2}$ of hydrogenic lines: the Doppler broadening (both thermal and non-thermal), the Stark broadening, and a self-absorption. Indeed, if the source function S_λ does not depend on the wavelength λ, then the observed profile $I(\Delta\lambda)$ of a spectral line is given by (see, e.g., [1–3]),

$$I[\tau(\Delta\lambda)] = \{1 - \exp[-\tau(\Delta\lambda)]\}S, \tag{1.6}$$

where $\tau(\Delta\lambda)$ is the optical depth and $\Delta\lambda$ is the detuning from the unperturbed wavelength λ_0 of the spectral line. In the case of the Doppler broadening, one has,

$$\tau(\Delta\lambda) = \tau_0 \exp[-(\Delta\lambda/\delta\lambda_D)^2]. \tag{1.7}$$

Here, τ_0 is the optical depth in the center of the line and $\delta\lambda_D = \lambda V_0/c$, $V_0 = (V_t^2 + 2kT/M)^{1/2}$, where V_t is a characteristic non-thermal velocity and $\lambda = \lambda(n, n_0)$ is the unperturbed wavelength of the spectral line. The FWHM $\Delta\lambda_{1/2\mathrm{DO}}$ of the combined Doppler-opacity profile is the solution of the following equation:

$$\begin{aligned} 1 - \exp(-\tau_0 \exp\{-[\Delta\lambda_{1/2\mathrm{DO}}/(2\delta\lambda_D)]^2\}) \\ = [1 - \exp(-\tau_0)]/2. \end{aligned} \tag{1.8}$$

The solution of this equation is

$$\begin{aligned} \Delta\lambda_{1/2\mathrm{DO}} &= (2V_0\lambda/c)[\ln\, y(\tau_0)]^{1/2}, \\ y(\tau_0) &= \tau_0(\ln\{2/[1 + \exp(-\tau_0)]\})^{-1}. \end{aligned} \tag{1.9}$$

By combining $\Delta\lambda_{1/2\mathrm{DO}}$ with the theoretical Stark FWHM $\Delta\lambda_{1/2\mathrm{S}}$ (due to electron and ion microfields) and by dividing the widths by λ (to proceed to the scaled FWHM, like in Eq. (1.3)), we obtain:

$$[\delta\lambda_{1/2}(n)]^{3/2} = (2V_0/c)^{3/2}[\ln\, y(\tau_0)]^{3/4} + (\delta\lambda_{S,1/2})^{3/2}. \tag{1.10}$$

In the model, where the source function does not depend on the wavelength, the ratios of optical depths at the center of different hydrogenic lines are well-known constants. For example, for the low-n of the Balmer series, these ratios are as follows: $\tau_{0\alpha}:\tau_{0\beta}:\tau_{0\gamma}:\tau_{0\delta} = 7.25:1:0.334:0.156$. Thus, by measuring the experimental FWHM of three hydrogenic lines in the series and using Eq. (1.10) three times — once for each line — it is possible to exclude parameters V_0 and τ_0, and to extract the scaled Stark width $\delta\lambda_{1/2\mathrm{S}}(N_\mathrm{e})$, and to determine the electron density N_e. Finally, substituting the obtained value of N_e in Eq. (1.10) written for any two of the three lines, one also finds the Doppler width and thus the radiators temperature using Eq. (1.5).

As for the theoretical Stark width of hydrogenic spectral lines, required for determining the electron density N_e from the experimental Stark widths (extracted from the observed widths of lines), the situation is as follows. Theories of the SB of spectral lines in plasmas were developed for over 70 years by now. The most user-friendly analytical theories are *semiclassical* theories. In semiclassical theories, the radiator is described quantally, while the perturbing charges are described classically — see Appendix A.

The simplest semiclassical theory of the SB of hydrogen lines is a so-called Conventional Theory (CT) — sometimes called a standard theory. The CT, in particular:

(A) employs the impact approximation and the perturbation theory for all components of the electron microfield; the impact approximation considers a sequence of binary collisions of the perturbing electrons with the radiator and these collisions are considered to be completed; (B) neglects the ion dynamics, i.e., treats the ion microfield in the quasistatic approximation; (C) neglects any coupling between the electron and ion microfields. (Different versions of the CT are described in detail in Appendix B). While the above approximations (A) and (B) were gradually removed in the course of a further development of the theory (as described in Appendix A), for a very long time — until mid-1990s — it was still considered that there is zero or little coupling between the electron and ion microfields.

In reality, the coupling between the electron and ion microfields can be strong and can significantly affect profiles of hydrogen and hydrogen-like spectral lines. This was shown with the development of the most advanced analytical theory of the SB called the *Generalized Theory* (GT). The coupling, facilitated by the radiator, becomes stronger and stronger with the increase of the electron density N_e and/or the principal quantum number n, as well as with the decrease of the temperature T. The GT achieved this result by going beyond the fully-perturbative description of the electron microfield used in the CT. All the details are presented in Appendices C and D.

In parallel to the development of more sophisticated analytical theories, there were developed various simulation models — see Appendix A. Simulation models could be useful to some extent, though they have problems in calculating Stark shifts (especially in dense plasmas of relatively low temperatures) and they lack the physical insight.

From the practical point of view, it is beneficial to know some formulas, relating in a simple manner the Stark FWHMs of certain hydrogenic spectral lines to the electron density N_e. Some of these formulas were extracted from two types of SB tables, one type of the tables being based on the GT (the tables presented in book [4]), while the other type — on simulations [5, 6]. Some other formulas were obtained from experiments. All kinds of these formulas are presented below.

1.2.1 H_β *line*

The H_β line is usually the most suitable for measuring the electron density N_e because its Stark width practically depends only on N_e, being practically independent of the plasma temperature and of the mass of perturbing ions — in distinction, e.g., from the H_α line. While the H_α line is more intense than the H_β line, its Stark width significantly depends on the plasma temperature and on the mass of perturbing ions (more rigorously, on the ratio of the reduced mass of the ion-radiator pair to the proton mass) — because of a strong central Stark component in the H_α line (the H_β line does not have the central Stark component). Gigosos *et al.* [6] suggested the following formula for the theoretical Stark FWHM of the H_β line:

$$\Delta\lambda_{1/2\mathrm{S}}\ (\mathrm{nm}) = 4.800\ [\mathrm{N_e}\ (\mathrm{cm}^{-3})/10^{17}]^{0.68116}, \quad 10^{14}\,\mathrm{cm}^{-3} < N_\mathrm{e} < 10^{19}\,\mathrm{cm}^{-3}. \qquad (1.11)$$

A slightly different formula was suggested by Surmick and Parigger [7]:

$$\Delta\lambda_{1/2\mathrm{S}}\ (\mathrm{nm}) = 4.50\ [N_\mathrm{e}\ (\mathrm{cm}^{-3})/10^{17}]^{0.71\pm0.03}, \quad 10^{14}\,\mathrm{cm}^{-3} < \mathrm{N_e} < 10^{19}\,\mathrm{cm}^{-3}. \qquad (1.12)$$

Since, over a broad range of electron densities, the theoretical Stark profile of the H_β line has a shape of a doublet, it could be also possible deducing N_e from the *separation* $\Delta\lambda_{\rm peaks}$ *between the two peaks.* In particular, Ivkovic *et al.* [8], by analyzing simulated Stark profiles by Gigosos *et al.* [6], suggested the following formula for the situations where the plasma composition (and thus the reduced mass of the perturbing ion-radiator pair) is not known,

$$\log[\Delta\lambda_{\rm peaks}\ ({\rm nm})] = \{\log[N_e({\rm cm}^{-3})] - 16.618\}/1.457,$$
$$10^{14}\,{\rm cm}^{-3} < N_{\rm e} < 10^{19}\,{\rm cm}^{-3}, \quad (1.13)$$

where $\log[\cdots]$ is the decimal logarithm. Alternatively, if there are no perturbing ions, for which the reduced mass is no more than two in units of the proton mass, the suggested formula [8] was:

$$\log[\Delta\lambda_{\rm peaks}\ ({\rm nm})] = \{\log[N_e\ ({\rm cm}^{-3})] - 16.621\}/1.452,$$
$$10^{14}\,{\rm cm}^{-3} < N_e < 10^{19}\,{\rm cm}^{-3}, \quad (1.14)$$

For pure hydrogen plasmas (reduced mass of the perturbing ion-radiator is 1/2 of the proton mass), the suggested formula [8] was:

$$\log[\Delta\lambda_{\rm peaks}\ ({\rm nm})] = \{\log[N_e\ ({\rm cm}^{-3})] - 16.661\}/1.416,$$
$$10^{14}\,{\rm cm}^{-3} < N_e < 10^{19}\,{\rm cm}^{-3}. \quad (1.15)$$

1.2.2 *H_α line*

Pardini *et al.* [9] by analyzing the simulated Stark FWHM presented by Gigosos *et al.* [7], suggested the following formula,

$$\Delta\lambda_{1/2{\rm S}}\ [{\rm nm}] = [0.1/(a + b/N_{\rm e}^{1/2})][N_{\rm e}\ ({\rm cm}^{-3})/(8.02 \times 10^{12})]^{2/3},$$
$$3 \times 10^{14}\,{\rm cm}^{-3} < N_{\rm e} < 10^{19}\,{\rm cm}^{-3}, \quad (1.16)$$

$$a = (4033.8 - 24.45\{\ln[T(K)]\}^2)^{1/2},$$
$$b = 1.028 \times 10^9 + 174576.3\ T(K).$$

Parigger *et al.* [10] by analyzing the table of theoretical Stark FWHM based on the GT, Parigger suggested the following simpler formula (averaged over temperatures and over the ratio of the reduced mass of the ion-radiator pair to the proton mass),

$$\Delta\lambda_{1/2\mathrm{S}}\ [\mathrm{nm}] = 5.68[N_e\ (\mathrm{cm}^{-3})/10^{18}]^{0.64\pm0.03}. \tag{1.17}$$

1.2.3 *He II Balmer-alpha line* 468.6 *nm*

Büscher *et al.* [11] deduced from their benchmark experiment at the gas-liner pinch the following relation:

$$\Delta\lambda_{1/2\mathrm{S}}\ [\mathrm{nm}] = 2.74 \times 10^{-20}(N_\mathrm{e}[\mathrm{m}^{-3}])^{0.831}. \tag{1.18}$$

1.2.4 *He II Balmer-beta line* 320.3 *nm*

Büscher *et al.* [11] deduced from their benchmark experiment at the gas-liner pinch the following relation:

$$\Delta\lambda_{1/2\mathrm{S}}\ [\mathrm{nm}] = 9.78 \times 10^{-18}(N_e[\mathrm{m}^{-3}])^{0.74}. \tag{1.19}$$

1.2.5 *Modifications in strongly-magnetized plasmas*

In many types of plasmas — such as, tokamak plasmas (see, e.g., [12]), laser-produced plasmas (see, e.g., [13]), capacitor-produced plasmas [14], astrophysical plasmas [15, 16] — there are strong magnetic fields. There are two major effects of a strong magnetic field on the SB of low-n hydrogenic spectral series. We present these effects here using hydrogen/deuterium spectral lines.

The first effect is the inhibition of the dynamical Stark broadening by a strong magnetic field.

Physically, this is because a relatively large Zeeman splitting of hydrogen energy levels diminishes the range of ion impact parameters ρ, for which the characteristic frequency of the variation v_i/ρ of the electric field of perturbing charged particles exceeds the Zeeman splitting.

This effect was described analytically in year 1994 by Derevianko and Oks [17] using an advanced SB formalism — the GT. Derevianko and Oks applied their results to the edge plasmas of magnetic fusion devices (e.g., in the divertor region of tokamaks), where the homogeneous Stark width of hydrogen/deuterium spectral lines is

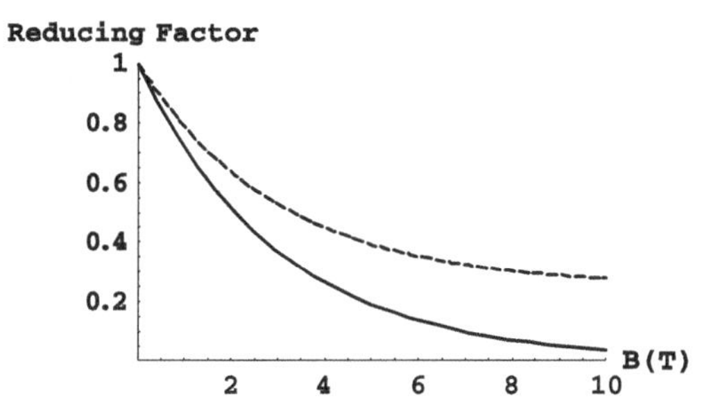

Fig. 1.1. Stark width reducing factor for the Ly_β line at $N_e = 10^{14}\,\mathrm{cm}^{-3}$ and $T = 5\,\mathrm{eV}$ versus the magnetic field B: for σ-components (solid line) and for the π-component (dotted line). The reducing factor is defined as the ratio of the Stark width at a particular value of the magnetic field B to the corresponding Stark width at $B = 0$.

controlled by the dynamic part of the *ion microfield.* (For definitions of homogeneous and inhomogeneous Stark widths see the first paragraph of Appendix E.1.). As an example, Fig. 1.1 presents the plot of the Stark width reducing factor for the Ly_β line at $N_e = 10^{14}\,\mathrm{cm}^{-3}$ and $T = 5\,\mathrm{eV}$: for σ-components (solid line) and for the π- component (dotted line). It is seen that the strong magnetic field (which in magnetic fusion devices reaches up to $\sim 10\,\mathrm{T}$) very significantly diminishes the dynamical Stark width.

For explaining the inhibition effect more accurately, the following details are necessary. In the semiclassical theories of the dynamical SB of *non-hydrogenic* spectral lines by electron or ion microfields, the broadening is controlled by *virtual transitions* between different sublevels characterized by the same principal quantum number n (see, e.g., book [18]). This is a *non-adiabatic* contribution — since it involves different sublevels. For hydrogenic spectral lines, under the usual assumption that the fine structure can be neglected, in addition to the non-adiabatic contribution there is also a significant *adiabatic* contribution — see, e.g., review [19] and book [5]. Physically, the adiabatic contribution is due to the fact that for hydrogenic spectral lines, the overwhelming majority of sublevels have permanent electric dipole moments (which is especially clear in the parabolic quantization). Even an isolated sublevel having a permanent electric dipole moment would produce some adiabatic contribution — in other words, the latter does not require virtual transitions between different sublevels.

A strong magnetic field, by splitting hydrogen/deuterium energy levels, very significantly diminishes the non-adiabatic contribution to the Stark width, so that practically only the adiabatic contribution remains. Fortunately, the adiabatic contribution to the dynamical Stark width has been calculated in frames of the GT *exactly* (without using the perturbation theory) [17]. The result is presented in Appendix E.1 (Eqs. (E.2)–(E.7)), where all the further details can be also found and used for measuring the electron density in strongly magnetized plasmas.

The second effect of strong magnetic fields on the SB of the headlines of hydrogenic spectral series has to do with the following. In a strong magnetic field **B**, perturbing electrons basically spiral along magnetic field lines. Therefore, their trajectories are not rectilinear in the case of neutral radiators or not hyperbolic in the case of charged radiators. However, until the year 2016, all semiclassical theoretical calculations of the SB of hydrogenic spectral lines considered only rectilinear trajectories of perturbing electrons in the case of neutral radiators or hyperbolic trajectories in the case of charged radiators. In 2016, in the paper [20], the spiraling trajectories of the perturbing electrons have been taken into account. It turned out that the allowance for the spiraling trajectories affects the SB in the following two ways.

The primary effect is a very significant change in the ratio of intensities of the three Zeeman components of hydrogen/deuterium spectral lines (we remind that in a strong magnetic field, these lines have a shape of a triplet consisting of two lateral peaks of the equal intensity and one central peak). The allowance for the spiraling trajectories increases the ratio of the intensity of the central peak to the intensity of either one of the two lateral peaks by up to a factor of two. The secondary effect is an additional shift of the lateral components: the red one — further to the red side, the blue one — further to the blue side (by the same amount as the additional shift of the red component).

Examples analyzed in the paper [20], show that the effect of spiraling trajectories on the SB of hydrogen/deuterium spectral lines plays a significant role especially in magnetic fusion devices (electron densities 10^{13}–10^{15} cm^{-3}) and in astrophysical objects such as DA white dwarfs (electron densities $\sim 10^{17}$ cm^{-3}). All the further details can be found in Appendix E.3 — the details that can be used for measuring the electron density in strongly magnetized plasmas.

1.3 Using Highly-Excited Hydrogen/Deuterium Lines

Highly-excited hydrogen/deuterium spectral lines can be used for measuring the electron density in relatively low-density plasmas. The three most frequent applications are edge plasmas of magnetic fusion devices [12,21], where $N_e \sim (3 \times 10^{13}$–$10^{15})$ cm^{-3}, plasmas of radio-frequency discharges [22], where $N_e \sim 10^{13}$ cm^{-3}, as well as the plasma of the solar chromosphere, where in the quiet Sun $N_e \sim 10^{11}$ cm^{-3} [23] (in solar flares typically $N_e \sim 10^{13}$ cm^{-3}, but the SB can be controlled by the electrostatic plasma turbulence — as presented in Chapter 6.1). While the absolute intensity of the lines decreases along the spectral series, the SB (and thus the sensitivity to plasma microfields) increases.

In frames of the CT of the SB described in Appendix B, the FWHM of a highly excited hydrogen lines is typically presented in the form,

$$\Delta\lambda_{1/2}^{\mathrm{high}} \approx A_e(n,n')N_e/T_e^{1/2} + B_i(n,n')N_e^{2/3}, \qquad (1.20)$$

where the term $A_e(n,n')N_e/T_e^{1/2} \equiv \Delta\lambda_e$ is due to electrons (i.e., represents the impact broadening by the electron microfield) and the term $B_i(n,n')N_e^{2/3} \equiv \Delta\lambda_i$ is due to the ions (i.e., represents the quasistatic broadening by the ion microfield). Here, the coefficient B does not depend on the temperature or on the density, while the coefficient A has only a very weak, logarithmic dependence on N_e and T_e.

Ferri *et al.* [24] rewrote Eq. (1.20) in the form (following Bengtson *et al.* [25]),

$$\Delta\lambda_{1/2}^{\mathrm{high}}(A) \approx 2.5 \times 10^{-13}(\beta N_e + \alpha N_e^{2/3}), \qquad (1.21)$$

Table 1.1. Numerical values of coefficients α and β in Eq. (1.21) for some Balmer and Paschen lines at $T = 4\,\text{eV}$.

	Balmer		Paschen	
n	α	$\beta(10^{-9})$	α	$\beta(10^{-9})$
8	0.216	5.65	1.158	3.57
9	0.288	6.37	1.691	3.76
10	0.392	8.95	2.042	5.26
11	0.453	12.3	2.655	5.52
12	0.597	11.9	2.786	9.52
13	0.736	12.3	4.016	6.28

where N_e is in m^{-3}, and used simulations to tabulate the coefficients α and β for several Balmer and Paschen lines, but only for one value of the temperature: $T = 4\,\text{eV}$. The resulting values of α and β are shown in Table 1.1.

Stehle and Hutcheon used a different simulation method (compared to [24]) to produced table of Stark profiles of Lyman, Balmer, and Pasche lines. In their paper [26], Stehle and Hutcheon announced the availability of these tables.

Stambulchik and Maron [27] made some advances into approximate analytical calculations of the Stark widths of highly-excited hydrogen/deuterium lines. (They started within the CT, where the ion-dynamical effects are neglected, but then tried to take into account the ion dynamics in a rough approximation). Their main point was to approximate intensities of Stark components of a particular spectral line by some common value (instead of the true intensities that differ from each other). Within this approximation, they obtained some useful analytical results but did not offer any practical formulas like Eq. (1.21).

Practical analytical formulas for the Stark widths of highly-excited hydrogen/deuterium lines were obtained within the frames of the more advanced, GT (described in Appendix C), as follows [28] (see also book [5], Chapter 7):

$$\text{FWHM}(A) = \Delta\lambda_L(A) + [0.94761 \times 10^{-19} Z_i^{3/2} N_e n^6 n'^6 / (n^2 - n'^2)^{3/2}] / [\Delta\lambda_L(A)]^{1/2}, \quad (1.22)$$

where

$$\Delta\lambda_L = \Delta\lambda_e + \Delta\lambda_i, \tag{1.23}$$

$$\begin{aligned}\Delta\lambda_{\rm e}(A) = 2.1200 \times 10^{-20} n^4 n'^4 (n^2 - n'^2)^{-2} {\rm g_{nn'}} N_e {\rm T}^{-1/2}\{1 \\ + \ln[1.5192 \times 10^{15} {\rm T g'^{-1/2}_{nn}} (3.1826 \times 10^9 N_e \\ + 53.2 n^2 Z_i^{2/3} N_e^{4/3})^{-1/2}]\},\end{aligned} \tag{1.24}$$

$$\begin{aligned}\Delta\lambda_i(A) = f(T)(N_e/N_{cr})^{1/3}\{1 - \exp\{-9\pi Z_{\rm i} {\rm g_{nn'}} [2{\rm f(T)} n^2_{\rm eff}]^{-1}[1 \\ + \ln[1 + 6.2847 \times 10^8 (T/Z_i)(N_{\rm e} {\rm g_{nn'}} \mu/M_{\rm p})^{-1/2}]] \\ \times (N_e/N_{cr})^{2/3}\}\},\end{aligned} \tag{1.25}$$

$$f({\rm T}) = 0.34941[n^4 n'^4/(n^2 - n'^2)^2]{\rm T M}_p/[\mu Z_i n^2_{\rm eff}], \tag{1.26}$$

$$\begin{aligned}N_{cr} = (9/\pi) Z_{\rm i}^{-2} n_{\rm eff}^{-6} (m_e/\hbar)^3 (T_i/\mu)^{3/2} \\ \approx 1.7309 \times 10^{18}\,{\rm cm}^{-3} Z_i^{-2} n_{\rm eff}^{-6} [T({\rm eV}) M_p/\mu]^{3/2},\end{aligned} \tag{1.27}$$

$$n_{\rm eff} \equiv (n^2 + n - {\rm n}'^2 - {\rm n}')^{1/2}. \tag{1.28}$$

$${\rm g_{nn'}} \equiv (n^2 - n'^2)^2 - n^2 - n'^2. \tag{1.29}$$

Here, Z_i is the charge of perturbing ions, μ is the reduced mass of the perturber-radiator pair, M_p is the proton mass, n and n' are the principal quantum numbers of the upper and lower levels (involved in the radiative transition), respectively. *In the above practical formulas* (1.22)–(1.27), *the electron density* $\mathrm{N_e}$ *is in* cm^{-3}, *the temperature* T *is in* eV.

The validity condition of the above formulas consists of the requirement that the second term in Eq. (1.22) should remain smaller than the first term.

As an example, Fig. 1.2 shows the experimental FWHM of five highly-excited Balmer lines observed in the divertor region of tokamak DIII-D [21]. It is seen that the GT more accurately reproduces the shape of the experimental FWHM dependence than the CT. The shape of the FWHM dependency in the GT is more

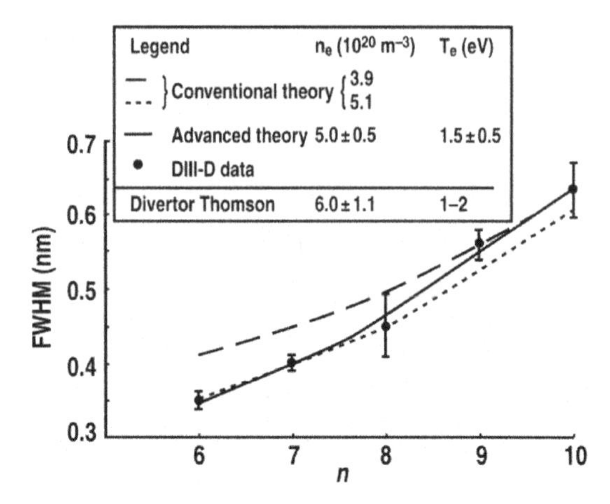

Fig. 1.2. FWHM versus the principal quantum number, n, of deuterium high-n Balmer lines, as measured in the DIII-D divertor. Fits for SB theories are shown: the GT (labeled "advanced theory" in the legend) with $N_e = 5.0 \times 10^{20}\text{m}^{-3}$ and $T = 1.5\,\text{eV}$ (solid line), and the CT with $N_e = 3.9 \times 10^{20}\text{m}^{-3}$ (dashed line) and $N_e = 5.1 \times 10^{20}\text{m}^{-3}$ (dotted line).

complicated than that in the CT because of the competition of two additional effects included in the GT, but neglected in the CT, namely, the coupling of electron and ion microfields (via the radiator) and the ion-dynamical broadening. The SB due to coupling increases with n, while that due to ion dynamics decreases with n.

The theoretical results presented in Fig. 1.2 were obtained from the best fit to the shapes of the experimental Balmer lines. The best fit for the CT yielded $N_e = (4.5 \pm 0.6) \times 10^{20}\text{m}^{-3}$, but no information about the temperature. The most probable values of the densities, deduced from individual line profiles fitted by the CT, varied by over 15%. The best fit for the GT yielded $N_e = (5.0 \pm 0.5) \times 10^{20}\text{m}^{-3}$ and the temperature $T = 1.5 \pm 0.5\,\text{eV}$. The most probable values of the densities deduced from individual line profiles fitted by the GT, varied by less than 5%.

1.3.1 *Modifications in strongly-magnetized plasmas*

All the simulations and analytical approaches mentioned above in this Sec. 1.3 did not take into account a very important broadening mechanism that becomes effective in strongly-magnetized plasmas:

the broadening by the Lorentz field (hereafter, Lorentz broadening). Radiating hydrogen atoms moving with the velocity $\mathbf{v}$ across the magnetic field $\mathbf{B}$ experience a Lorentz electric field $\mathbf{E}_\mathrm{L} = \mathbf{v} \times \mathbf{B}/c$ in addition to other electric fields. In plasmas, the atomic velocity $\mathbf{v}$ has a distribution. Things get complicated by the fact that Lorentz and Doppler broadenings cannot be accounted for via a convolution, but rather they intertwine in a more complicated manner — see Appendix E.2 for details and analytical results for Lorentz–Doppler profiles.

There are practically important situations where the Lorentz broadening can predominate over other broadening mechanisms for highly-excited hydrogen lines. In Appendix E.2, it is shown that the Lorentz broadening can significantly exceed both the SB by the plasma microfield and the Zeeman splitting for high-n hydrogen lines. Also in Appendix E.2, it is shown that the ratio of the FWHM $(\Delta\omega_L)_{1/2}$ caused by the Lorentz broadening to the FWHM $(\Delta\omega_D)_{1/2}$ caused by the Doppler broadenings is

$$(\Delta\omega_L)_{1/2}/(\Delta\omega_D)_{1/2} = n_\alpha^2 n_\beta^2 B\ (\mathrm{Tesla})/526. \tag{1.30}$$

We note that this ratio does not depend on the temperature.

For Balmer lines ($n_\beta = 2$) Eq. (1.30), becomes

$$(\Delta\omega_L)_{1/2}/(\Delta\omega_D)_{1/2} = n_\alpha^2 B\ (\mathrm{Tesla})/131. \tag{1.31}$$

So, as stated in Appendix E.2, e.g., for the edge plasmas of tokamaks, where Balmer lines of $n_\alpha \sim (10\text{–}16)$ have been observed, the Lorentz broadening dominates over the Doppler broadening when the magnetic field exceeds the critical value $B_c \sim 1$ Tesla. This condition is fulfilled in the modern tokamaks and will be fulfilled also in the future tokamaks.

Another example presented in Appendix E.2: in solar chromosphere, where Balmer lines of $n_\alpha \sim (25\text{–}30)$ have been observed, the Lorentz broadening dominates over the Doppler broadening when the magnetic field exceeds the critical value $B_c \sim (0.15\text{–}0.2)$ Tesla. This condition can be fulfilled in sunspots where B can be as high as 0.4 Tesla.

Therefore, it is practically useful to calculate pure Lorentz-broadened profiles of highly-excited Balmer lines. The corresponding calculations are presented in Appendix E.2 in detail. For the practical diagnostic of the electron density, there is an important result derived from the calculated Lorentz-broadened profiles of highly-excited Balmer lines, as follows.

For any two adjacent high-n Balmer lines (such as, H_{16} and H_{17}, or H_{17} and H_{18}), the sum of their Half Widths at Half Maximum (HWHM) in the frequency scale, the sum being denoted here simply as $\Delta\omega_{1/2}$, turned out to be

$$\Delta\omega_{1/2} = A[3n^2\hbar B v_T/(2m_e ec)], \tag{1.32}$$

where the constant A depends on the direction of observation as follows:

$$A = 0.80 \text{ (observation perpendicular to } \mathbf{B}), \tag{1.33}$$

$$A = 1.00 \text{ (observation parallel to } \mathbf{B}), \tag{1.34}$$

$$A = 0.86 \text{ ("isotropic" observation)}. \tag{1.35}$$

Here by the "isotropic" observation, we describe the situation where along the line of sight there are regions with various directions of the magnetic field, which could be sometimes the case in astrophysics.

The results presented in Eqs. (1.32)–(1.35) lead to a revision of the simple diagnostic method based on the principal quantum number $n_{\max}$ of the last observed line in the spectral series of hydrogen/deuterium lines, such as, Lyman, or Balmer, or Paschen lines (though typically Balmer lines are used). This simple method was first proposed by Inglis and Teller [29]. The idea of the method was that the SB of hydrogen lines by the ion microfield (in case it is quasistatic) in the spectral series scales as $\sim n^2$. Therefore, at some value $n = n_{\max}$, the sum of the Stark HWHM of the two adjacent lines becomes equal to the unperturbed separation of these two lines, so that they (and the higher lines) merge into a quasicontinuum. Since, the SB is controlled by the ion density N_i (equal to the electron density N_e for hydrogen plasmas), this had led previously to the following simple reasoning.

At the electric field E, for the multiplet of the principal quantum number $n \gg 1$, the separation $\Delta\omega(n)$ of the most shifted Stark sublevel from the unperturbed frequency $\omega_0(n)$ is $\Delta\omega(n) = 3n^2\hbar E/(2m_e\mathrm{e})$. Then the sum of the "halfwidths" of the two adjacent Stark multiplets of the principal quantum numbers n and $n+1$ is

$$\Delta\omega_{1/2}(n) = 3n^2\hbar E/(m_e e). \tag{1.36}$$

The unperturbed separation (in the frequency scale) between the hydrogen spectral lines, originating from the highly-excited levels n and $n+1$ is

$$\omega_0(n+1) - \omega_0(n) = m_e \mathrm{e}^4/(n^3\hbar^3). \tag{1.37}$$

By equating (1.36) and (1.37), one finds

$$n_{\mathrm{max}}^5 E = E_{\mathrm{at}}/3 = 5.71 \times 10^6 \mathrm{CGS}, \quad E_{\mathrm{at}} = m^2e^5/\hbar^4 \tag{1.38}$$

($E_{\mathrm{at}} = 1.714 \times 10^7$ CGS $= 5.142 \times 10^9$ V/cm is the atomic unit of electric field). For the field E, Inglis and Teller [29] used the most probable field of the Holtsmark distribution, which they estimated as $E_{\mathrm{imax.}} = 3.7eN_i^{2/3} = 3.7eN_e^{2/3}$, and obtained from Eq. (1.38) the following relation:

$$N_e \mathrm{n}_{\mathrm{max}}^{15/2} = 0.027/a_0^3 = 1.8 \times 10^{23}\,\mathrm{cm}^{-3}, \tag{1.39}$$

where a_0 is the Bohr radius. We note that Hey [30], by using a more accurate value of the most probable Holtsmark field $E_{\mathrm{imax}} = 4.18eN_e^{2/3}$, obtained a slightly more accurate numerical constant in the right side of Eq. (1.39), namely, $0.0225/a_0^3$, while Griem [31] suggested this constant to be even twice smaller.

Thus, Inglis–Teller relation (39) constituted a simple method for measuring the electron density by the number n_{max} of the observed lines of a hydrogen spectral series. The simplicity of this method is the reason why, despite the existence of more sophisticated (but more demanding experimentally) spectroscopic methods for measuring N_e, this method is still used in both laboratory and astrophysical plasmas. For example, Welch *et al.* [12] used it (with the constant in the right side of Eq. (1.39) suggested by Griem [31]) for determining the electron density in the low-density discharge at Alcator C-Mod.

However, in magnetized plasmas, the Lorentz field E_L can significantly exceed the most probable Holtsmark field $E_{\max}$, as shown in Appendix E.2. In this situation, the number $n_{\max}$ of the last observable hydrogen line will not be controlled by the electron density, but rather by different parameters, as shown below. Let us first conduct a simplified reasoning along the approach of Inglis and Teller [29]. By substituting $E = E_L$ in the left side of Eq. (1.38), we obtain the following relation:

$$n_{\max}^{10} B^2 T(K) = 1.78 \times 10^{18} M/M_p \text{ or}$$
$$n_{\max}^{10} B^2 T(\text{eV}) = 1.54 \times 10^{14} M/M_p, \tag{1.40}$$

where B is the magnetic field in Tesla; M and M_p are the atomic and proton masses, respectively.

More accurate relations can be derived using the results of our calculations of Lorentz broadened profiles of high-n Balmer lines and the corresponding formulas (1.32)–(1.35) for the sum of the HWHM of two adjacent Balmer lines. In this more accurate way, we obtained the following relations.

For the observation perpendicular to $\mathbf{B}$:

$$n_{\max}^{10} B^2 T(K) = 2.79 \times 10^{18} M/M_p \quad \text{or}$$
$$n_{\max}^{10} B^2 T(\text{eV}) = 2.40 \times 10^{14} M/M_p. \tag{1.41}$$

For the observation parallel to $\mathbf{B}$:

$$n_{\max}^{10} B^2 T(K) = 1.78 \times 10^{18} M/M_p \quad \text{or}$$
$$n_{\max}^{10} B^2 T(\text{eV}) = 1.54 \times 10^{14} M/M_p. \tag{1.42}$$

For the "isotropic" case (the meaning of which was explained after Eq. (1.35)):

$$n_{\max}^{10} B^2 T(K) = 2.38 \times 10^{18} M/M_p \quad \text{or}$$
$$n_{\max}^{10} B^2 \mathrm{T}(\text{eV}) = 2.05 \times 10^{14} M/M_p. \tag{1.43}$$

Thus, the above formulas, by using the observable quantity $n_{\max}$, allow to measure the atomic temperature T, if the magnetic field is known, or the magnetic field B, if the temperature is known.

The accuracy can be increased by taking into account additional broadening mechanisms as follows:

$$[(\Delta\omega_{1/2}(n))^2 + \Delta\omega_Z^2 + (\Delta\omega_D)_{1/2}^2]^{1/2} = m_e e^4/(n^3\hbar^3), \qquad (1.44)$$

where the sum $\Delta\omega_{1/2}(n)$ of Lorentz halfwidths of the two adjacent high-n Balmer lines is given by Eq. (1.32). Here, the Zeeman width $\Delta\omega_Z$, and the Doppler width $(\Delta\omega_D)_{1/2}$ are,

$$\Delta\omega_Z = eB/(2m_e c),$$
$$(\Delta\omega_D)_{1/2} = 2(\ln 2)^{1/2}\omega_0 v_T/c = 1.665\omega_0 v_T/c, \qquad (1.45)$$

where v_T is the thermal velocity of the radiators and ω_0 is the unperturbed frequency of the spectral line. For highly-excited hydrogen lines, where $n_\alpha \gg n_\beta$, one can use the expression $\omega_0 = m_e e^4/(2n_\beta^2\hbar^3)$.

We also note that the Lorentz and Zeeman mechanisms can be combined together "exactly" using the fact that the problem of a hydrogen atom in the crossed electric and magnetic fields allows an analytical solution which is exact within the subspace spanned on the states of the principal quantum number n. This fact, being the consequence of the O(4) symmetry of the hydrogen atom (and of the corresponding Kepler problem), was presented already in 1927 by Born in frames of the "old quantum theory" [32] and was later elaborated in more detail in the contemporary quantum theory by Demkov *et al.* [33]. The characteristic frequencies, arising in this solution, are $\xi\Omega$, where,

$$\Omega = \{[\Delta\omega_L/(n-1)]^2 + \Delta\omega_Z^2\}^{1/2},$$
$$\Delta\omega_L = 3n(n-1)\hbar B v_T/(2m_e ec),$$
$$\xi = -(n-1), -(n-2), \ldots, (n-1), \qquad (1.46)$$

It is seen that the characteristic frequencies $\xi\Omega$ are similar to the left side of Eq. (1.44) if in the latter the Doppler broadening would be disregarded.

We emphasize that in the expression for Ω in Eq. (1.46), the second term under the square root is just a small correction to the first term if the atomic temperature $T_a \gg T_{cr}(n) = (104/n^2)\,\mathrm{eV}$. This is the case for the edge plasmas of tokamaks (n ~10–16), as well as for the solar chromosphere ($n \sim 30$). Indeed, for example, for $n = 13$ (which is the middle of the interval of $n = 10$–16 observed at Alcator C-Mod [12]), we have $T_{cr}(n) = 0.6\,\mathrm{eV}$, while the actual atomic temperature was about 6 eV. For the solar chromosphere ($n \sim 30$) we have $\mathrm{T}_{cr}(n) \sim 0.1\,\mathrm{eV}$, while the actual atomic temperature is ~1 eV.

Coming back to the diagnostic application of Eq. (1.44) and using it for the conditions of the low-density discharge at Alcator C-Mod [12], where deuterium Balmer lines up to $n_{\max} = 16$ were observed and the magnetic field was $B = 8$ Tesla, we obtain $T = 6\,\mathrm{eV}$.

For the solar chromosphere, where Balmer lines up to $n_{\max} \sim 30$ have been observed [23], assuming $T \sim 10^4\,\mathrm{K}$ and the average non-thermal velocity $v_{\mathrm{nonth}} \sim 20\,\mathrm{km/s}$, we obtain from Eq. (1.44): $B \sim 0.2$ Tesla.

We also note that when in the above two examples, instead of using Eq. (1.44), we employed the approach based in part on Eq. (1.46), we obtained essentially the same results.

The bottom line is that in strongly-magnetized plasmas, the principal quantum number $n_{\max}$ of the last observed line in the spectral series of hydrogen/deuterium lines can no longer be used for determining the electron density, but rather for determining the product $T^{1/2}B$ — according to Eqs. (1.41)–(1.43). Then if the temperature T of the radiators is known, one can determine the magnetic field B; inversely, if the magnetic field B is known, one can determine the temperature T of the radiators.

1.4 Using Langmuir-Waves-Cause Dips in Hydrogenic Line Profiles as the Most Accurate Passive Spectroscopic Method for Measuring the Electron Density

Let us state upfront that if Langmuir-wave-caused dips (hereafter, L-dips) can be identified in the experimental profile of a hydrogenic spectral line, then from the separation between the L-dips, it is possible to determine the electron density N_e with an accuracy ~1%. All other methods for measuring N_e from the spectral line profiles have the relative error of at least ~10%. Here are the details.

According to the theory of L-dips presented in Appendix G, under a combined action of a Langmuir wave $\mathbf{E}(t) = \mathbf{E}_0 \cos\omega_{\mathrm{pe}}t$ at the plasma electron frequency and a quasistatic field $\mathbf{F}$ (representing the quasistatic part of the ion microfield and/or a low-frequency electrostatic plasma turbulence), L-dips can appear in hydrogenic line profiles at the following positions measured from the unperturbed wavelength λ_0 of the line:

$$\Delta\lambda_{\mathrm{dip}}(k) = [\lambda_0^2/(2\pi c)]n^{-1}(nq - n'q')k\omega_{\mathrm{pe}}. \tag{1.47}$$

Here, $q = n_1 - n_2$, and $q' = n_1' - n_2'$, where n_1, n_2, m and n_1', n_2', m' are the parabolic quantum numbers of the upper and lower Stark sublevels, respectively; n and n' are the corresponding principal quantum numbers; k is the number of quanta of the Langmuir field $\mathbf{E}(\mathrm{t})$ involved in the resonance with Stark sublevels split by the quasistatic field $\mathbf{F}$ (typically, $k = 1, 2$). For the spectral lines of the Lyman series, one has $q' = 0$, so that Eq. (1.47) simplifies.

The L-dips appear in pairs: an L-dip, corresponding to a particular set of the quantum numbers (q, q'), has a partner L-dip, corresponding to the set $(-q, -q')$. This means that one L-dip in the pair is in the red side of the line profile, while the other L-dip in the pair is in the blue side of the line profile. The separation $\Delta\lambda_{\mathrm{pair}}(k)$ between the two L-dips in the pair is

$$\Delta\lambda_{\mathrm{pair}}(k) = 2|\Delta\lambda_{\mathrm{dip}}(k)|. \tag{1.48}$$

Therefore, by identifying a pair (or several pairs) of the L-dips in the experimental profile, one can deduce the value of the plasma electron frequency ω_{pe} and then determine the electron density N_e from the definition of the plasma electron frequency,

$$\omega_{\mathrm{pe}} = (4\pi e^2 N_{\mathrm{e}}/\mathrm{m_e})^{1/2} = 5.641 \times 10^4[N_{\mathrm{e}}(\mathrm{cm}^{-3})]^{1/2}. \tag{1.49}$$

In high-density plasmas, each pair of the L-dips can be shifted a little to the red — according to Eq. (G.7) of Appendix G. (Physically this is due to the spatial non-uniformity of the ion microfield at the location of the radiator). However, the separation between the two L-dips in the pair is still given by Eqs. (1.48), (1.47). Therefore, the common shift of the two L-dips in the pair does not affect the method for measuring the electron density N_e from the separation of the two L-dips in any pair.

Here are just a couple of experimental examples — all other experimental examples are in Chapter 8 devoted to measuring the amplitude and the degree of anisotropy of Langmuir waves in plasmas. This is because L-dips serve simultaneously two different diagnostic purposes: to measure both the electron density and the parameters of Langmuir waves.

The first example is from the benchmark experimental study of L-dips performed by Kunze's group at the gas-liner pinch [34]. The gas-liner pinch operates as a modified z-pinch with a special gas inlet system, creating a very uniform plasma column. The implosion time was 3.0 μs, the decay time was 4.0 μs, the maximum electron density was $2.9 \times 10^{18}\,\mathrm{cm}^{-3}$, and the maximum electron temperature was 12.5 eV. The purpose of the experiment was to study possible L-dips in the profile of the hydrogen Ly-alpha line at different electron densities. The benchmark nature of the study was ensured by the fact that at each shot plasma parameters, including the electron density, were measured independent of the spectral line profiles by the coherent Thomson scattering.

Figure 1.3 shows the shot-by-shot comparison of the electron density N_e^{dips} obtained from the separation between the observed L-dips with the electron density N_e^{Thomson} measured by the coherent Thomson scattering. It is seen that there is an excellent agreement between N_e^{dips} and N_e^{Thomson} over the entire range of the experimental electron densities. Thus, the passive spectroscopic method for measuring N_e from L-dips is just as highly-accurate as the active method (much more complicated experimentally) using coherent Thomson scattering.

Figure 1.4 presents an example of the experimental spectrum from [34], showing two pairs of the L-dips. These L-dips are in the profiles of the lateral components of the Ly-alpha line, originating from the Stark sublevels (100) and (010), the sublevels being labeled

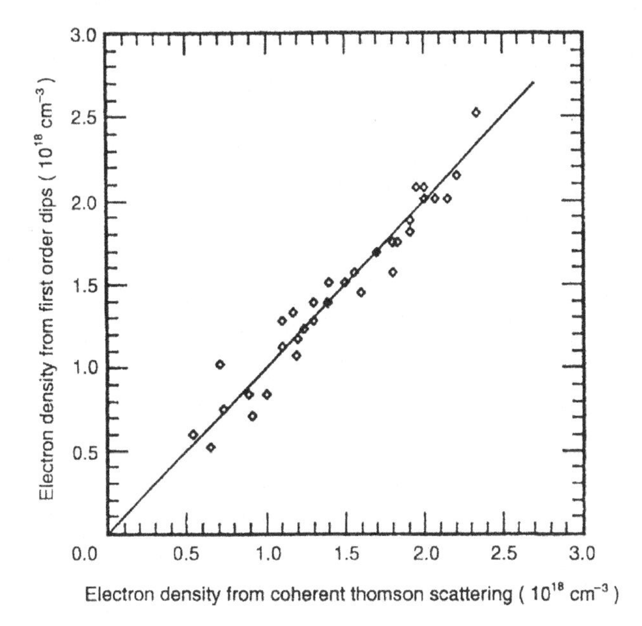

Fig. 1.3. Comparison of the electron density N_e, measured in each shot (at various N_e) from the coherent Thomson scattering with N_e from the L-dips observed in the experimental profiles of the hydrogen Ly-alpha line observed at the gas-liner pinch.

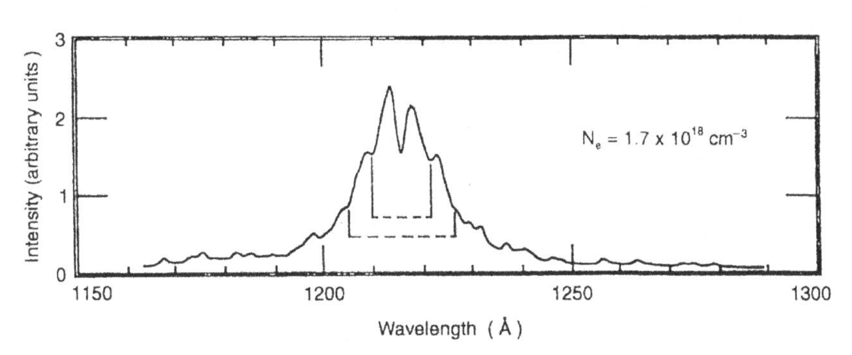

Fig. 1.4. Four experimental L-dips in the profile of the hydrogen Ly-alpha line at $N_e = 1.7 \times 10^{18}$ cm^{-3} observed at the gas-liner pinch. The theoretically expected positions of the L-dips, calculated using N_e obtained by the coherent Thomson scattering, are shown by pairs of the vertical lines connected by dashed horizontal lines.

by the parabolic quantum numbers. The first pair of the L-dips corresponds to the one-quantum resonance ($k = 1$ in Eq. (1.48)) and the second pair of the L-dips corresponds to the two-quantum resonance ($k = 2$ in Eq. (1.48)). The separation between the experimental L-dips was found to scale with the electron density N_e as $N_e^{1/2}$ — just as it was expected from the theory of the L-dips.

Another experimental example deals with a femtosecond laser-driven cluster-based plasma [35]. The experiments have been performed at Kansai Photon Science Institute, Japan Atomic Energy Agency (KPSI, JAEA, Japan) where two Ti:sapphire laser facilities (wavelength approximately 800 nm) were used. In the first experiment, the JLITE-X laser generated 40 fs pulses of the energy of 160 mJ, a contrast of 10^5. The laser beam was focused about 1.5 mm above the nozzle orifice by an off-axis parabola with the spot size of around 50 μm, which yielded the laser intensity of 4×10^{17} W/cm^2 in a vacuum. In the second experiment, the J-KAREN laser provided the laser pulses with a high contrast of 10^8–10^{10}, achieved using an additional saturable absorber and an additional Pokkels cell switch. The pulse duration was 40 fs at the pulse energy of about 800 mJ and the intensity of laser radiation in the focal spot with a diameter of 30 μm reached 3×10^{18} W/cm^2 in a vacuum. The clusters were created when a gas of a high initial pressure was expanded into vacuum through a specially designed supersonic nozzle, which consisted of three coaxial conical surfaces. CO_2 clusters with a diameter of about 0.5 μm for pure CO_2 (gas pressure 20 bar) and 0.22 μm for the mixed gas of 90% He + 10% CO_2 (gas pressure 60 bar) were produced in both experiments. The spatially resolved X-ray spectra have been obtained by employing a focusing spectrometer with spatial resolution FSSR-1D.

Figure 1.5 shows a typical experimental spectrum of the O VIII Ly-epsilon line obtained at the laser intensity 3×10^{18} W/cm^2 [35]. Two solid vertical lines correspond to the positions of two L-dips: one dip in the blue wing at −20 mA from the slightly shifted center of the line, another L-dip in the red wing at 37 mA from the slightly shifted position of the center line, as shown in Fig. 1.3 inset. The center of gravity of the two L-dips is shifted to the red by 9 mA, which

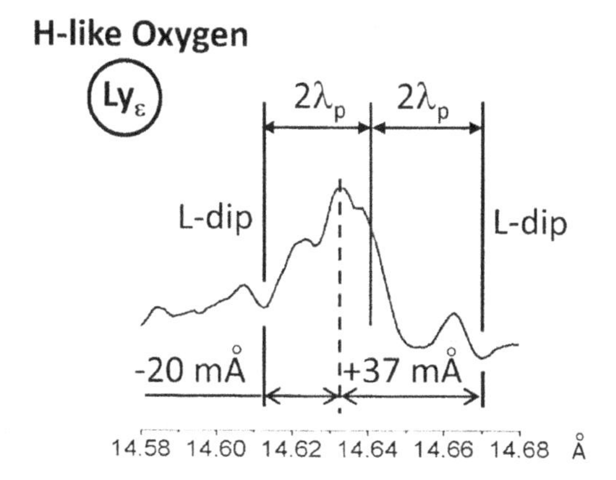

Fig. 1.5. The typical experimental spectrum of the O VIII Ly-epsilon line observed in the femtosecond laser-driven cluster-based plasma (the laser intensity 3×10^{18} W/cm^2).

according to the theory presented in Appendix G is because at high electron densities, the spatial non-uniformity of the ion microfield becomes significant.

The superposition of a bump–dip–bump structure in the blue wing with a significantly inclined spectral profile created a secondary minimum at about 14.63 Å of no physical significance. As for the L-dip in the red wing, its near bump is clearly visible, but the far bump is only faintly outlined because it practically merged with the noise. Here and below the "near" (or "far") bump means the bump closer to (or further from) the line center with respect to the central minimum of the dip.

The two L-dips are separated from each other by $4\lambda_p$, where $\lambda_{pe} = \omega_{pe}\lambda_0^2/(2\pi c)$. They are one-quantum resonance dips in the profiles of the two most intense lateral components of the Ly_ε line, originating from the Stark sublevels (311) and (131), the sublevels being labeled by the parabolic quantum numbers. The electron density deduced from the separation of the two L-dips was $N_e = 5.0 \times 10^{20}$ cm^{-3}.

Finally, we note that in the entire scope of experiments at various plasma machines, the observed L-dips allowed measuring the electron density in a huge range of $N_e = (3 \times 10^{13} - 4 \times 10^{22}\,\text{cm}^{-3})$. Some of these experiments are presented in Chapter 8.

1.5 Combining Measurements of Stark Widths and Stark Shifts

Stark shifts of hydrogenic spectral lines are about one order of magnitude smaller than the corresponding Stark widths. Therefore, measuring the Stark shifts only cannot be recommended as a reliable diagnostic of the electron density. However, in high-density plasmas, simultaneous measurements of both the Stark width and the Stark shift enhances the reliability of the diagnostic of the electron density compared to measuring only the Stark width. The underlying theory of the Stark shifts can be found, e.g., in Secs. 2.6, 5.2 and 6.1 of book [5].

We illustrate this by an example of the experiment [36], which was one of the applications of the Laser-Induced Breakdown Spectroscopy (commonly abbreviated as LIBS). In this experiment, a continuum YG680S-10 Nd:YAG laser, operated at 1064 nm and 10 Hz, was used to generate the plasma. The laser beam, with 150-mJ energy per pulse and 7.5-ns pulse duration, was focused typically to 1400 Gw/cm^2 in a pressure cell that was filled to a pressure of 810 ± 25 Torr subsequent to an evacuation of the cell with a diffusion pump. Stark broadened profiles of the H-beta line, as well as Stark broadened and red-shifted profiles of the H-alpha line, were measured at different time delays after the laser shot. The temperatures were determined from Boltzmann plots (i.e., from the ratios of the absolute intensities of the hydrogen spectral lines in the range of 6600–100,000 K). Large uncertainties resulted from the extreme broadening and a partial overlap of the H-alpha, H-beta, and H-gamma spectral lines early in the plasma decay. In fact, the temperatures were determined only for time delays longer than 0.25 μs following the optical breakdown.

Figure 1.6 shows the comparison of the electron densities obtained from the H-alpha and H-beta widths with the electron densities obtained from the H-alpha shifts using the GT. It is seen that the measurements of the Stark shifts effectively complements the measurements of the Stark widths and enhances the accuracy of determining N_e: at each time delay, the actual N_e should be at the overlap of the error bars for the width and shift measurements.

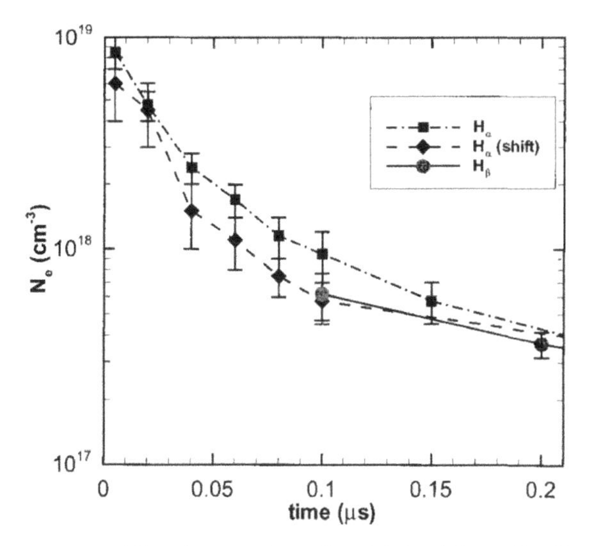

Fig. 1.6. Comparison of the electron densities obtained from the H_α and H_β line widths and from the H_α line shifts, following laser-induced optical breakdown in a cell containing H_2 gas at a pressure of 810 Torr. Electron densities are inferred by using the GT. The Nd:YAG laser radiation was focused to typically 1400 Gw/cm^2, which was more than 10 times larger than the nanosecond breakdown threshold of hydrogen gas for 1064-nm radiation.

It should be emphasized that at the time delays (after the laser shot) $\Delta t < 0.1\mu$s, corresponding to the electron densities up to 10^{19} cm^{-3}, it was impossible to measure the width of the H-beta line because it was so broad that it practically merged with the H-gamma line.

We note that attempts to deduce the electron density $N_{e,\alpha}$ from the Stark widths of the H-alpha line and simultaneously the electron density $N_{e,\beta}$ from the Stark widths of the H-beta line using the CT — sometimes called Griem's theory — yielded a huge inconsistency for time delays $\Delta t > 0.5\mu$s. Namely, the ratio $N_{e,\alpha}/N_{e,\beta}$, obtained by using Griem's theory, reaches 2.0 and 5.4 for time delays Δt of 1.5 μs and 2.5 μs, respectively. In distinction, the ratio $N_{e,\alpha}/N_{e,\beta}$, obtained by using the GT, is close to unity even for the time delay $\Delta t = 1.5\mu s$ (as well as for all smaller time delays) and reaches the value of 2 only for the time delay $\Delta t = 2.5\mu s$. This is illustrated in Fig. 1.7.

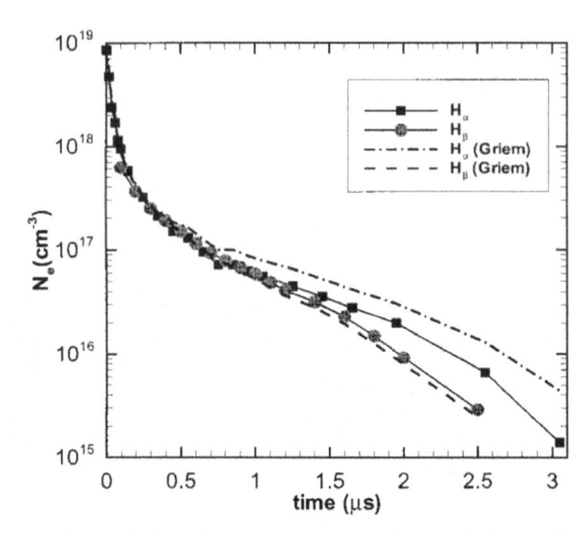

Fig. 1.7. Comparison of the electron densities obtained from H_α and H_β line profiles, following laser-induced optical breakdown in a cell containing H_2 gas at a pressure of 810 Torr. Electron densities infer, inferred by using the GT are shown by solid lines, while the electron densities deduced by using the CT (labeled Griem's theory) are shown by dashed and dashed-dotted lines.

The fact that in this experiment, the GT yielded self-consistent results for N_e ranging from $10^{19}\,\mathrm{cm}^{-3}$ (for $\Delta t = 0.005\,\mu s$) down to $3 \times 10^{16}\ \mathrm{cm}^{-3}$ (for $\Delta t = 1.5\,\mu s$) is primarily due to its accurate treatment of the ion dynamics, which was totally disregarded in Griem's theory. At the lower end of the experimental density range, the ion dynamics contributes significantly to the width of the H_α line without affecting much the width of the H_β line.

As the density further decreased by an order of magnitude, the experimental width of the H_α line has been affected by an optical depth τ of this line. The values of $\tau > 1$ for $\Delta t \geq 2.5\,\mu s$ could be explained by the expansion of the plasma column and thus by the increase of the optical path at these time delays.

1.6 Using the Asymmetry of Hydrogen Spectral Lines — Especially for Bypassing the Optical Thickness Problem

The asymmetry of the hydrogen-like lines in dense plasmas is primarily caused by the spatial non-uniformity of the ion microfield at the location of the radiator [37, 38]. The intensity of the blue maximum is greater than the red, and the positions of the intensity maxima are asymmetrical with respect to the unperturbed line center.

For the theoretical description of the effects of the non-uniformity of the ion microfield on spectral line profiles, it is necessary to go beyond the dipole approximation (in the multipole expansion of the potential of the perturbing ion at the location of the atomic electron) and to take into account at least the quadrupole term.

In the dipole approximation, any hydrogenic spectral line splits up into pairs of Stark components. Within each pair, the components have equal intensities and are symmetrically located with respect to the unperturbed position of the spectral line. The allowance for the quadrupole interaction results in a slight redistribution of intensities in the pair and in an additional shift of both components in the pair by the same amount to the same side.

The theory behind this asymmetry-based method for measuring the electron density was first briefly presented in the paper [39] and later in more detail in the paper [40]. The main practical results are presented below.

The degree of asymmetry ρ is defined as follows:

$$\rho = (I_B - I_R)/[0.5(I_B + I_R)], \tag{1.50}$$

where I_B and I_R are integrated intensities of the blue and red parts of the spectral line profile, respectively. The boundary between the blue of red parts of the experimental profile (i.e., the reference point, from which the integrals I_B and I_R are calculated) is the center of gravity of the profile. In the paper [41], it has been proven analytically that the quadrupole interaction, despite shifting Stark components, does not shift the center of gravity of the line profile.

The practical formula for deducing the electron density N_e from the experimental degree of asymmetry ρ is the following,

$$\rho = 0.46204(N_e/10^{21}\,\mathrm{cm}^{-3})^{1/3}(Z_p^{2/3}/Z_r^2)\sum_{k>0} I_k^{(0)}\varepsilon_k^{(1)}. \tag{1.51}$$

Here, Z_p and Z_r are charges of the perturbing ions and of the radiator, respectively; the quantities $I_k^{(0)}$ and $\varepsilon_k^{(1)}$ for various hydrogenic spectral lines were tabulated in the paper [38].

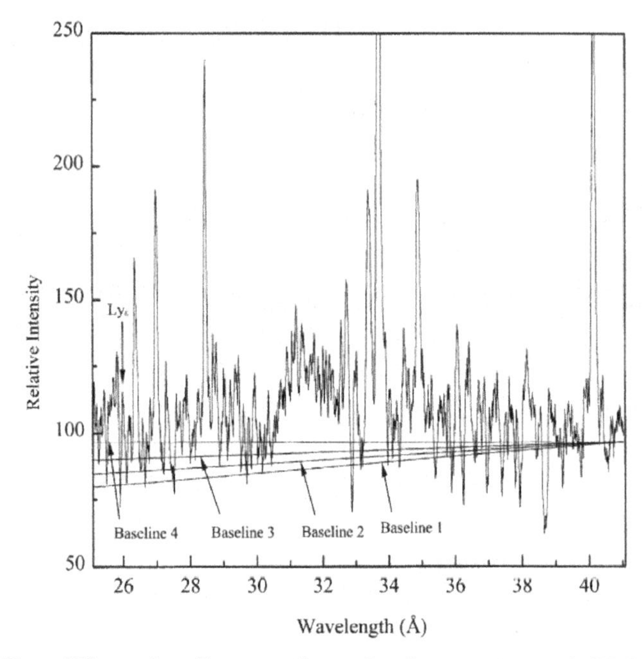

Fig. 1.8. Four different baselines are shown for the experimental C VI spectrum, observed at the vacuum spark device, starting from the bottom of the Ly-epsilon line and then incremented forward.

As an example, Fig. 1.8 shows the experimental spectrum of C VI observed in the vacuum spark device [40]. For the analysis of asymmetry, four different baselines were chosen starting from the bottom of the Ly-epsilon line and then incremented forward.

Figure 1.9 shows the experimental profile of the C VI Ly-delta line observed at the vacuum spark device [40] (with a choice of baseline 2). The center of gravity is indicated by the dashed line. There is a significant asymmetry: the integrated intensity of the blue part of the profile is about 16% greater than the integrated intensity of the red part of the profile.

The asymmetry parameter ρ varied from 0.147 to 0.183 for the four baselines shown in Fig. 1.8. With the help of Eq. (1.51), the electron density range of $(1.8–3.8)\times 10^{20}$ cm^{-3} was determined.

Another example of using the asymmetry for determining the electron density relates to the experiment, described in the paper

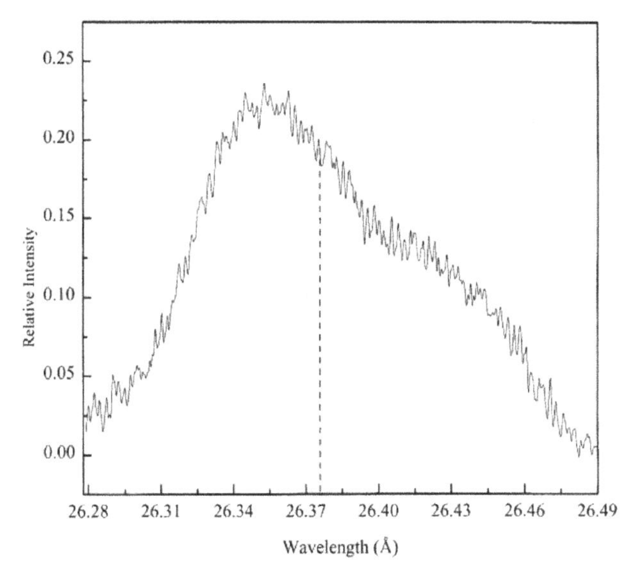

Fig. 1.9. The experimental line profile of the Ly-delta line for baseline 2 observed at the vacuum spark device. The asymmetry between the blue and red wings is very noticeable. The integrated intensity of the blue part of the profile is about 16% greater than the red one. The center of gravity of the profile is given by the dashed line.

[42]. In this experiment, performed in frames of the Laser-Induced Breakdown Spetroscopy (LIBS), a Nd-YAG laser beam was focused into a spot size of about 50 μm in diameter, directly in the laboratory air. A small concentration of naturally occurring water in the air caused the appearance of the observed hydrogen lines of the Balmer series — specifically H-alpha and H-beta lines were recorded. The asymmetry was studied using the experimental H-beta profiles.

Figure 1.10 presents the profile of the H-beta line observed 5 μs after the laser shot, gate width 1.0 μs. (Also shown is fitting the experimental profile by a superposition of two Voigt profiles.) The experimental profile shows a significant asymmetry.

Table 1.2 shows the experimental asymmetry parameter ρ and the inferred electron densities N_e for time delays between 5 and 10 μs (gate width 1.0 μs). The column "N_e from ρ" shows the most probable values of N_e for each time delay. The column "N_e range" shows

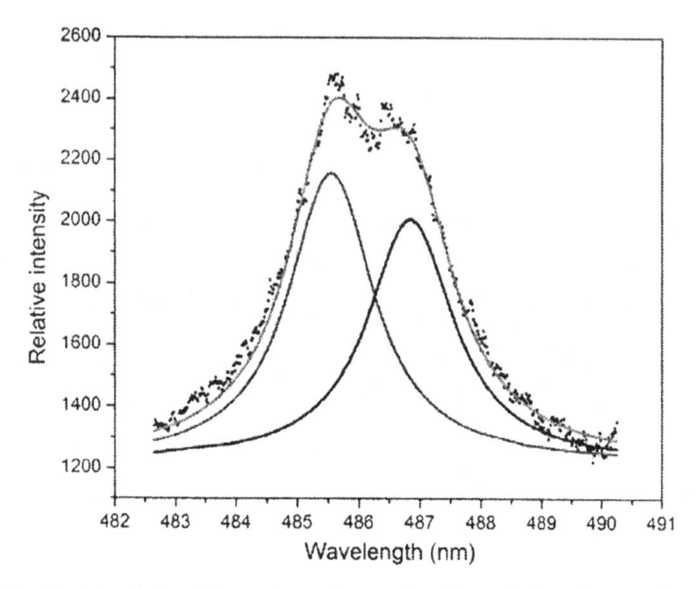

Fig. 1.10. Profile of the H-beta line observed in laser-induced breakdown spectroscopy experiment. The spectrum was observed 5 μs after the laser shot. Also shown is fitting the experimental profile by a superposition of two Voigt profiles.

Table 1.2. Asymmetry parameter ρ and inferred electron densities for time delays between 5 and 10 μs.

$\Delta\tau\mu$s	ρ %	Ne from $\rho 10^{17}$/ccm	Ne range 10^{17}/ccm
5	2.9 ± 1.5	3.3	0.38–11
6	2.5 ± 1.3	2.2	0.24–7.7
7	2.9 ± 1.5	3.2	0.38–11
8	1.7 ± 0.8	0.67	0.10–2.2
9	0.6 ± 0.5	0.038	0.0004–0.19
10	1.0 ± 0.6	0.13	0.009–0.57

the range of the N_e values (for each time delay) corresponding the experimental uncertainty in measuring the asymmetry parameter ρ.

Finally, we note that the primary advantage of this method of measuring the electron density is that it works well even when the central part of the experimental profile is affected by self-absorption [39]. This is because the asymmetry parameter ρ is controlled primarily by the line wings and the line wings are optically thin.

1.7 Using Helium and Lithium Spectral Lines

Among helium and lithium spectral lines, the most appropriate for measuring the electron density are the lines having a dipole-forbidden component (F) — sometimes several dipole-forbidden components — next to the dipole-allowed line (A). The physics behind this phenomenon can be explained by the example of the most frequently used radiative transition 2P – 4D. Relatively close to the level 4D lies the level 4F (the separation between levels 4D and 4F is much smaller than the separation between the levels 4D and 4P). At the absence of an electric field, the radiative transition 2P – 4F is dipole-forbidden, so that only the allowed line 2P – 4D is observed. Under a static (or quasistatic) electric field, such as the quasistatic part of the ion microfield in plasmas, the wave function of the level 4F acquires an admixture of the wave function of the level 4D. As a result, the forbidden component, corresponding to the radiative transition 2P – 4F, starts showing up. The higher is the electron density, the greater is the ratio F/A of the intensities of the forbidden and allowed components, thus serving as a measure of the electron density.

Two other parameters can be also used for measuring the electron density: the FWHM of the allowed component and the separation(s) between the peaks of the forbidden and allowed components. The reason why the separation between the peaks increases with the growth of the electron density is because the electric field shifts the A- and F-components away from each other.

In dense plasmas, the allowed component could be optically thick. This would compromise measurements of the electron density based of the FWHM or on the F/A ratio. However, the separation(s) between the peaks would not be affected by the self-absorption. Therefore, the separation between the peaks is the most reliable tool for measuring the electron density.

Spectral lines corresponding to the radiative transition $2P - 4D$, such as He I 447.1 nm (triplet), He I 492.2 nm (singlet), and Li I 460.3 nm, are the most frequently used. They allow measuring the electron density starting from $N_e \sim 10^{15}\,\mathrm{cm}^{-3}$ and higher. For measuring lower electron densities, one should use spectral lines corresponding to the radiative transition $2P-5D$, such as He I 402.6 nm (triplet), He I 438.8 nm (singlet), and Li I 413.3 nm. In this case, the ion microfield would intermix the wave functions of three closely lying levels $5D$, $5F$, and $5G$, so that two forbidden components ($2P-5F$ and $2P-5G$) would show up.

Below for the spectral lines He I 447.1 nm, He I 492.2, and Li I 460.3 nm, we present practical formulas (derived by various authors) connecting the peaks separation s with the electron density N_e.

1.7.1 *He I 447.1 nm*

Czernichowski and Chapelle [43] suggested the following relation for the range of $N_e = (0.3\text{–}30) \times 10^{15}\,\text{cm}^{-3}$:

$$\log N_e[\text{cm}^{-3}] = 17.056 + 1.586(\log s[\text{nm}] - 0.156) + 0.225(\log s[\text{nm}] - 0.156)^2. \quad (1.52)$$

Perez *et al.* [44] suggested the following formula for the range of $N_e = (1.5\text{–}15) \times 10^{16}\,\text{cm}^{-3}$:

$$\log N_e[\text{cm}^{-3}] = 15.790 + 1.0436(\log s[\text{nm}] - 0.1557)^{1/2}. \quad (1.53)$$

For the range that includes higher electron densities $N_e = (10^{15}\text{–}10^{18})\,\text{cm}^{-3}$ and specifically for He perturbed by He^+ ions, Ivkovic *et al.* [45] produced the following relation:

$$\log N_e[\text{cm}^{-3}] = 15.5 + \log\{(\text{s}[\text{nm}]/0.149)^{b(T)} - 1\}, \quad b(T) = 1.46 + 8380/\text{T}^{1/2}, \quad (1.54)$$

where $T[K]$ is the electron temperature. In Eqs. (1.52)–(1.54), log(...) stands for the decimal logarithm.

1.7.2 *He I 492.2 nm*

For the range of electron densities $N_e = (3 \times 10^{14}\text{–}10^{18})\,\text{cm}^{-3}$ and specifically for He perturbed by He^+ ions, Ivkovic *et al.* [46] suggested the following formula:

$$\log N_e[\mathrm{cm}^{-3}] = 15.3065 + 1.141\ \log\{(s[\mathrm{nm}]/0.13187)^{b(T)} - 1\},$$
$$b(T) = 1.25 + 994/T^{1/2}, \tag{1.55}$$

where $T[K]$ is the electron temperature.

In Eqs. (1.52)–(1.55), $\log(\cdots)$ stands for the decimal logarithm.

1.7.3 *Li I 460.3 nm*

For the range of electron densities $N_e = (0.5-11)\times10^{18}\ \mathrm{cm}^{-3}$ and the electron temperature $T = 5800\,\mathrm{K}$, and specifically for Li perturbed by Li^+ ions, Cvejic *et al.* [47] suggested the following relation:

$$s[\mathrm{nm}] = [0.1059^2 + 0.036(N_e/10^{16}\ \mathrm{cm}^{-3})^{2/3} + 0.041(N_e/10^{16}\ \mathrm{cm}^{-3})^{4/3}]^{1/2}. \tag{1.56}$$

References

[1] V.V. Sobolev, *Course in Theoretical Astrophysics*, NASA, Washington, DC (1969).

[2] D. Mihalas, *Stellar atmospheres*, W.H. Freeman, San Francisco (1970).

[3] E.G. Gibson, *The Quiet Sun*, NASA, Washignton, DC (1973).

[4] E. Oks, *Stark Broadening of Hydrogen and Hydrogenlike Spectral Lines in Plasmas: The Physical Insight*, Alpha Science International, Oxford, United Kingdom (2006).

[5] M.A. Gigosos and V. Cardenoso, *J. Phys. B* **29** 4795 (1996).

[6] M.A. Gigosos, M.A. Gonzalez and V. Cardenoso, *Spectrochimica Acta B* **58** 1489 (2003).

[7] D.M. Surmick and C.G. Parigger, *Intern. Review Atom. Mol. Phys.* **5** 73 (2014).

[8] M. Ivkovic, N. Konjevic and Z. Pavlovic, *J. Quant. Spectr. Rad. Transfer* **154** 1 (2015).

[9] L. Pardini, S. Legnaioli, G. Lorenzetti *et al.*, *Spectrochimica Acta B* **88** 98 (2013).

[10] C.G. Parigger, D.M. Surmick, G. Gautam and A.M. El Sherbini, *Optics Lett.* **40** 3436 (2015).

[11] S. Büscher, S. Glenzer, T. Wrubel and H.-J. Kunze, *J. Phys. B: Atom. Mol. Opt. Phys.* **29** 4107 (1996).

[12] B.L. Welch, H.R. Griem, J. Terry, C. Kurz, B. LaBombard, B. Lipschultz, E. Marmar and J. McCracken, *Phys. Plasmas* **2** 4246 (1995).
[13] U. Wagner, M. Tatavakis, A. Gopal *et al.*, *Phys. Rev. E* **70** 026401 (2004).
[14] S. Fujioka, Z. Zhang, K. Ishihara *et al.*, *Scientific Reports* **3** 1170 (2013).
[15] L.J. Silvers, *Phil. Trans. R. Soc. A* **366** 4453 (2008).
[16] A.A. Schekochihin and S.C. Cowley, in: S. Molokov, R. Moreau and H.K. Moffett (Eds.) *Magnetohydrodynamics — Historical Evolution and Trends*, Springer, Berlin, p. 85 (2007).
[17] A. Derevianko and E. Oks, *Phys. Rev. Lett.* **73** 2059 (1994).
[18] H.R. Griem, *Spectral Line Broadening by Plasmas*, Academic, New York (1974).
[19] V.S. Lisitsa, *Sov. Phys. Uspekhi* **122** 603 (1977).
[20] E. Oks, *J. Quant. Spectrosc. Rad. Transfer* **171** 15 (2016).
[21] N.H. Brooks, S. Lisgo, E. Oks, D. Volodko, M. Groth, A.W. Leonard and the DIII-D Team, *Plasma Phys. Reports* **35** 112 (2009).
[22] E. Oks, R.D. Bengtson and J. Touma, *Contributions to Plasma Phys.* **40** 158 (2000).
[23] U. Feldman and G.A. Doschek, *Astrophys. J.* **212** 913 (1977).
[24] S. Ferri, A. Calisti, R. Stamm, B. Talin and R.W. Lee, *AIP Conf. Proc.* **467** 115 (1999).
[25] R.D. Bengtson, J.D. Tannich and P. Kepple, *Phys Rev. A* **1** 532 (1970).
[26] C. Stehle and R. Hutcheon, *Astron. Astrophys. Suppl. Ser.* **140** 93 (1999).
[27] E. Stambulchik and Y. Maron, *J. Phys. B: Atom. Mol. Opt. Phys.* **41** 095703 (2008).
[28] E. Oks, *Phys. Rev. E* (*Rapid Communications*) **60** 2480 (1999).
[29] D.R. Inglis and E. Teller, *Astrophys. J.* **90** 439 (1939).
[30] J. Hey, *J. Phys. B: Atom. Mol. Opt. Phys.* **46** 175702 (2013).
[31] H.R. Griem, *Plasma Spectroscopy*, McGraw-Hill, New York (1964).
[32] M. Born, *The Mechanics of the Atom*, Bell and Sons, London (1927).
[33] Yu.N. Demkov, B.S. Monozon and V.N. Ostrovskii, *Sov. Phys. JETP* **30** 775 (1970).
[34] E. Oks, St. Böddeker and H.-J. Kunze, *Phys. Rev. A* **44** 8338 (1991).
[35] E. Oks, E. Dalimier, A.Ya. Faenov *et al.*, *J. Phys. B: Atom. Mol. Opt. Phys.* (*Rapid Communications*) **47** 221001 (2014).
[36] C.G. Parigger, D.H. Plemmons and E. Oks, *Appl. Optic.* **42** 30 (2003).
[37] L.P. Kudrin and G.V. Sholin, *Sov. Phys. Doklady* **7** 1015 (1963).
[38] G.V. Sholin, *Opt. Spectrosc.* (USSR) **26** 275 (1969).

[39] L.G. Golubchikov, E. Oks and G.V. Sholin, in Eleventh Intern. Conf. on Phenomena in Ionized Gases, I. Stoll, L. Pekarek, & L. Laska Prague (Eds.) (1973).

[40] N.K. Podder, E.J. Clothiaux and E. Oks, *Contrib. Plasma Phys.* **39** 529 (1999).

[41] E. Oks, *J. Quant. Spectr. Rad. Transfer* **58** 821 (1997).

[42] C.G. Parigger, L.D. Swafford, A.C. Woods, D.M. Surmick and M.J. Witte, *Spectrochimica Acta Part B* **99** 28 (2014).

[43] A. Czernichowski and J. Chapelle, *J. Quant. Spectrosc. Rad. Transfer* **33** 427 (1985).

[44] C. Perez, I. de la Rosa, J.A. Aparicio, S. Mar and M.A. Gigosos, *Jpn. J. Appl. Phys.* **35** 4073 (1996).

[45] M. Ivkovic, M.A. Gonzalez, S. Jovicevic, M.A. Gigosos and N. Konjevic, *Spectrochimica Acta Part B* **65** 234 (2010).

[46] M. Ivkovic, M.A. Gonzalez, N. Lara, M.A. Gigosos and N. Konjevic, *J. Quant. Spectr. Rad. Transfer* **127** 82 (2013).

[47] M. Cvejic, E. Stambulchik, M.R. Gavrilovic, S. Jovicevic and N. Konjevic, *Spectrochimica Acta Part B* **100** 86 (2014).

Chapter 2

Temperatures

2.1 Electron Temperature

The electron temperature could be measured by methods that are based on various intensity ratios, rather than on lineshapes. These methods are described, e.g., in Sec. 10.3.2 of the book [1], such as:

— ratios of integrated intensities of several spectral lines,
— line-to-continuum ratios,
— ratios of intensities of recombination continua,
— short- and long-wavelength continuum,
— ratio of ionization stages.

However, these methods depend on model assumptions about the plasma state (such as, e.g., local thermodynamic equilibrium, or partial local thermodynamic equilibrium, or the corona model, or the collisional-radiative model, etc.) — the assumptions that not always are reliable — in distinction to the diagnostic methods using lineshapes.

For measuring the electron temperature based on shapes of hydrogenic spectral lines, one can use the fact that hydrogenic lines having the central Stark components (such as, alpha- and gamma-lines) are much more affected by the electron temperature than hydrogenic lines that do not have the central Stark component (such as, beta- and delta-lines). Let us consider, for example, the situation where the H-alpha and H-beta lines are observed simultaneously.

In this situation, by using one of the Eqs. (1.11)–(1.15) related to the H-beta line, it is possible to determine the electron density N_e (the electron temperature does not enter Eqs. (1.11)–(1.15)). Then by substituting the obtained N_e value in Eq. (1.16) related to the H-alpha line, it is possible to determine the electron temperature — since the FWHM of the H-alpha line in Eq. (1.16) depends both on the electron density and the electron temperature.

Another possibility concerns helium lines having forbidden components. The separation between the peaks of allowed and forbidden components depends both on the electron density and the electron temperature, but those dependencies are different for different helium lines. For example by observing simultaneously the lines He I 447.1 nm and He I 492.2 nm, and using Eq. (1.54) for the line He I 447.1 nm and Eq. (1.55) for the line He I 492.2 nm, it is possible to determine both the electron density and the electron temperature.

The above methods for measuring the electron temperature based on the lineshapes are adequate for high-density plasmas. In low-density plasmas, for measuring both the electron density and the electron temperature, one could observe a series of highly-excited hydrogen/deuterium spectral lines and measure their FWHM — as, e.g., in the experiments described in the papers [12, 21] and discussed in Sec. 1.3 (see, e.g., Fig. 1.2).

2.2 Ion and Atomic Temperatures

In non/weakly magnetized plasmas, ion and atomic temperatures can be determined from Doppler-broadened profiles of selectively chosen spectral lines of ions and atoms, respectively. One should choose non-hydrogenic spectral lines because they have smaller contribution of the Stark broadening (SB) than hydrogenic spectral lines. For non-hydrogenic spectral lines, the SB usually results into the Lorentzian component of the profile while the Doppler broadening — into the Gaussian component of the profile. Therefore, the observed profiles can be represented by the Voigt function, which is a convolution of the Gaussian and Lorentzian profiles. An approximate, but highly-accurate formula relating the FWHM of the Voigt profile $\Delta\lambda_{1/2\mathrm{V}}$ to both the FWHM of the Gaussian component $\Delta\lambda_{1/2\text{G}}$ and the FWHM of the Lorentzian component $\Delta\lambda_{1/2\text{L}}$, was suggested by Olivero and Longbothum [2] as follows:

$$\Delta\lambda_{1/2\text{V}} = 0.5346 + (0.2166\Delta\lambda_{1/2\text{L}}^2 + \Delta\lambda_{1/2\text{G}}^2)^{1/2}. \qquad (2.1)$$

The Lorentzian FWHM $\Delta\lambda_{1/2\text{L}}$ is controlled primarily by the electron density, while the Gaussian FWHM $\Delta\lambda_{1/2\text{G}}$ is controlled by the ion temperature (in the case of radiating ions) or by the atomic temperature (in the case of the radiating atoms):

$$\begin{aligned}\Delta\lambda_{1/2\text{G}} &= \lambda_0[2(\ln 2)^{1/2} v_\text{T}/c] \\ &= \lambda_0[7.715 \times 10^{-5}(T[\text{eV}]/M[\text{amu}])^{1/2}. \qquad (2.2)\end{aligned}$$

Here, v_T, T and M are the mean thermal velocity, the temperature, and the mass of the radiators, respectively; amu stands for the atomic mass unit, which is practically the same as the proton mass.

The formula (2.1) for the FWHM of the Voigt profile has the accuracy of 0.02%. By measuring the FWHM of two different non-hydrogenic lines and by using formula (2.1) for each of the line, it is possible to determine both the Lorentzian FWHM (and thus the electron density) and the Gaussian FWHM (and thus the corresponding temperature).

In strongly magnetized plasmas, the situation becomes more complicated, as described in detail in Appendix E.2. Radiators moving with the velocity $\mathbf{v}$ across the magnetic field $\mathbf{B}$ experience a Lorentz electric field $\mathbf{E}_\text{L} = \mathbf{v}\times\mathbf{B}/\text{c}$ in addition to other electric fields. In plasmas, the radiator velocity $\mathbf{v}$ has a distribution — therefore, so does the Lorentz field, thus making an additional contribution to the broadening of spectral lines. Lorentz and Doppler broadenings cannot be taken into account by a convolution: they entangle in a more complicated way.

Analytical results for Lorentz–Doppler profiles of hydrogen/deuterium spectral lines, obtained in papers [3, 4] and presented in Appendix E.2, demonstrate the complexity of the profiles — as illustrated, e.g., in Figs. E.10, E.11, E.15–E.17. In view of this complexity, it seems advisable to use a simple diagnostic method based on the principal quantum number $n_{\max}$ of the last observed line in the spectral series of hydrogen/deuterium lines, such as, Lyman, or Balmer, or Paschen lines (though typically Balmer lines are used). For the applicability of the method to measuring the temperature, several conditions should be met. First, the magnetic field should exceed the following critical value,

$$B_{\mathrm{c}}\ (\mathrm{Tesla}) = 4.69 \times 10^{-7} N_e^{2/3}/[T(K)]^{1/2}, \tag{2.3}$$

so that the average Lorentz field exceeds the most probable ion microfield. Second, the ratio of the Zeeman width of hydrogen lines $\Delta\omega_Z$ to the corresponding "halfwidth" of the n-multiplet due to the Lorentz broadening $\Delta\omega_L$,

$$\Delta\omega_Z/\Delta\omega_L = 5680/[n(n-1)T(K)^{1/2}], \tag{2.4}$$

should be smaller than unity. Third, the ratio of the FWHM $(\Delta\omega_L)_{1/2}$ caused by the Lorentz broadening to the FWHM $(\Delta\omega_D)_{1/2}$ caused by the Doppler broadenings,

$$(\Delta\omega_L)_{1/2}/(\Delta\omega_D)_{1/2} = n_\alpha^2 n_\beta^2 B\ (\mathrm{Tesla})/526, \tag{2.5}$$

should be greater than unity.

As discussed in detail in Sec. 3 of Chapter 1, in this situation, the value of $n_{\max}$ is controlled by the product B^2T, where B is the magnetic field and T is the temperature of radiators. Here are the practical formulas (presented also in Sec. 3 of Chapter 1).

For the observation perpendicular to **B**:

$$n_{\max}^{10} B^2 T(\mathrm{K}) = 2.79 \times 10^{18} M/M_{\mathrm{p}} \quad \text{or}$$
$$n_{\max}^{10} B^2 T(\mathrm{eV}) = 2.40 \times 10^{14} M/M_{\mathrm{p}}. \tag{2.6}$$

For the observation parallel to **B**:

$$n_{\max}^{10} B^2 T(K) = 1.78 \times 10^{18} M/M_{\mathrm{p}} \quad \text{or}$$
$$n_{\max}^{10} B^2 T\ (\mathrm{eV}) = 1.54 \times 10^{14} M/M_{\mathrm{p}}. \tag{2.7}$$

For the "isotropic" case,

$$n_{\max}^{10} B^2 T(K) = 2.38 \times 10^{18} M/M_p \quad \text{or}$$
$$n_{\max}^{10} B^2 T(eV) = 2.05 \times 10^{14} M/M_p, \tag{2.8}$$

the "isotropic" observation corresponding to the situation where along the line of sight there are regions with various directions of the magnetic field, which could be sometimes the case in astrophysics. In Eqs. (2.6)–(2.8), M and M_p are the atomic and proton masses, respectively.

Thus, if the magnetic field is known (being measured, e.g., by methods described in Chapter 3), then Eqs. (2.6)–(2.8) allow determining the radiators temperature T.

References

[1] H.-J. Kunze, *Introduction to Plasma Spectroscopy*, Springer, Berlin (2009).
[2] J.J. Olivero and R.L. Longbothum, *J. Quant. Spectrosc. Rad. Transfer* **17** 233 (1977).
[3] E. Oks, *Intern. Review of Atom. and Mol. Phys.* **4** 105 (2013).
[4] E. Oks, *J. Quant. Spectrosc. Rad. Transfer* **156** 24 (2015).

Chapter 3

Magnetic Field

Typically, in plasmas, the magnetic-field-caused Zeeman splitting of spectral lines is smaller than the Stark broadening (SB) and/or the Doppler broadening. Specifically, in dense plasmas, the Stark splitting in the most probable ion field F',

$$\omega_{F'} = 3n\hbar F'/(2Z_r m_e e), \quad F' \sim 4Z_p^{1/3} e N_e^{2/3}, \tag{3.1}$$

is much greater than the Zeeman splitting,

$$\omega_B = eB/(m_e c). \tag{3.2}$$

Here, n is the principal quantum number, Z_r is the nuclear charge of the radiator, and Z_p is the charge of perturbing ions.

In these situations, measurements of the magnetic field require a polarization analysis. One of the effective methods for this purpose was developed in Ref. [1]. The authors of Ref. [1] suggested using the polarization difference profile

$$P(\Delta\omega) = I_{\text{par}}(\Delta\omega) - I_{\text{perp}}(\Delta\omega) \tag{3.3}$$

of hydrogenic spectral lines having an intense *central* Stark component, such as alpha- and gamma-lines. (In Eq. (3.3), $I_{\text{par}}(\Delta\omega)$ and $I_{\text{perp}}(\Delta\omega)$ are profiles with the polarization parallel and perpendicular to the magnetic field **B**, respectively.) This was a significant advancement compared to the earlier work by Demura and Lisitsa [2], who suggested using the polarization profile $P(\Delta\omega)$ of hydrogenic spectral lines having only lateral Stark components, but not a central Stark component, such as beta- and delta-lines. The physics behind the advance from Ref. [1] is as follows.

For lateral Stark components, the polarization effect is relatively small:

$$\max P(\Delta\omega) \sim \omega_B^2/\omega_{F'}^3, \quad \omega_B/\omega_{F'} = \varepsilon \ll 1. \tag{3.4}$$

However, for the central Stark components,

$$\max P_c(\Delta\omega) \sim 1/\omega_B, \tag{3.5}$$

so that the polarization effect is greater than for lateral stark components by a factor of $1/\varepsilon^3 \gg 1$. In dense plasmas, the typical value of this factor is $1/\varepsilon^3 \sim 10^3$ or even higher.

Physically, this result can be understood as follows. First, if we would calculate the frequency-integrated absolute values, of the polarization difference profiles

$$p_c = \int_0^\infty d(\Delta\omega)|P_c(\Delta\omega)|,$$
$$p = \int_0^\infty d(\Delta\omega)|P(\Delta\omega)| \tag{3.6}$$

for a central and a lateral component, respectively, and compare them, we could find that

$$p_c \sim 1, \quad p \sim \varepsilon^2 \ll 1. \tag{3.7}$$

This is because for the central component, the quantity p_c is controlled only by the magnetic field, which has a highly anisotropic (δ-function-type) angular distribution. However, for the lateral component, the quantity p is controlled, in the first approximation, by the angular distribution of the electric field, which is isotropic. Therefore, in this approximation, the quantity p vanishes. For the lateral component, the magnetic field enters the game only in the next approximation and results into a small (compared to unity) contribution which is quadratic with respect to B. The property that p is proportional to B^2 could be expected in advance since the quantity p should be invariant with respect to the reversal of the direction of the magnetic field.

Second, we should take into account that the polarization difference profile of the central component $P_c(\Delta\omega)$ has the width of the order of ω_B, while the polarization difference profile of the lateral component $P(\Delta\omega)$ has the width of the order of $\omega_{F'}$. This circumstance contributes the extra factor of $1/\varepsilon$ into the ratio $\max P_c(\Delta\omega)/\max P(\Delta\omega)$, bringing the total to the order of $1/\varepsilon^3$.

Detailed analytical expressions for the polarization difference profiles with the allowance for the Doppler and electron dynamical broadenings can be found in Ref. [1]. Figure 3.1 presents an example of the theoretically calculated polarization difference profile of the Brackett-alpha line of Ne X 40.48 nm ($n = 5, n_0 = 4$) corresponding to the following plasma parameters: $N_e = 10^{20}\,\text{cm}^{-3}$, $T = 400\,\text{eV}$, $B = 400\,\text{T}$ (these are plasma parameters that can be found in Z-pinches). It is seen that the degree of polarization reaches its maximum at the very center of the line and that $P_{\max} = 17\%$. This degree of polarization is large enough to be observed in experiments.

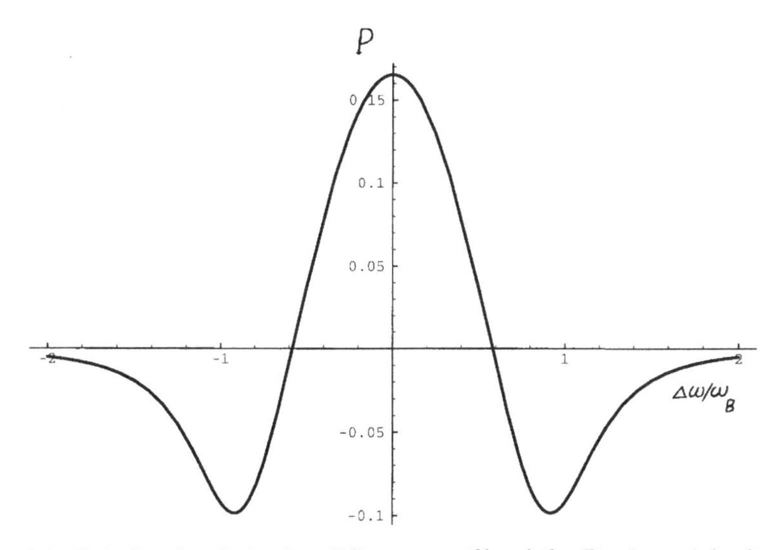

Fig. 3.1. Calculated polarization difference profile of the Brackett-alpha line of Ne X 40.48 nm ($n = 5$, $n' = 4$) corresponding to the following plasma parameters: $N_e = 10^{20}\,\text{cm}^{-3}$, $T = 400\,\text{eV}$, $B = 400$ Tesla.

It should be emphasized that by scaling plasma parameters to significantly lower values of N_e, T and B, one would obtain the same order of magnitude of the polarization effect. Thus, this method for measuring the magnetic field can be used in a broad range of plasma parameters.

A simple alternative is using the diagnostic method based on the principal quantum number $n_{\max}$ of the last observed line in the spectral series of hydrogen/deuterium lines, such as, Lyman, or Balmer, or Paschen lines (though typically Balmer lines are used). For the applicability of the method to measuring the magnetic field B, several conditions should be met. First, the magnetic field should exceed the following critical value

$$B_c(\text{Tesla}) = 4.69 \times 10^{-7} N_e^{2/3}/[T(\text{K})]^{1/2}, \tag{3.8}$$

so that the average Lorentz field exceeds the most probable ion microfield. Second, the ratio of the Zeeman width of hydrogen lines $\Delta\omega_Z$ to the corresponding "halfwidth" of the n-multiplet due to the Lorentz broadening $\Delta\omega_L$

$$\Delta\omega_Z/\Delta\omega_L = 5680/[n(n-1)T(\text{K})^{1/2}] \tag{3.9}$$

should be smaller than unity. Third, the ratio of the FWHM $(\Delta\omega_L)_{1/2}$ caused by the Lorentz broadening to the FWHM $(\Delta\omega_D)_{1/2}$ caused by the Doppler broadenings

$$(\Delta\omega_L)_{1/2}/(\Delta\omega_D)_{1/2} = n_\alpha^2 n_\beta^2 B(\text{Tesla})/526 \tag{3.10}$$

should be greater than unity.

As discussed in detail in Sec. 3 of Chapter 1, in this situation, the value of $n_{\max}$ is controlled by the product B^2T, where B is the magnetic field and T is the temperature of radiators. Here are the practical formulas (also presented in Sec. 3 of Chapter 1).

For the observation perpendicular to $\mathbf{B}$:

$$n_{\max}^{10} B^2 T(\text{K}) = 2.79 \times 10^{18} M/M_p \quad \text{or}$$
$$n_{\max}^{10} B^2 T(\text{eV}) = 2.40 \times 10^{14} M/M_p. \tag{3.11}$$

For the observation parallel to **B**:

$$n_{\text{max}}^{10}B^2T(\text{K}) = 1.78 \times 10^{18}M/M_p \quad \text{or}$$
$$n_{\text{max}}^{10}B^2T(\text{eV}) = 1.54 \times 10^{14}M/M_p. \tag{3.12}$$

For the "isotropic" case:

$$n_{\text{max}}^{10}B^2T(\text{K}) = 2.38 \times 10^{18}M/M_p \quad \text{or}$$
$$n_{\text{max}}^{10}B^2\text{T (eV)} = 2.05 \times 10^{14}M/M_p, \tag{3.13}$$

the "isotropic" observation corresponding to the situation where along the line of sight there are regions with various directions of the magnetic field, which could be sometimes the case in astrophysics. In Eqs. (3.11)–(3.13), M and M_p are the atomic and proton masses, respectively.

Thus, if the temperature is known (being measured, e.g., by methods described in Chapter 2), then Eqs. (3.11)–(3.13) allow determining the magnetic field.

In some magnetic fusion devices, such as tokamaks, it is important to measure not only the total magnetic field, but also a so-called magnetic pitch angle.

$$\gamma_{\text{pitch}} = \tan^{-1}(B_P/B_T), \tag{3.14}$$

where B_P is the poloidal field and B_T is the toroidal field. This can be done in a typical geometry of beam spectroscopy experiments, where the beam velocity $\mathbf{v}$ constitutes a known angle α with the toroidal magnetic field $\mathbf{B}_T$ at the point of the observation. Below, we follow Ref. [3] that presented a method for measuring simultaneously the magnetic pitch angle γ_{pitch} and the effective charge Z_{eff} of ions in tokamaks; details on measuring Z_{eff} are presented in the next chapter. The results from Ref. [3] are based on the Generalized Theory (GT) of the Stark broadening of hydrogen/deuterium lines from Ref. [4].

The beam travels in the horizontal mid-plane of a tokamak, so that the poloidal magnetic field $\mathbf{B}_P$ is vertical. Therefore, the angle θ between the vectors $\mathbf{v}$ and $\mathbf{B} = \mathbf{B}_T + \mathbf{B}_P$, the magnetic pitch angle

γ_{pitch}, and the angle α are related by the formula:

$$\cos\gamma_{\text{pitch}} = (\cos\theta)/\cos\alpha. \tag{3.15}$$

The direction of observation $\mathbf{k}$ constitutes a known angle Ω with the toroidal magnetic field $\mathbf{B}_T$. Typically, the vector $\mathbf{k}$ is in the mid-plane or very close to it. Therefore, the tilt angle γ_{tilt} between the vectors $\mathbf{k}$ and $\mathbf{B}$ (which enters expressions for relative intensities of Zeeman components), the magnetic pitch angle γ_{pitch}, and the angle Ω are related by a similar formula:

$$\cos\gamma_{\text{pitch}} = (\cos\gamma_{\text{tilt}})/\cos\Omega. \tag{3.16}$$

Frequency-integrated intensities I_π of the Zeeman π-component (i.e., the central component of the Zeeman triplet) should be compared at two orthogonal orientations of the polarizer: $I_{\pi 1} = I_{\max}\sin^2\gamma_{\text{pitch}}$ (polarizer is vertical) and $I_{\pi 2} = I_{\max}\sin^2\Omega\cos^2\gamma_{\text{pitch}}$ (polarizer is perpendicular to the previous orientation). From the measured ratio $I_{\pi 1}/I_{\pi 2}$, the magnetic pitch angle is determined by the relation:

$$\tan^2\gamma_{\text{pitch}} = (I_{\pi 1}/I_{\pi 2})\sin^2\Omega. \tag{3.17}$$

For the applicability of the above method, the beam energy should satisfy the following condition:

$$E(\text{keV}) \ll E_0(\text{keV}) = 11.12(M/M_p)/(n\sin\theta)^2. \tag{3.18}$$

Finally, we note that in some situations, there is another way to bypass the obstacle caused by the fact that the magnetic-field-caused Zeeman splitting of spectral lines is smaller than the SB — the way that does not require the polarization analysis. This can be achieved by observing forbidden magnetic-dipole transitions between the sublevels of the ground state of ions possessing the configurations $2s^2 2p^k$ and $3s^2 2p^k$; $k = 1, 2, 3, 4, 5$. (Such ions are not hydrogen-like, or He-like, or Li-like, except for $k = 1$ corresponding to Li-like ions.) This was suggested by Sealy *et al.* [5] and implemented by Wroblewski *et al.* [6] and by Shpitalnik *et al.* [7] in fusion devices. However, the applications of this idea are limited to plasmas having heavy multi-charged ions.

References

[1] A.V. Demura and E. Oks, *IEEE Trans. Plasma Sci.* **26** 1251 (1998).
[2] A.V. Demura and V.S. Lisitsa, *Sov. Phys. JETP* **35** 1130 (1972).
[3] A. Derevianko and E. Oks, *Rev. Sci. Instrum.* **68** 998 (1997).
[4] A. Derevianko and E. Oks, *Phys. Rev. Lett.* **73** 2059 (1994).
[5] J.F. Seely, U. Feldman, N.R. Sheehy Jr., S. Suckewer and A.M. Title, *Rev. Sci. Instrum.* **56** 855 (1985).
[6] D. Wroblewski, H.W. Moos and W.L. Rowan, *Appl. Phys. Lett.* **48** 21 (1986).
[7] R. Shpitalnik, A. Weingarten, K. Gomberoff, Y. Krasik and Y. Maron, *Phys. Plasma* **5** 792 (1998).

Chapter 4

Effective Charge of Ions

In plasmas containing several sorts of ions, the effective charge of ions is defined as

$$Z_{\text{eff}} = \sum Z_{\text{i}}^2 N_{\text{i}} / N_{\text{e}}, \tag{4.1}$$

where the summation over all sorts of ions. It is a very important parameter controlling, e.g., the "vitality" of magnetic fusion machines.

The effective charge of ions can be in principle determined experimentally from the measured Stark width. This is due to the fact that in plasmas of relatively low densities and/or relatively high temperatures (such as, e.g., the edge plasmas of tokamaks), the ion-caused Stark broadening (SB), of intense, low-n lines of hydrogen/deuterium spectral series (Ly-alpha, Ly-beta, H-alpha, H-beta, etc.) can be treated dynamically on the same footing as the SB by electrons. Under these conditions, the dynamical SB by ions predominates over the dynamical SB by electrons — because both contributions are inversely proportional to the velocity of perturbers and ions move significantly slower than electrons.

Roughly speaking, the ion dynamical Stark width γ_s is proportional to $\sum_{\text{i}} Z_i^2 N_i / \langle v_i \rangle$. Therefore, at the first glance, it looks like γ_s is a linear function of the effective charge Z_{eff}, so that Z_{eff} could be determined from the experimental Stark width approximately equal to γ_s. This idea was brought up in the paper [1] and was attempted to be practically implemented in the paper [2].

However, in the paper [3] it was shown that for almost all hydrogen/deuterium spectral lines, this idea would not work for the following reasons. Contributions to the dynamical Stark width due to ions can be separated into adiabatic and non-adibatic — see Appendix E.1 for details. The adiabatic contribution to the dynamical Stark width γ_{ad}^{i} due to ions of the sort "i" has the following form, which is the *exact, non-perturbative* analytical result obtained by the Generalized Theory (GT):

$$\gamma_{\mathrm{ad}}^{i} = 18(\hbar/m_e)^2(X_{\alpha\beta}/Z_r)^2 Z_{\mathrm{i}}^2 N_{\mathrm{i}}(2\pi M_{\mathrm{i}}/T_{\mathrm{i}})^{1/2} I(R_{\mathrm{i}}),$$
$$X_{\alpha\beta} = |n_\alpha q_\alpha - n_\beta q_\beta|. \tag{4.2}$$

Here, Z_i, N_i and T_i are the charge, the density and the temperature of the plasma ions, respectively; Z_r is the nuclear charge of the radiator; M_i is the reduced mass of the pair "radiator — perturbing ion"; n_γ and $q_\gamma = (n_1 - n_2)_\gamma$ are, respectively, the principal quantum number and the electric quantum number, the latter being expressed via the parabolic quantum numbers n_1 and n_2 ($\gamma = \alpha$ or β). The function $I(R_i)$ in Eq. (4.2) is defined as follows:

$$I(R_i) = \{R_i^2[3 - \cos(1/R_i)]$$
$$+(R_i - 2R_i^3)\sin(1/R_i) - ci(1/R_i)\}/6. \tag{4.3}$$

In Eq. (4.3), $ci(1/R_i)$ is the cosine integral function, the quantity R_i being

$$R_{\mathrm{i}} = r_{\mathrm{D}}/r_{\mathrm{Wa}}, \tag{4.4}$$

where

$$r_D = [T_e/(4\pi e^2 N_e)]^{1/2} = 743.40[T_e(\mathrm{eV})/N_e(\mathrm{cm}^{-3})]^{1/2}\mathrm{cm}, \tag{4.5}$$

is the Debye radius and

$$r_{\mathrm{Wa}} = 3X_{\alpha\beta}\hbar/(Z_{\mathrm{r}} m_e v_{\mathrm{i}})$$
$$= 3.5486 \times 10^{-6}(X_{\alpha\beta}/Z_r)(M_i/M_p)^{1/2}/[T_i(\mathrm{eV})]^{1/2}\mathrm{cm}, \tag{4.6}$$

is the adiabatic Weisskopf radius (M_p is the proton mass).

In the typical situation, where $R_i \gg 1$, the function $I(R_i)$ from Eq. (4.3) reduces approximately to

$$I(R_i) = 0.209 + (1/6)\ln R_i. \quad (4.7)$$

If in this situation one would disregard the weak logarithmic dependence of the function I on $R_i(M_i)$, then the total adiabatic Stark width,

$$\gamma_{\text{ad}} = \sum_i \gamma_{\text{ad}}^i \quad (4.8)$$

would be proportional to

$$\sum_i Z_i^2 N_i M_i^{1/2}. \quad (4.9)$$

For heavy impurity ions in tokamaks, the reduced mass M_i of the atom-ion pair is equal to the proton mass M_p and thus can be factored out of the summation in Eq. (4.9), so that the remaining sum would be equal to $N_e Z_{\text{eff}}$ and the total adiabatic width γ_{na} would be proportional to Z_{eff}. For the primary perturbers — protons or deuterons — the reduced mass of the perturber–radiator pair differs significantly from M_p: for example, in a hydrogen plasma for proton perturbers one has $M_i = M_p/2$. However, this circumstance might be easily accounted for: instead of γ_{ad} to be proportional to Z_{eff}, one would have γ_{ad} to be proportional to $(Z_{\text{eff}} - 1 + 1/2^{1/2})$, so that it would be still possible to deduce Z_{eff} from the experimental Stark width if it were not for the non-adiabatic contribution γ_{na} discussed below.

According to the GT, presented in detail in Appendices C and E.1, the non-adiabatic contribution is controlled by a generalized width function $a(Y)$ depending on a dimensionless parameter $Y = 0.160 n Z_i B(\text{Tesla})/T_i(\text{eV})$. Physically, Y is the ratio of a magnetic splitting $\Delta\omega_B$ to the ion Weisskopf frequency $\Omega_{Wi} = v_i/\rho_{Wa}$. Thus, the total ion dynamical width γ_s has the form $\gamma_s = \gamma_{\text{na}} + \gamma_{\text{ad}}$, where,

$$\gamma_{\text{na}} \propto \sum \frac{Z_i^2 N_i a(Y(Z_i))}{\langle v_a \rangle}, \quad \gamma_{\text{ad}} \propto \sum \frac{Z_i^2 N_i}{\langle v_a \rangle}. \quad (4.10)$$

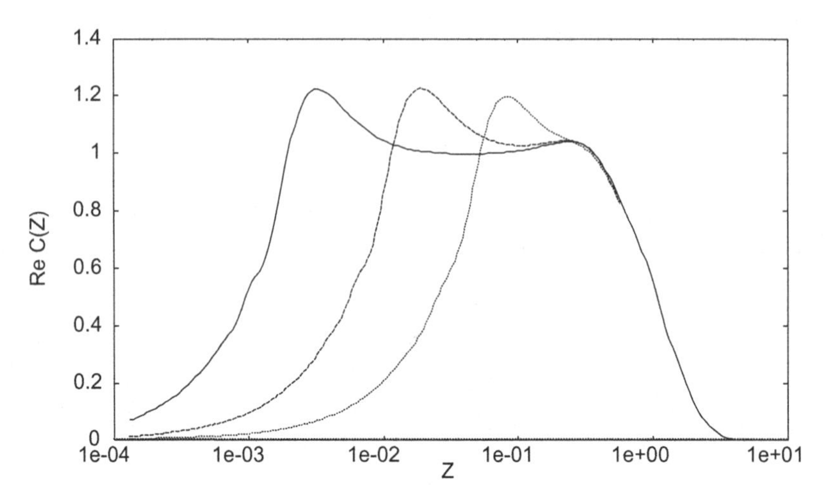

Fig. 4.1. The broadening function $ReC(Y(Z_i), Z)$ versus the dimensionless impact parameter $Z = \rho\mu_B B(m_\alpha - m_{\alpha''})/(\hbar \mathrm{v})$ for three different values of the perturbing ion charge: $Z_i = 1$ (solid curve); $Z_i = 6$ (dashed curve); $Z_i = 26$ (dotted curve). The calculation is for the σ-components of the hydrogen L_α line for plasma parameters $N_e = 10^{13}$ cm^{-3}, $T = 100$ eV, $B = 2$T.

In paper [3], it was shown that ion perturbers of different charges $Z_i = 1$–30 result in *significantly different* width functions $a(Y(Z_i))$ making it impossible to express γ_{na} as a function of Z_{eff}. In other words, the same width γ_s may correspond to values of Z_{eff} differing by 30–40%. This is illustrated in Fig. 4.1.

Thus, practically all hydrogen/deuterium spectral lines cannot be employed for deducing Z_{eff} from the Stark width. Luckily, there exists one exception to this rule. For the unshifted (π) component of the Ly-alpha line, the non-adiabatic width vanishes as a result of a mutual cancellation of diagonal and non-diagonal matrix elements of the ion dynamical operator [3]. This unique phenomenon allows measuring Z_{eff} by employing formulas for the (purely adiabatic) impact width of the unshifted (π) component of the Ly-alpha line as follows:

$$\gamma^i_{\pi,\mathrm{ad}} = Z_i^2 N_i \gamma_0 / N_e, \quad \gamma_0 \equiv 72\left(\frac{\hbar}{m_e}\right)^2 N_e \sqrt{\frac{2\pi m_p}{T}} I(R)$$

$$\approx 2.45 \times 10^4 N_e(\mathrm{cm}^3)/\sqrt{T(\mathrm{eV})} I(R)(1/\mathrm{s}),$$

$$I(R) \approx 0.209 + 1/6 \ln R, \quad R \equiv \frac{m_e v_p}{6\hbar Z_i}\sqrt{\frac{T_e}{4\pi e^2 Z_{\text{eff}}, N_e}}$$

$$\approx 1.1 \times 10^8 T(\text{eV})/\sqrt{Z_{\text{eff}}^3 N_e(\text{cm}^3)}. \tag{4.11}$$

A weak logarithmic dependence of γ_0 on Z_i in (4.11) allows us to factor γ_0 out of the summation over i and to express the impact width of the central π-component of the L_α as a function of Z_{eff}:

$$\gamma_s^\pi = \sum_i \gamma_{\pi,\text{ad}}^i = (Z_{\text{eff}} - 1 + 1/\sqrt{2})\gamma_0. \tag{4.12}$$

As mentioned in Chapter 3, it is possible to use the same experimental line for measuring simultaneously the effective charge Z_{eff} and the magnetic pitch angle γ_{pitch} in tokamaks — following paper [4]. For this purpose, one could employ a beam spectroscopy experiments including a polarization analysis as follows.

Frequency-integrated intensities I_π of the Zeeman π-component (i.e., the central component of the Zeeman triplet) should be compared at two orthogonal orientations of the polarizer: $I_{\pi 1} = I_{\max} \sin^2 \gamma_{\text{pitch}}$ (polarizer is vertical) and $I_{\pi 2} = I_{\max} \sin^2 \Omega \cos^2 \gamma_{\text{pitch}}$ (polarizer is perpendicular to the previous orientation). Here, Ω is a known angle between the direction of observation $\mathbf{k}$ and the toroidal magnetic field $\mathbf{B}_T$. From the measured ratio $I_{\pi 1}/I_{\pi 2}$, the magnetic pitch angle is determined by the relation:

$$\tan^2 \gamma_{\text{pitch}} = (I_{\pi 1}/I_{\pi 2})\sin^2 \Omega. \tag{4.13}$$

The beam velocity $\mathbf{v}$ constitutes a known angle α with the toroidal magnetic field $\mathbf{B}_T$. The angle θ between $\mathbf{v}$ and the total magnetic field $\mathbf{B} = \mathbf{B}_T + \mathbf{B}_P$, where $\mathbf{B}_P$ is the poloidal magnetic field, is related to the above angles as follows

$$\cos \theta = (\cos \gamma_{\text{pitch}})\cos \alpha. \tag{4.14}$$

So, after the experimental determination of the magnetic pitch angle γ_{pitch} from the measured ratio $I_{\pi 1}/I_{\pi 2}$ via Eq. (4.13), it is possible to find the value of

$$\sin^2 \theta = 1 - (\cos^2 \gamma_{\text{pitch}})\cos^2 \alpha. \tag{4.15}$$

Finally, the ion dynamical width γ_π of the π-component of the Ly-alpha line is measured and the effective charge Z_{eff} is deduced from it by using the following formula with the known value of $\sin^2\theta$:

$$\begin{aligned}\gamma_\pi(\text{s}^{-1}) &= 2.17\times10^{-4}Z_{\text{eff}}\{M/[M_{\text{p}}E(\text{eV})]\}^{1/2}N_e(\text{cm}^{-3})\\ &\quad\times[0.279+(\ln \text{R})/4]\sin^2\theta,\\ R &= 1.49\times10^{8}Z_{\text{eff}}^{-3/2}\{E(\text{eV})T(\text{eV})M_p/\\ &\quad[N_e(\text{cm}^{-3})M]\}^{1/2}/\sin\theta.\end{aligned} \tag{4.16}$$

There is also an alternative recipe. First, the impact width $\gamma_\pi^{(0)}$ should be measured at the absence of the beam. Then the effective charge Z_{eff} is determined from the value of $\gamma_\pi^{(0)}$ using Eq. (4.11). Thereafter, the ion dynamical width γ_π of the π-component of the Ly-alpha line should be measured at the presence of the beam. Then the value of $\sin^2\theta$ is deduced from the measured ratio $\gamma_\pi/\gamma_\pi^{(0)}$ using Eq. (4.16). Finally, the magnetic pitch angle γ_{pitch} is determined from the value of θ by using Eq. (4.14). For this second recipe to yield a reasonably accurate value of the magnetic pitch angle, the angle α should be chosen small enough ($\alpha \ll 1$), preferably $\alpha < \gamma_{\text{pitch}}$ or $\alpha \sim \gamma_{\text{pitch}}$.

While applying either one of the above methods, if it would be necessary to bypass the Doppler broadening, one of the possible experimental techniques can be based on the Doppler-free two-photon excited laser fluorescence. Under the two-photon excitation from the ground state, selection rules ensure that only the π-component of the Ly-alpha line would be excited.

References

[1] V.A. Abramov and V.S. Lisitsa, *Sov. J. Plasma Phys.* **3** 451 (1977).
[2] S. Bychkov, R.S. Ivanov and G.I. Stotskii, *Sov. J. Plasma Phys.* **13** 769 (1987).
[3] A. Derevianko and E. Oks, *Phys. Rev. Lett.* **73** 2059 (1994).
[4] A. Derevianko and E. Oks, *Rev. Sci. Instrum.* **68** 998 (1997).
[5] A. Derevianko and E. Oks, *J. Quant. Spectrosc. Rad. Transfer* **54** 137 (1995).

Part II

Plasmas Containing Oscillatory Electric Fields

Chapter 5

Low-frequency Electrostatic Turbulence

5.1 Non/Weakly-Magnetized Plasmas (Including Solar Flares and Flare Stars)

At the absence of a magnetic field, there is only one type of a Low-frequency Electrostatic Turbulence (LET): ion acoustic waves — frequently called *ionic sound*. It is a broadband oscillatory electric field, whose frequency spectrum is below or of the order of the ion plasma frequency ω_{pi} — see Appendix F and Eq. (F.1).

In magnetized plasmas (even in weakly-magnetized plasmas) in addition to the ionic sound, propagating along the magnetic field **B**, two other types of the LET are possible. One is *electrostatic ion cyclotron wave*, whose wave vector is nearly perpendicular to **B** and whose frequency is close to the ion cyclotron frequency ω_{ci} — see Appendix F and Eq. (F.2). Another type is *lower hybrid oscillations* having the wave vector perpendicular to **B** and whose frequency is $\omega = 1/[(\omega_{\text{ci}}\,\omega_{\text{ce}})^{-1} + (\omega_{\text{pi}})^{-2}]^{1/2}$, where ω_{ce} is the electron cyclotron frequency — see Appendix F and Eqs. (F.3), (F.4). Frequencies of both ionic sound and lower hybrid oscillations are below or of the order of the ion plasma frequency ω_{pi} .

It is usually assumed that hydrogenic radiators perceive Oscillatory Electric Fields (OEFs), associated with a low-frequency plasma turbulence as *quasistatic*. This assumption is discussed in Appendix F in all details.

At the absence of a magnetic field, the broadening of hydrogenic spectral lines by quasistatic electric field yields the lineshape that reproduces the shape of the distribution $W(\mathbf{F})$ of the quasistatic fields — see, e.g., books [1, 2]. If a magnetic field is present, then from the spectroscopic point of view, one deals here with a hydrogenic atom/ion in crossed *static* electric ($\mathbf{F}$) and magnetic ($\mathbf{B}$) fields — the problem allowing an exact solution for each value of $\mathbf{F}$ and $\mathbf{B}$ [3].

If the emission is observed from a relatively small volume, within which the magnetic field can be considered homogeneous, then the spectral line profiles can be obtained by averaging the solution from [3] over the ensemble distribution $W(\mathbf{F})$ of the quasistatic field $\mathbf{F}$. In other words, the key part of the problem becomes the calculation of $W(\mathbf{F})$.

The distribution $W(\mathbf{F})$ of the quasistatic fields is either the distribution $W_t(\mathbf{E}_t)$ of the turbulent fields (the fields of the LET) or a convolution of $W_t(\mathbf{E}_t)$ with the distribution $W_i(\mathbf{F}_i)$ of the ion microfield, if the latter is also quasistatic. It should be noted that there are situations, where the turbulent field is quasistatic, but the ion microfield is not quasistatic, and that this is possible even if their frequencies are of the same order of magnitude — as long as the average turbulent field E_0 is much greater than the characteristic ion microfield $\langle F_i \rangle$ — see Appendix F for details.

However, the typical situation is where both the LET and the ion microfield are perceived by the radiator as quasistatic, so that the distribution $W(\mathbf{F})$ of the total quasistatic field is a convolution of $W_t(\mathbf{E}_t)$ with the distribution $W_i(\mathbf{F}_i)$. Details on the distribution $W_t(\mathbf{E}_t)$ of the LET and on its convolution with the distribution $W_i(\mathbf{F}_i)$ are given in Appendix F.

There are basically two methods for diagnosing the LET using hydrogenic spectral lines. The first method requires measuring at least two lines of a particular (e.g., Balmer or Paschen) spectral series, or at least three lines if some of them are affected by a self-absorption. The details of this method are as follows.

The method consists of two parts. The first part is the analysis of the dependence of the reduced halfwidths $\Delta\lambda_{1/2}(n)/\lambda_n$ for two or more lines of a particular series on the principal quantum number

n of the upper level involved in the radiative transition (here λ_n is the unperturbed wavelength). Here and below, the halfwidth means the full width at the half maximum. There are typically three competing mechanisms: Stark, Doppler (thermal and non-thermal) and opacity broadenings. Opacity: $\Delta\lambda_{1/2}(n)/\lambda_n$ decreases as n increases. Doppler: $\Delta\lambda_{1/2}(n)/\lambda_n$ does not depend on n. Stark: $\Delta\lambda_{1/2}(n)/\lambda_n$ increases roughly as n^2. Therefore, $\Delta\lambda_{1/2}(n)/\lambda_n$ dependence either starts flat for low-n lines (if they are optically thin) and then goes up for higher-n lines, or it first goes down for low-n lines (if they are optically thick), then goes through a minimum, and then goes up. In either case, from the observed $\Delta\lambda_{1/2}(n)/\lambda_n$ dependence, it is possible to determine the contributions of all three broadening mechanisms *separately*, including the Stark broadening (SB) — see more details in Chapter 2, Sec 2.

The second part of the method is the analysis of the shapes of each hydrogenic line in the wings. This allows determining whether the SB is primarily due to the ion and electron microfields or due to the LET. The shape of the wings due to the former is a power law $\sim 1/(\Delta\lambda)^\alpha$, where $2 \leq \alpha \leq 3$, while due to the latter the shape of the wings is $\exp[-k(\Delta\lambda)^\gamma]$, where $\gamma \approx 1$, as shown in paper [4] (for the Doppler broadening it would be $\gamma = 2$).

Here is an example of the application of this method to two powerful solar flares [5]. The hypothesis of the development of the LET was brought up as the cause of the anomalous resistivity of plasmas, which helps explaining, e.g., a huge energy release — up to 10^{25} J or higher — in a relatively short time in powerful solar flares. The observations and the analysis presented in the paper [5] was the first proof of that hypothesis and of the corresponding mechanism of solar flares.

Figure 5.1 shows the dependence of the reduced halfwidths $\Delta\lambda_{1/2}(n)/\lambda_n$ on the principal quantum number n of the upper level for Balmer lines from H_5 to H_{11} observed in the flare of 08.18.59 and from H_4 to H_{15} observed in the flare of 09.26.63.

In the flare of 08.18.59 the entire data correspond to the rising branch of this dependence. This is a manifestation of a strong contribution by the Stark broadening.

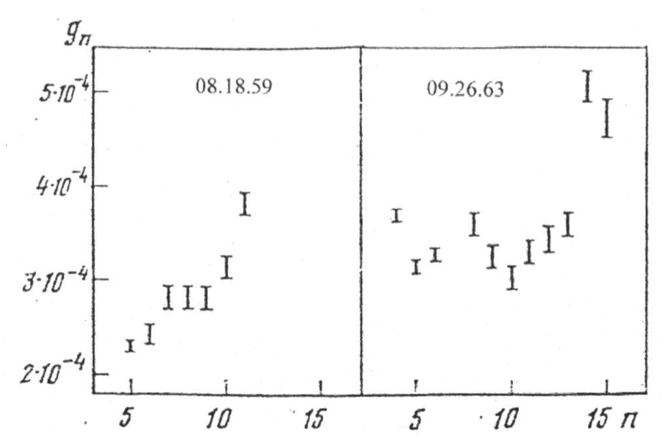

Fig. 5.1. Dependence of the reduced halfwidths $\Delta\lambda_{1/2}(n)/\lambda_n = g_n$ on the principal quantum number n of the upper level for Balmer lines observed in the solar flares of 08.18.59 and of 09.26.63.

In the flare of 09.26.63, it starts from a falling branch, reaches a minimum, and finally switches to a rising branch. This means that lines H_4–H_6 were optically thick, lines H_8–H_{10} were dominated by the Doppler broadening, but lines H_{11}–H_{15} were dominated by the Stark broadening.

The next step for both flares was to find out whether the SB was primarily due to the ion and electron microfields or due to the LET. Figure 5.2 shows the shape of the wings of lines H_5–H_{11} in the flare of 08.18.59 presented in the double logarithmic scale: Log(Intensity) versus Log($\Delta\lambda$). If the shapes would follow a power law dictated by the broadening due to the ion and electron microfields, then the plots would be straight lines, but in reality, they are not.

Figure 5.3 shows the shape of the wings of the same lines presented as Log(Intensity) versus $\Delta\lambda$. It is seen that in this scale, they look practically as straight lines.

The combination of these findings was the proof that the SB was due to the LET. From the detailed analysis of the observed widths and observed shapes in the flare of 18.08.59, it was deduced that:

— the LET field strength was 1.5 kV/cm, which was several times greater than the typical strength of electron and ion microfields,

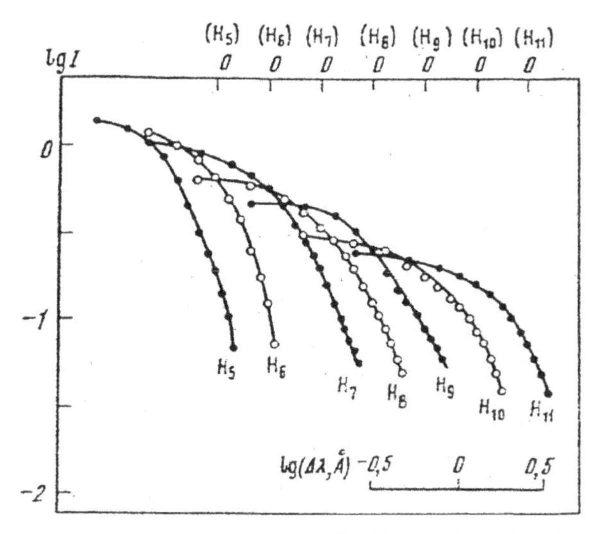

Fig. 5.2. The shape of the wings of lines H_5–H_{11} in the solar flare of 08.18.59 presented in the double logarithmic scale.

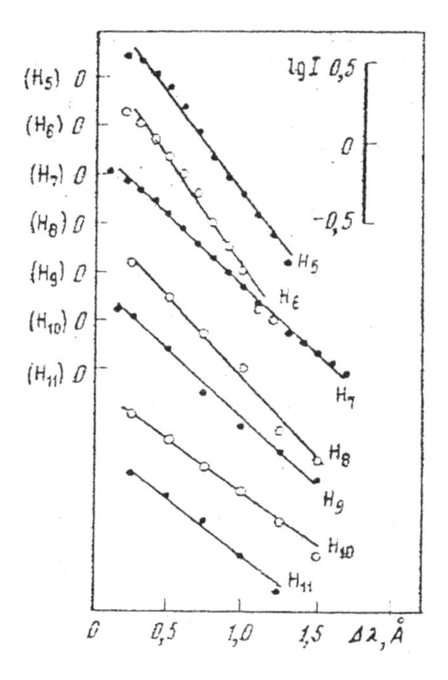

Fig. 5.3. Same as in Fig. 2, but presented as Log(Intensity) versus $\Delta\lambda$.

— the electric field of the LET most probably had the circular polarization.[1] If so, they should have been Bernstein modes, which is one of the LET causing the anomalous resistivity of the flare plasma.

Figures 5.4 and 5.5 show a similar analysis of the shapes in the solar flare of 09.26.63. Once again, in the double logarithmic scale, the Stark-broadening-controlled shapes do not become straight lines, but in the scale Log(Intensity) versus $\Delta\lambda$ they do. Again, this is the proof that the SB was due to LET.

From the detailed analysis of the observed widths and observed shapes in the solar flare of 09.26.63, it was deduced that:

— the LET field strength was 1.0 kV/cm, which was several times greater than the typical strength of electron and ion microfields,
— the electric field of the LET most probably had the circular polarization. If so, they should have been Bernstein modes, which

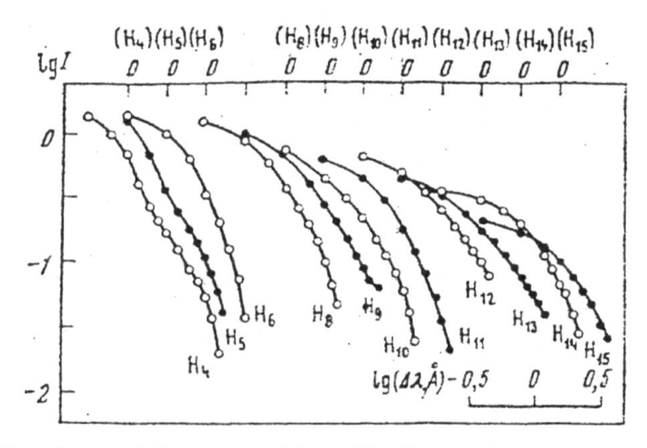

Fig. 5.4. The shape of the wings of lines H_4–H_{15} in the solar flare of 09.26.63 presented in the double logarithmic scale.

[1]In paper [6], not only the isotropic distribution of the LET (see Appendix F, Eq. (F.13)), but also a variety of axially-symmetric distributions (Appendix F, Eq. (F.12)), including a very prolate one (corresponding to the linear polarization) and a very oblate one (corresponding to the circular polarization) were derived. The latter provided the best fit to the observed shapes of the spectral lines.

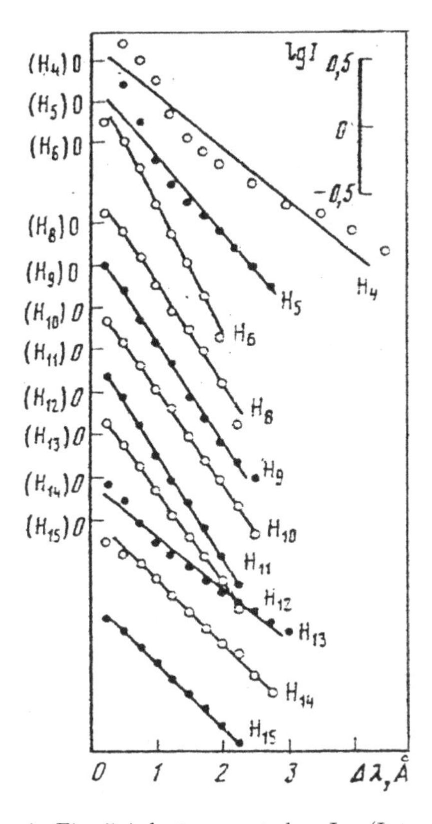

Fig. 5.5. Same as in Fig. 5.4, but presented as Log(Intensity) versus $\Delta\lambda$.

is one of the LET causing the anomalous resistivity of the flare plasma.

A slight modification of the above method was applied for diagnosing the LET in another type of astrophysical objects — in flare stars. It is commonly accepted that the activity of these stars is similar to the solar activity. In paper [7], the LET was diagnosed in many flare stars. Below are few examples from the paper [7].

Figures 5.6–5.8 shows the dependence of the reduced halfwidths $\Delta\lambda_{1/2}(n)/\lambda_n$ on the principal quantum number n of the upper level for Balmer lines observed in two flares of AD Leo [8, 9]. It is seen that for both flares, the reduced halfwidths $\Delta\lambda_{1/2}(n)/\lambda_n$ increase as the principal quantum number n of the upper level increases for

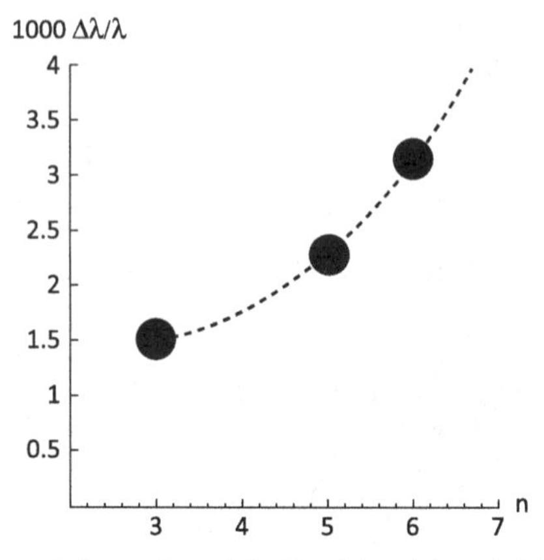

Fig. 5.6. Dependence of the reduced halfwidths $\Delta\lambda_{1/2}(n)/\lambda_n$ on the principal quantum number n of the upper level for Balmer lines observed by Gershberg and Shakhovskaja [8] in the flare of AD Leo of 03.02.70 in spectrum #1 (dots) from [8]. The average trend is shown by the dashed line.

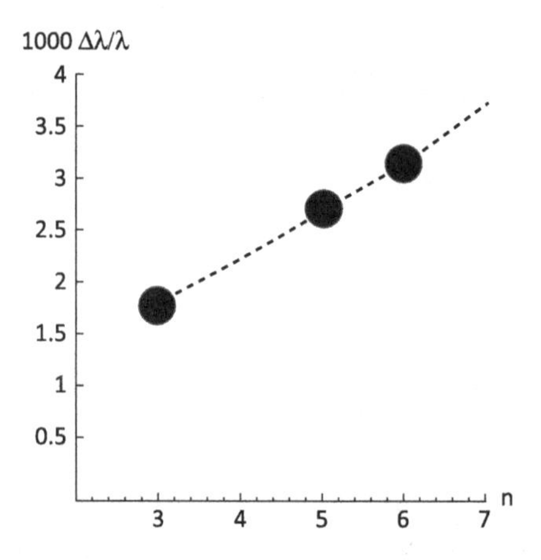

Fig. 5.7. The same as in Fig. 5.6, but in spectrum #2 from [8].

$n > 4$. Therefore, for both flares, there is a significant contribution by the SB.

In distinction to the hydrogen lines in solar flares, the analysis of the shapes of the line wings in flare stares was not possible because of the insufficient spectral resolution. Therefore, there were *initially* two alternative interpretations of the observed SB in flare stars: either

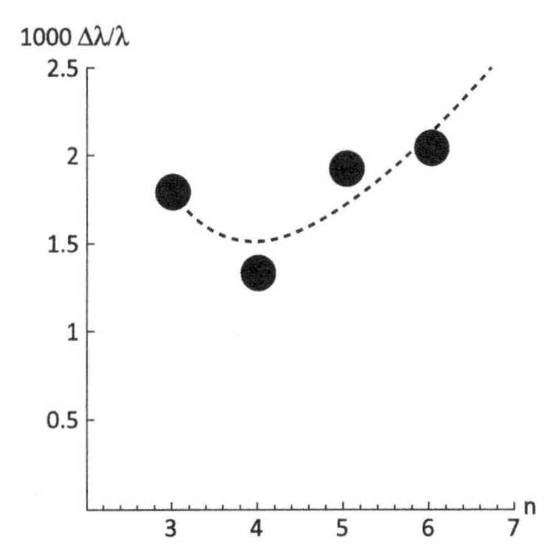

Fig. 5.8. The same as in Fig. 5.6, but observed by Kulapova and Shakhovkaja [9] in the flare of AD Leo of 02.18.71.

by the LET dominating the ion microfield or by the ion microfield at the absence of the LET. Additional considerations based on the analysis of Balmer decrements helped to discriminate between these two interpretations as follows.

The would-be values of $N_e = (2-6) \times 10^{15}$ cm^{-3}, deduced above under the assumption that the LET is absent, exceeded by more than two orders of magnitude the values of $N_e^{\text{decr}} \sim 10^{13}$ cm^{-3}, deduced by Gershberg [10] from the analysis of the Balmer decrements in the same spectrograms, and by more than one order of magnitude the values of $N_e^{\text{decr}} \sim 10^{14}$ cm^{-3}, deduced by Katsova [11] from the analysis of the Balmer decrements in the same spectrograms. Therefore, it was very likely that the LET was developed in those flares, since in this case the N_e values deduced from the analysis of the halfwidths of the Balmer lines, being by one or even one and a half orders of magnitude lower than at the absence of the LEPT, are much closer to the values of N_e^{decr}.

Another example deals with more recent observations by Crespo-Chacon *et al.* [12] concerning flares of AD Leo on April 2–5, 2001. Figure 5.9 shows the dependence of the reduced halfwidths $\Delta\lambda_{1/2}(n)/\lambda_n$ on the principal quantum number n of the upper level for Balmer lines H_4–H_{11} observed in spectrum #2 from [12].

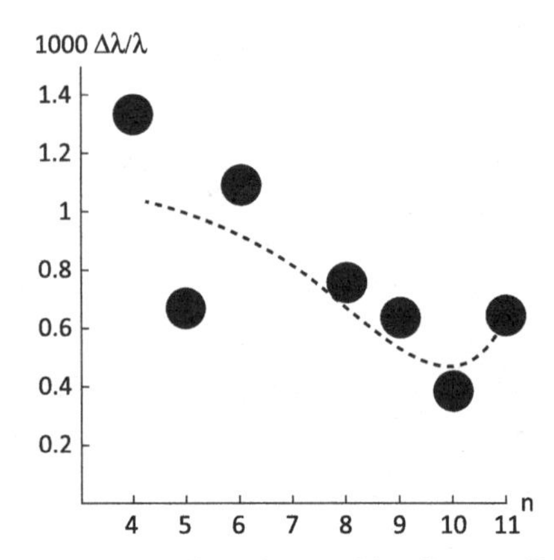

Fig. 5.9. The same as in Fig. 5.6, but observed by Crespo-Chacon [12] in the flare of AD Leo of 04.2–5.2001 in spectrum/flare #2 in its maximum.

The analysis of these reduced widths without engaging the LET does not yield self-consistent results, especially concerning the ratio of the widths of the lines H_{11} and H_{10}. In distinction, engaging the LET of the root-mean-square strength $E_0 = 4\,\mathrm{kV/cm}$ allows the self-consistent explanation of these reduced widths, leading also to the conclusion that the electron density N_e was several times smaller than 6×10^{13} cm^{-3}. Being combined with the analysis of the Balmer decrements by Crespo-Chacon *et al.* [12], it allowed in paper [7] to significantly narrow the relevant interval of N_e: 1.3×10^{13} cm^{-3} $< N_e < N_{e,\max}$, where $N_{e,\max} \sim 2 \times 10^{13}$ cm^{-3}.

All of the above was just about one method (and also about its slight modification) for diagnosing the LET in plasmas. For the second method based on hydrogenic spectral lines, it is sufficient to study just one line, but in two orthogonal linear polarizations. The theory behind this method was developed in paper [13].

In paper [13], it was shown that, for axially-symmetric distributions of the LET, both the average field of the LET and the degree of its anisotropy $\eta = (2\langle E^2_{\mathrm{par}}\rangle/\langle E^2_{\mathrm{perp}}\rangle)^{1/2}$ can be determined from the polarization difference profile, obtained by subtracting the experimental profile, observed with a polarizer axis perpendicular to the axis of symmetry of the LET, from the experimental profile, observed with a polarizer axis parallel to the axis of symmetry

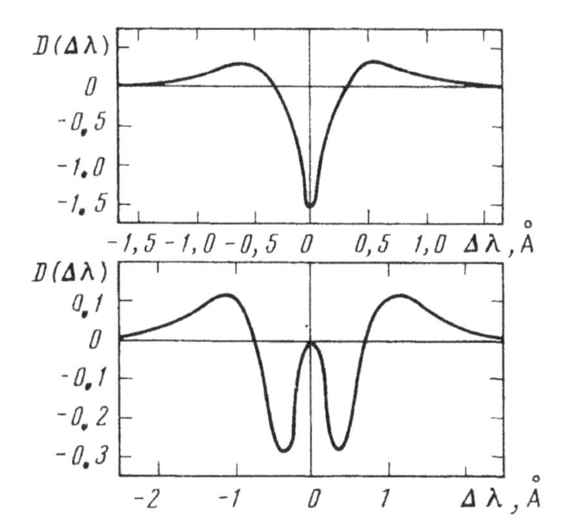

Fig. 5.10. Theoretical polarization difference profiles for the H-alpha (top) and H-beta (bottom) spectral lines for $\eta = 3, (\langle F^2_{\text{perp}} \rangle)^{1/2} = 1.4\,\text{kV/cm}$, and the Doppler width parameter $d = 0.01\,\text{nm}$ ($d = \lambda_0(2T/M)^{1/2}/c$).

of the LET. Both the experimental polarization profiles should be normalized to unity before the subtraction.

Figure 5.10 shows theoretical polarization difference profiles for the H-alpha (top) and H-beta (bottom) spectral lines for $\eta = 3$, $(\langle E^2_{\text{perp}} \rangle)^{1/2} = 1.4\,\text{kV/cm}$, and the Doppler width parameter $d = 0.01\,\text{nm}$ ($d = \lambda_0(2\text{T}/\text{M})^{1/2}/\text{c}$). Since, the polarization profiles were normalized to unity before the subtraction, the areas of the top and bottom parts of any polarization difference profiles are equal to each other. Any of these two equal areas denoted as S_D are controlled by a universal function $S_F(\eta)$ that depends only on the degree of anisotropy η:

$$\begin{aligned} S_F(\eta) &= (\eta^2 + 1/2)(\eta^2 - 1) - (3/2)(\eta^2/|\eta^2 - 1|^{3/2}) \\ &\quad \arcsin(1 - 1/\eta^2)^{1/2}, \quad \eta > 1, \\ S_F(\eta) &= (\eta^2 + 1/2)(\eta^2 - 1) - (3/2)(\eta^2/|\eta^2 - 1|^{3/2}) \\ &\quad \ln[1/\eta + (1/\eta^2 - 1)^{1/2}]^{-1}, \quad \eta < 1. \end{aligned} \tag{5.1}$$

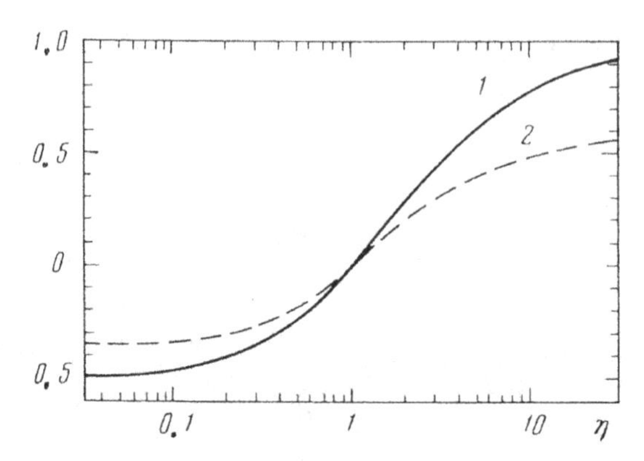

Fig. 5.11. The dependence of the universal function S_F on the anisotropy degree η (solid line) and the dependence of the area S_D of the top or the bottom of the polarization difference profile for the H-alpha line on the anisotropy degree η (dashed line).

Figure 5.11 shows the dependence of the universal function S_F on the anisotropy degree η (solid line) and the dependence of the area S_D of the top or the bottom of the polarization difference profile for the H-alpha line on the anisotropy degree η (dashed line). Thus, the experimental polarization difference profile allows determining the degree of the anisotropy η of the LET.

The experimental polarization difference profile also allows determining the average filed of the LET in each of the two orthogonal directions $(\langle E^2_{\text{par}}\rangle)^{1/2}$ and $(\langle E^2_{\text{perp}}\rangle)^{1/2}$. The quantity $E_{\text{max}} = \max[(\langle E^2_{\text{par}}\rangle)^{1/2}, (\langle E^2_{\text{perp}}\rangle)^{1/2}]$ can be found from the shape of the wing $D_{\text{As}}(x), (x = \lambda - \lambda_0)$, of the polarization difference profile using the following formulas:

$$
\begin{aligned}
D_{\text{As}}(x) &= \exp[-x^2/(2F^2_{\text{max}}\Delta^2_\pi)], \quad D_{\text{As}}(x) > 0, \\
D_{\text{As}}(x) &= -x\,\exp[-x^2/(2F^2_{\text{max}}\Delta^2_\pi)], \quad D_{\text{As}}(x) < 0.
\end{aligned} \tag{5.2}
$$

Here,

$$
\Delta^2_\pi = (\Sigma_{\text{k}} I_{\text{k}\pi}/\Delta^{\text{p}}_{\text{k}})/(\Sigma_k I_{\text{k}\pi}/\Delta^{p+2}_k), \quad p = 1 \quad \text{for } \eta > 1,
$$

$$
p = 2 \quad \text{for } \eta < 1, \tag{5.3}
$$

where $I_{k\pi}$ are the relative intensities of the π-components of the hydrogenic spectral line and Δ_k are the Stark constants of the Stark components:

$$\Delta_k = 3ea_0\lambda_0^2(nq - n_0q_0)/(4\pi\hbar c), \quad (5.4)$$

a_0 being the Bohr radius and (n, q) and (n_0, q_0) being the principal and electric quantum numbers of the upper and lower Stark sublevels, respectively.

This polarization method was practically implemented for the first time in the experiment presented in paper [14]. The experiment was performed at the "Dimpol" plasma machine, which was a magnetic mirror trap. The initial plasma was created by a Penning-type discharge in a DC magnetic field. Then an AC magnetic field of a high amplitude was imposed. Thus, the opposing magnetic fields configuration arose in a θ-pinch geometry.

Figure 5.12 shows the experimental profiles of the H-alpha line observed in two orthogonal polarizations, as well as the polarization difference profile. The direction of the observation was perpendicular to the magnetic field. Profile I was observed with the polarizer perpendicular to the magnetic field, profile II — with the polarizer parallel to the magnetic field.

By analyzing the experimental polarization difference profile, it was found that the LET had an axially-symmetric distribution (the axis of the symmetry coinciding with the magnetic field direction, the latter being denoted as z-axis) of an oblate shape: $E_{\mathrm{par}} = E_z = 8\,\mathrm{kV/cm}$, $E_{\mathrm{perp}} = (E_\varphi^2 + E_\mathrm{r}^2)^{1/2} = 30\,\mathrm{kV/cm}$. These experimental results allowed identifying the LET as Bernstein modes.

Non-hydrogenic spectral lines can also be used in some situations for diagnosing the LET. For example, in paper [15] this was done by the simultaneous observation and analysis of two helium spectral lines: He I 667.8 nm and He I 587.6 nm. Both spectral lines correspond to the transitions $2^{2S+1}P - 3^{2S+1}D$in helium atoms, where $S = 1$ for He I 587.6 nm and $S = 0$ for He I 667.8 nm.

In the experiment [15], a two-dimensional (2D) magnetic field was produced by the currents in the external conductors aligned along the axis of the cylindrical quartz vacuum chamber. The $\boldsymbol{X}$-type null line of the 2D magnetic field coincided with the chamber axis. An auxiliary Θ-discharge with strong preionization produced

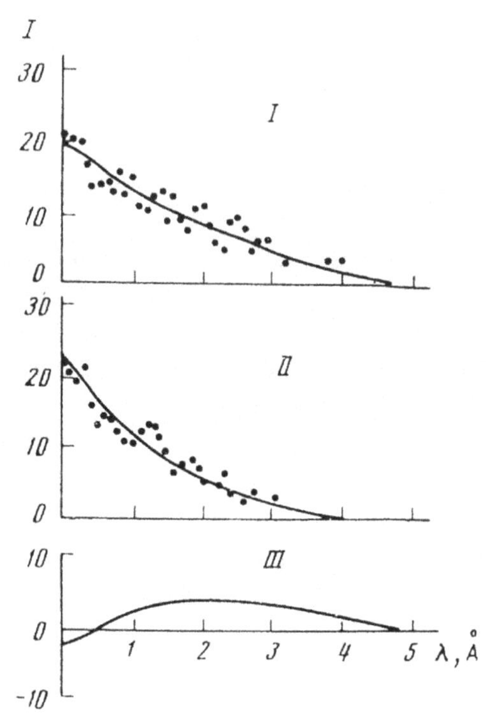

Fig. 5.12. The experimental profiles of the H-alpha line observed in experiment [14] in two orthogonal polarizations, as well as the polarization difference profile. The direction of the observation was perpendicular to the magnetic field. Profile I was observed with the polarizer perpendicular to the magnetic field, profile II — with the polarizer parallel to the magnetic field.

an initial plasma in the magnetic field: the electron density was close to the initial gas-atom density, $N_e \leq 10^{16}\,\text{cm}^{-3}$. The plasma current was excited by a pulsed voltage applied between two grid electrodes inserted into the chamber from both ends. The plasma current was directed parallel to the null line. The current generation initiated 2D plasma flows in the plane orthogonal to the chamber axis, which resulted in both formations of a current sheet and effective plasma compression into the flat sheet.

A noticeable asymmetry (the blue wing more intense than the red wing) of the experimental profiles of He I 667.8 nm spectral line was found, while the experimental profiles of He I 587.6 nm spectral line

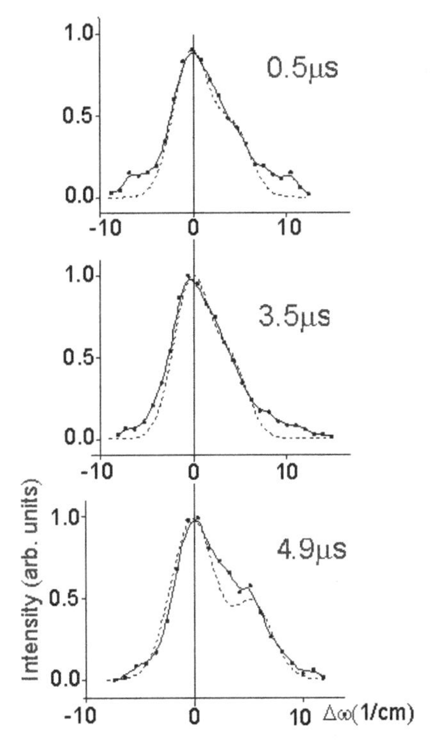

Fig. 5.13. Profiles of the spectral line He I 667.8 nm corresponding to various instants after the start of the plasma current. Dots and solid line — experimental profiles; dashed line — theoretical profiles, allowing in particular for the Stark effect in a quasistatic circularly-polarized electric field **F**. Theoretical profiles correspond to the direction of observation perpendicular to the polarization plane of the field **F**, which is perpendicular to the plasma current. They were calculated for the following field amplitudes F_0:F_0 = 105 kV/cm (τ = 0.5 μs), F_0 = 110 kV/cm (τ = 3.5 μs), and F_0 = 120 kV/cm (τ = 4.9 μs).

were practically symmetric with respect to the vertical lines crossing their maxima. Figure 5.13 shows by solid lines the experimental profiles of He I 667.8 nm recorded at the instants $\tau = 0.5, 3.5$, and 4.9 μs after the start of the plasma current. The observed asymmetry of He I 667.8 nm can be explained only by anisotropically-developed electric fields. It cannot be explained by plasma microfields (which are isotropic): the broadening by the ion microfield would lead to the opposite type of the asymmetry (the red wing more intense than the blue wing), while the broadening by the electron microfield would lead to the symmetric profile. So, the observed blue asymmetry should be attributed to the anisotropic LET dominating over the plasma microfields, as explained below.

The observed type of the asymmetry of He I 667.8 nm can be explained only by the development of the LET, having the elliptical or (as a particular case, circular) polarization of the amplitude F_0 and of the ellipticity degree μ, in the plane perpendicular to the plasma current. For the direction of observation perpendicular to the polarization plane and the LET frequency ω significantly smaller than the spacing ε_{21} between the levels 3^1P and 3^1D, the spectrum of He I 667.8 nm should be similar to the one presented in Fig. 5.14(a). Each of the three spectral components in Fig. 5.14(a) would be additionally broadened by the following three mechanisms: (1) instrumental broadening; (2) Doppler broadening; (3) Stark broadening by plasma microfields. Therefore, the profiles of the three spectral components could overlap, as shown in Fig. 5.14(a) by the dashed line. The resulting profile of He I 667.8 nm could have the maximum close to the position of the most intensive component I_1. The profile would be significantly asymmetric with respect to the vertical line crossing its maximum, the blue part of the profile being more intense than the red part — because $I_2 > I_0$. This is indeed the type of the asymmetry of the observed profiles of He I 667.8 nm.

For the three calculated profiles in Fig. 5.13 providing the best fit, the LET amplitude F_0 was in the range of (105–120) kV/cm. For these values of F_0, the SB of the other studied helium line, He I 587.6 nm, is relatively small. For example, for $F_0 = 120$ kV/cm, the spacing between the spectral components of He I 587.6 nm, that are the most distant from each other is 0.046 nm. This spacing is about three times smaller than the instrumental broadening. Therefore the Stark effect by the LET makes only a relatively insignificant contribution to the broadening of He I 587.6 nm. This explains why the line He I 587.6 nm did not show a noticeable asymmetry — in distinction to the line He I 667.8 nm.

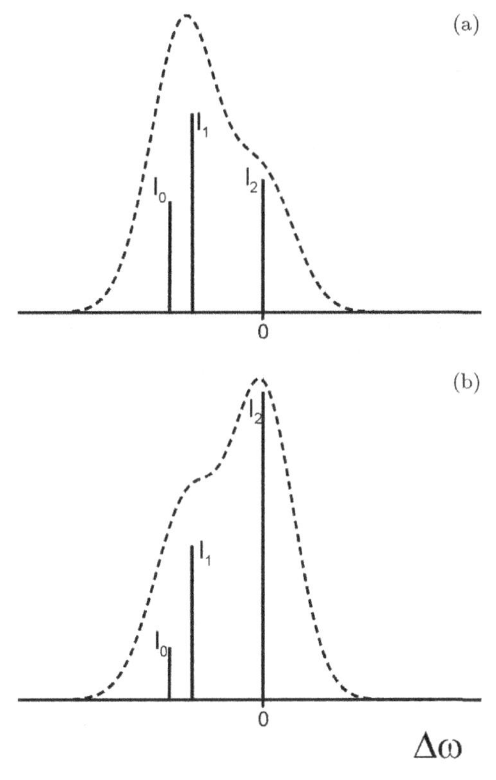

Fig. 5.14. Theoretical Stark splitting of the spectral line He I 667.8 nm in a quasistatic electric field **F** for the direction of observation perpendicular to **F** (Fig. 5.14(a) or parallel to **F** (Fig. 5.14b). The frequency detuning $\Delta\omega = 0$ corresponds to the unperturbed position of the spectral line He I 667.8 nm.

5.2 Strongly Magnetized Plasmas

The average turbulent electric field $\langle E_t \rangle$ can significantly exceed the most probable ion microfield in weakly coupled (a.k.a. ideal) plasmas. The energy density of the turbulent field $\langle E_t^2 \rangle/(16\pi N_e T_e)$ can reach values ~0.1, so that the ratio $\langle E_t \rangle / E_{i\,\max}$ can be as high as

$$\langle E_t \rangle / E_{\text{imax}} \sim \nu^{1/3} >> 1, \tag{5.5}$$

where ν is the number of ions in the sphere of the Debye radius,

$$\nu = [T_{\text{e}}^3/(36\pi N_e e^6)]^{1/2} = 1377[T_e^3/N_e e^6]^{1/2} >> 1, \tag{5.6}$$

(T_e is the electron temperature in Kelvin and N_e is the electron density in cm^{-3}), and $E_{i\max}$ is the most probable ion field. For relatively low-density hydrogen/deuterium plasmas,

$$E_{i\max} = 1.608E_0, \quad E_0 = 2.603eN_e^{2/3} = 1.25 \times 10^{-9} N_e^{2/3}. \quad (5.7)$$

For example, for edge plasmas of tokamaks, such as a low-density discharge in Alcator C-Mod [16], where $N_e \sim 3 \times 10^{13}$ cm^{-3} and $T \sim 5 \times 10^4$ K, one can get $\langle E_t \rangle / E_{i\max} \sim 20$. We note that such a strong LET had been detected via a spectroscopic diagnostic, e.g., in tokamak T-10 [17].

In non-turbulent magnetized plasmas, where either **B,** or n, or both are high enough, the average Lorentz field predominates over the plasma microfield, as shown in Appendix E. So, for turbulent magnetized plasmas we encounter the situations where the two primary broadening mechanisms of hydrogen lines are LET and the Lorentz broadening. In these situations, the shape of hydrogen lines is controlled by the distribution $W(\mathbf{E})$ of the total electric field $\mathbf{E} = \mathbf{E}_t + \mathbf{E}_L$.

In magnetized plasmas, the distribution of the turbulent field $W_t(\mathbf{E}_t)$ should be axially symmetric, the axis of symmetry being along the magnetic field. It is Sholin-Oks' distribution [13] of the following form,

$$\begin{aligned} &W_t(E_t, \cos\theta) dE_t d(\cos\theta) \\ &\quad = [4/(\pi E_{\text{par}}^2)]^{1/2} (E_t^2/E_{\text{perp}}^2) \exp[-E_t^2/E_{\text{perp}}^2 \\ &\qquad - \cos^2\theta (E_t^2/E_{\text{par}}^2 - E_t^2/E_{\text{perp}}^2)] \mathrm{d}E_t d(\cos\theta), \end{aligned} \quad (5.8)$$

where E_{par} and E_{perp} are the root-mean-square values of the turbulent field components parallel and perpendicular to **B**, respectively; θ is the angle between **E** and **B**.

The distribution $W(\mathbf{E})$ of the total electric field $\mathbf{E} = \mathbf{E}_t + \mathbf{E}_L$ is a convolution of the distribution from Eq. (5.8) with the distribution of the Lorentz field $\mathbf{E}_L$, the latter being confined in the plane perpendicular to **B**:

$$\begin{aligned} W_L(\mathrm{E}_L) \mathrm{d}E_L &= (2E_L/E_{LT}^2) \exp(-E_{\mathrm{L}}^2/E_{\mathrm{LT}}^2) dE_L, \\ E_{\mathrm{LT}} &= v_T B/c. \end{aligned} \quad (5.9)$$

Here, E_{LT} is the average Lorentz field expressed via the thermal velocity v_T of the radiating atoms of mass M. The distribution W_L actually reproduces the shape of the two-dimensional Maxwell distribution of atomic velocities in the plane perpendicular to **B**. This is because the absolute value of the Lorentz field $\mathbf{E}_L = \mathbf{v}\times\mathbf{B}/\mathrm{c}$ is $E_L = v_R\mathrm{B}/\mathrm{c}$, where v_R is the component of the atomic velocity perpendicular to **B**.

In the general case, the resulting formula for the convolution is too cumbersome. Therefore, below we focus at two practically important particular cases.

The first case is where the distribution of the turbulent field has the shape of a very *prolate* spheroid, so that essentially it is the following one-dimensional distribution along **B**:

$$W_{\mathrm{t}1}(E_t)dE_{\mathrm{t}} = (2/\pi)^{1/2}(1/E_{\mathrm{par}})\exp[-E_t^2/(2E_{\mathrm{par}}^2)]dE_t. \qquad (5.10)$$

This situation is typical for the ion-acoustic turbulence. The corresponding convolution with the Lorentz field distribution has the form

$$\begin{aligned} &W_1(E,\cos\theta)dE\ \mathrm{d}(\cos\theta)\\ &\quad = dE\ d(\cos\theta)\int_0^\infty dE_L\mathrm{W}_L(E_\mathrm{L})\int_0^\infty dE_t W_{t1}(\mathrm{E}_t)\delta[\cos\theta\\ &\qquad - E_t/(E_t^2+E_L^2)^{1/2}]\delta[E-(E_t^2+E_L^2)^{1/2}]. \qquad (5.11)\end{aligned}$$

After using the two δ-functions to perform the two integrations, the result reduces to the following:

$$\begin{aligned} &W_1(E,\cos\theta)dE\ d(\cos\theta)\\ &\quad = \mathrm{d}E\ d(\cos\theta)W_L(E|\sin\theta|)W_{t1}(E\cos\theta)E/|\sin\theta|. \qquad (5.12)\end{aligned}$$

Based on the total field distribution from Eq. (5.12), the profiles $\mathrm{S}(\Delta\omega)$ of hydrogen lines can be calculated as follows:

$$S(\Delta\omega) = S_\pi(\Delta\omega) + S_\sigma(\Delta\omega), \qquad (5.13)$$

$$S_\pi(\Delta\omega) = \sum_{(\alpha,\beta)\pi} [J_{\alpha\beta}/(k|X_{\alpha\beta}|)]$$
$$\times \int_0 d(\cos\theta) f_\pi(\cos\theta) W_1[\Delta\omega/(k|X_{\alpha\beta}|), \cos\theta], \quad (5.14)$$

$$S_\sigma(\Delta\omega) = \sum_{(\alpha,\beta)\pi} [J_{\alpha\beta}/(k|X_{\alpha\beta}|)] \int_0^1 \mathrm{d}(\cos\theta) f_\sigma(\cos\theta)$$
$$W_1[\Delta\omega/(k|X_{\alpha\beta}|), \cos\theta], \quad (5.15)$$

the quantity k being defined in Eq. (5.25).

The functions $f_\pi(\cos\theta)$ and $f_\sigma(\cos\theta)$ in Eqs. (5.14), (5.15) depend on the direction of observation. For the observation parallel to **B**:

$$f_\pi(\cos\theta) = (1 - \cos^2\theta), f_\sigma(\cos\theta) = (1 + \cos^2\theta)/2. \quad (5.16)$$

For the observation perpendicular to **B**:

$$f_\pi(\cos\theta) = (1 + \cos^2\theta)/2, \quad f_\sigma(\cos\theta) = (3 - \cos^2\theta)/4. \quad (5.17)$$

Now we consider the second practically important case where the distribution of the turbulent field has the shape of a very *oblate* spheroid, so that essentially it is the following two-dimensional distribution in the plane perpendicular to **B**:

$$W_{t2}(E_t)\mathrm{d}E_t = (2E_\mathrm{t}/E_\mathrm{perp})\exp[-E_t^2/E_\mathrm{perp}^2)]\mathrm{d}E_t. \quad (5.18)$$

This situation is typical for the Bernstein modes. Obviously, in this case, the total electric field $\mathbf{E} = \mathbf{E}_t + \mathbf{E}_L$ is confined in the plane perpendicular to **B**. Its distribution has the form:

$$W_2(E)\mathrm{d}E = \mathrm{d}E(1/\pi) \int_0^\infty \mathrm{d}E_L W_L(E_L) \int_0^\infty \mathrm{d}E_t W_{t2}(E_t)$$
$$\times \int_0^\pi \mathrm{d}\psi\delta[E - (E_t^2 + E_L^2 + 2E_tE_L\cos\psi)^{1/2}]. \quad (5.19)$$

After using the δ-function to perform the integration over the angle ψ (which is the angle between vectors $\mathbf{E}_t$ and $\mathbf{E}_L$), the result reduces to the following:

$$W_2(E)\mathrm{d}E = \mathrm{d}E(E/\pi)\int_0^{\infty} \mathrm{d}E_{\mathrm{t}} W_{\mathrm{t}2}(\mathrm{E_t})\int_{Emin}^{Emax} \mathrm{d}E_L W_L(E_L)/$$
$$\{[(E_t + E_L)^2 - E^2][E^2 - (E_t - E_L)^2]\}^{1/2}, \qquad (5.20)$$

where the lower and upper limits of the integration over E_L are

$$E_{\mathrm{min}} = E - E_{\mathrm{t}}, E_{\mathrm{max}} = E + E_{\mathrm{t}}. \qquad (5.21)$$

Based on the total field distribution from Eq. (5.20), the profiles $\mathrm{S}(\Delta\omega)$ of hydrogen lines can be calculated as follows:

$$\mathrm{S}(\Delta\omega) = \mathrm{S}_\pi(\Delta\omega) + \mathrm{S}_\sigma(\Delta\omega), \qquad (5.22)$$

$$S_\pi(\Delta\omega) = \sum_{(\alpha,\beta)\pi} [J_{\alpha\beta}/(k|X_{\alpha\beta}|)] f_\pi W_2[\Delta\omega/(k|X_{\alpha\beta}|)], \qquad (5.23)$$

$$S_\sigma(\Delta\omega) = \sum_{(\alpha,\beta)\sigma} [J_{\alpha\beta}/(k|X_{\alpha\beta}|)] f_\sigma W_2[\Delta\omega/(k|X_{\alpha\beta}|)]. \qquad (5.24)$$

As an example, Fig. 5.15 presents the profile of the Balmer line H_{18} calculated by Eqs. (5.23)–(5.25) for the case of $E_{\mathrm{LT}} = 2^{1/2}E_{\mathrm{par}}$, where $E_{\mathrm{LT}} = v_T B/c$ and E_{par} is the root-mean-square values of the turbulent field component parallel to **B**. The argument $\beta = \Delta\omega/(kE_{LT})$ is a scaled dimensionless detuning from the line center; the quantity k denotes the following:

$$k = 3\hbar/(2m_e e). \qquad (5.25)$$

It should be noted that while the argument of the line profiles in the above equations is the detuning $\Delta\omega$ from the unperturbed frequency ω_0 of the spectral line, the corresponding profiles in the wavelength scale $\mathrm{S}(\Delta\lambda)$, i.e., where the argument is the detuning from the unperturbed wavelength of the spectral line, can be obtained from the same equations by changing the constant $k = 3\hbar/(2m_e e)$ in Eqs. (5.23), (5.24) to

$$k_0 = [3\hbar/(2m_e e)]\lambda_0^2/(2\pi c), \qquad (5.26)$$

where λ_0 is the unperturbed wavelength.

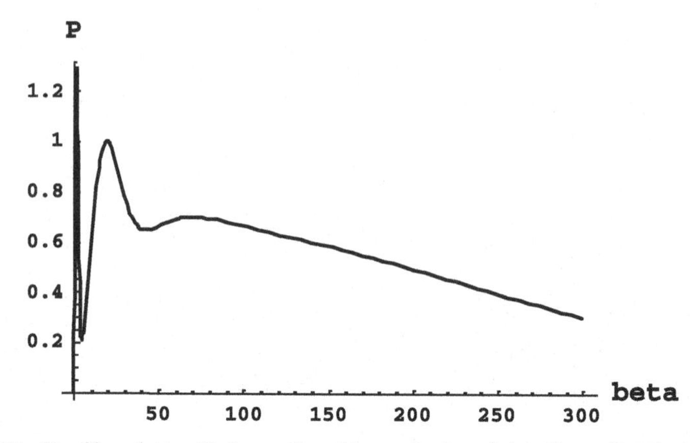

Fig. 5.15. Profile of the Balmer line H_{18} calculated by Eqs. (5.20)–(5.22) for the case of $E_{LT} = 2^{1/2}E_{par}$, where $E_{LT} = v_T B/c$ and E_{par} is the root-mean-square values of the turbulent field component parallel to **B**. The argument $\beta = \Delta\omega/(kE_{LT})$ is a scaled dimensionless detuning from the line center, k being defined in Eq. (5.25). The intensity is relative to the intensity of the 2nd maximum (whose intensity is set as unity). The 1st maximum would be smeared out by the Doppler broadening.

References

[1] H.R. Griem, *Spectral Line Broadening by Plasmas*, Academic, New York (1974).
[2] E. Oks, *Stark Broadening of Hydrogen and Hydrogenlike Spectral Lines in Plasmas: The Physical Insight*, Alpha Science International, Oxford, UK (2006).
[3] Yu. Demkov, B. Monozon and V. Ostrovsky, *Sov. Phys. JETP* **30** 775 (1970).
[4] E. Oks, *Sov. Astron. Lett.* **4** 223 (1978).
[5] A.N. Koval and E. Oks, *Bull. Crimean Astrophys. Observ.* **67** 78 (1983) (*Izv. Krym. Astrofiz. Obs.* **67** 90 (1983)).
[6] E. Oks and G.V. Sholin, *Sov. Phys. Tech. Phys.* **21** 144 (1976).
[7] E. Oks and R.E. Gershberg, *Astrophys. J.* **819** 16 (2016).
[8] R.E. Gershberg and N.I. Shakhovskaja, *Astron. J* (USSR) **48** 934 (1971).
[9] A.N. Kulapova and N.I. Shakhovskaja, *Izv. Krym. Astrofiz. Obs.* **48** 31 (1973).
[10] R.E. Gershberg, *Izv. Krym. Astrofiz. Obs.* **51** 117 (1974).
[11] M.M. Katsova, *Astron. J* (USSR) **67** 1219 (1990).
[12] I. Crespo-Chacon *et al.*, *Astron. Astrophys.* **452** 587 (2006).
[13] G.V. Sholin and E. Oks, *Sov. Phys.-Doklady*, **18** 254 (1973).

[14] M.V. Babykin, A.I. Zhuzhunashvili, E. Oks, V.V. Shapkin and G.V. Sholin, *Sov. Phys. JETP* **38** 86 (1974).
[15] A.G. Frank, V.P. Gavrilenko, N.P. Kyrie and E. Oks, *J. Phys. B: Atom. Mol. Opt. Phys.*, **39** 5119 (2006).
[16] B.L. Welch, H.R. Griem, J. Terry, C. Kurz, B. LaBombard, B. Lipschultz, E. Marmar and J. McCracken, *Phys. Plasmas* **2** 4246 (1995).
[17] V.P. Gavrilenko, E. Oks and V.A. Rantsev-Kartinov, *JETP Lett.* **44** 404 (1986).

Chapter 6

Principles of Spectroscopic Diagnostics of Plasmas Containing Quasimonochromatic Electric Fields (QEF)

Quasimonochromatic Electric Fields (QEF) in plasmas are single-frequency electric fields. They can be represented either by a *single-mode* field $\mathbf{E}(t) = \mathbf{E_0} \cos(\omega t + \varphi)$ or a *multi-mode* field $\mathbf{E}(t) = \Sigma_k \mathbf{E_{0k}} \cos(\omega t + \varphi_k)$; in the latter case the phase φ_k is uniformly distributed in the interval $(0, 2\pi)$.

Interactions of three subsystems are essential for diagnostics of QEF in plasmas: a radiator (R), QEF (F), and plasma (P). Their interactions are denoted below as V_{RP}, V_{RF}, and V_{FP}. Historically, the first theoretical works, which contributed to the diagnostics of QEF in plasmas, considered only the interaction V_{RF} of the subsystems R and F. The effect of QEF on hydrogenic lines was calculated analytically in 1933 by Blochinzew [1] for a single-mode QEF (see Appendix L for details) and in 1967 by Lifshits [2] for a multi-mode QEF (by "hydrogenic" lines we mean spectral lines of atomic hydrogen and of H-like ions). The effect of QEF on non-hydrogenic lines was calculated analytically in 1961 by Baranger and Mozer [3] for a multi-mode QEF and in 1969 by Cooper and Ringler [4] for a single-mode QEF. The outcome of the papers [1–4] was various structures of *satellites* at multiples of the QEF frequency. It should be emphasized that in [3, 4], the QEF was treated approximately by the standard perturbation theory, while in [1, 2] were found exact (non-perturbative) solutions. For non-hydrogenic lines, analytical results valid beyond the scope of the standard perturbation theory, i.e., valid for stronger QEF, were obtained later by Oks and Gavrilenko [5]. They also found non-perturbative analytical results for the effect of strong QEF on hydrogenic lines with the allowance for the fine structure [6] (the fine structure was disregarded in [1, 2]).

Other theoretical works subsequent to [1–4] considered a joint effect of the interaction V_{RF} and of the *quasistatic* part of the interaction V_{RP}, the latter being represented by the interaction of the radiator with a low-frequency ion microfield and/or low-frequency plasma turbulence. Those are numerical simulations by Kalman's group [7] and analytical results by Oks' group [8–10]. In particular, it was found in [8–10] that *resonances* between V_{RF} and the *quasistatic* part of the interaction V_{RP} yield *dips* in spectral line profiles. These dips were also found experimentally: first in [8] and then in a number of later experiments at various plasma sources (see, e.g. the experiments presented in book [11]). Gavrilenko and Oks calculated analytically also joint *non-resonant* effects of the interaction V_{RF} and of the *quasistatic* part of the interaction V_{RP}. The non-resonant effects are a partial anisotropic suppression of the quasistatic broadening by QEF for hydrogenic lines [12] (also confirmed experimentally in [12]) and a significant modification of intensities of satellites of dipole-forbidden non-hydrogenic lines [13].

The above theoretical works, while being important for the diagnostics of QEF in plasmas, dealt with a simplified picture. In 1984, Oks [14] presented a comprehensive analysis of the situation in all of its complexity: he developed general principles of the spectroscopic diagnostics of plasmas containing QEF and illustrated them by a number of applications. (In 1986, he also published a more detailed paper [15]).

The main idea of papers [14, 15] concerning the spectral line broadening was the following (the term "broadening" includes both the shape and the shift). *The spectral line broadening, caused by the interaction of the radiator with one of the subsystems* (P *or* F) *can be significantly affected by the interaction of the radiator with the other subsystem, as well as by the interaction of* P *and* F *with each other.* Generally, each of the interactions V_{RP} and V_{RF} should not be considered in isolation from the other one because their characteristic temporal parameters can "entangle" and thus couple these interactions. In other words, *the conventional approach, where one would first calculate two independent profiles of the same spectral line (one — caused by* V_{RP} *another — cause by* V_{RF}*) and then perform the convolution of the two profiles, would generally yield incorrect results.*

This idea and its applications were the central theme of book [11]. The book presented all of the results from [1–10, 12–15], as well as a large number of new, previously unpublished results on spectroscopy of plasmas containing QEF. In addition to the general principles, the book presented numerous applications to hydrogenic spectral lines (both with and without the allowance for the fine structure), to spectral lines of He, as well as to spectral lines of He-like and Li-like ions — in various regimes of interactions between, the subsystems R, P and F.

Here, we present a brief overview of the general principles of the spectroscopic diagnostics of plasmas containing QEF in the spirit of the papers [14, 15]. The physics of the spectral line broadening in plasmas containing QEF is very rich and complex due to the interplay of a large number of characteristic times and frequencies. There are seven following characteristic frequencies, which can be considered as "elementary" parameters.

1. $\Delta\omega$ — detuning from the unperturbed position of a given spectral line of the radiator. It affects the characteristic value of the argument τ of the correlation function.
2. $\omega_{pe}(N_e)$ — plasma electron frequency, which is also the inverse characteristic time of the formation of the screening by electrons (N_e is the electron density).
3. $\Omega_e(N_e, T_e) = v_{Te}/\min(\rho_{N_e}, \rho_{We})$ — characteristic frequency of the variation of the electron microfield, which is responsible

for the *homogeneous* broadening by electrons.[1] Here, T_e is the electron temperature, $v_{T_e} = (T_e/m_e)^{1/2}$ is the electron thermal velocity, $\rho_{N_e} \sim 1/N_e^{1/3}$ is the mean interelectronic distance, $\rho_{W_e} \sim n^2\hbar/(m_e v_{T_e})$ is the electron Weisskopf radius (n is the principal quantum number).

4. $\Omega_i(N_i, T_i) = v_{Ti}/\min(\rho_{N_i}, \rho_{W_i})$ — characteristic frequency of the variation of the *dynamic part* of the ion microfield, which is responsible for the *homogeneous* broadening by ions. Here, T_i is the ion temperature, $v_{Ti} = (T_i/m_i)^{1/2}$ is the ion thermal velocity, $\rho_{Ni} \sim 1/N_i^{1/3}$ is the mean interionic distance, $\rho_{Wi} \sim n^2\hbar/(m_e v_{Ti})$ is the ion Weisskopf radius.
5. ω — QEF frequency.
6. γ — homogeneous width of the power spectrum of QEF, which is also the inverse of the QEF *coherence time* τ_F .
7. $\delta_s(E_0)$ — instantaneous Stark shift at the amplitude value of QEF. For example, $\delta_s(E_0) = a_1 E_0$ in the case of the linear Stark effect or $\delta_s(E_0) = a_2 E_0^2$ in the case of the quadratic Stark effect; $a_1(k)$, $a_2(k)$ are Stark constants that depend on the set of quantum numbers of the particular states of the radiator. *Here and below, the set of quantum numbers is denoted by k.*

We note that the frequencies from items 2–4 characterize the subsystem P and its interaction with the subsystem R, while the frequencies from items 5–7 characterize the subsystem F and its interaction with the subsystem R.

On the basis of the above seven "elementary" frequencies, there occur four composite parameters that have various characteristic times as follows.

1. $\tau_e(k, N_e, T_e, \Delta\omega) \sim \min(1/\Omega_e, 1/\omega_{pe}, 1/\Delta\omega)$ — characteristic time of the formation of the homogeneous broadening by electrons.

[1]The Stark broadening (SB) of a spectral line is *homogeneous* when it is the same for all radiators. A typical example is the SB by the electron microfield. In distinction, the SB by the quasistatic part F_{qs} of the ion microfield is *inhomogeneous* because different radiators are subjected to generally different values of F_{qs}.

2. $\tau_i(k, N_i, T_i, N_e, \Delta\omega) \sim \min(1/\Omega_i, 1/\omega_{pe}, 1/\Delta\omega)$ — characteristic time of the formation of the homogeneous broadening by dynamic part of ions.
3. $\tau_{QS}(k, E_0, \omega)$ — characteristic time of the formation of Quasienergy States (QS):

$$\tau_{QS}(k, E_0, \omega) \sim \min(1/(\omega^2\delta_s)^{1/3}, 1/\omega). \tag{6.1}$$

Being subjected to QEF, the states of the radiator can oscillate with the QEF frequency ω. This effect is described as the emergence of QS, which were introduced in 1967 by Zel'dovich [16] and Ritus [17] (independently of each other). The above formula for $\tau_{QS}(k, E_0, \omega)$ was derived by Oks in [14]. So, for relatively weak QEF, the QS are formed at the timescale of the order of the period of the QEF $1/\omega$. However, for relatively strong QEF, the QS are formed at a much shorter time scale proportional to $1/E_0^{1/3}$ or to $1/E_0^{2/3}$ in the cases of the linear or quadratic Stark effect, respectively.
4. $\tau_{\text{life}}(k, N_e, T_e, N_i, T_i, \gamma, \omega, E_0, \Delta\omega)$ — the lifetime of the exited state of the radiator:

$$\tau_{\text{life}}(k, N_e, T_e, N_i, T_i, \gamma, \omega, E_0, \Delta\omega) \sim 1/\Gamma, \tag{6.2}$$

$$\Gamma = \gamma_e(k, N_e, T_e, \Delta\omega) + \gamma_i(k, N_i, T_i, N_e, \Delta\omega) + \gamma_F(k, \gamma, \omega, E_0). \tag{6.3}$$

Here, Γ is the sum of the *homogeneous* Stark widths due to electrons, dynamic part of ions, and QEF. The contribution $\gamma_F(k, \gamma, \omega, E_0)$ from QEF was calculated by Oks and Sholin in 1975 [18].

We emphasize that τ_e and τ_i entangle parameters of the subsystems P and R, τ_{QS} entangles parameters of the subsystems F and R, and τ_{life} entangles parameters of all three subsystems: P, F and R.

Below we provide several examples showing how the interplay of the above four characteristic times controls the physics of the SB in plasmas containing QEF.

First, we consider the profile of a hydrogenic line subjected to QEF $\mathbf{E}(t) = \Sigma_k \mathbf{E_{0k}} \cos(\omega t + \varphi_k)$ in a plasma, the phase φ_k being uniformly distributed in the interval (0, 2π). In this situation, each

Stark component of the line splits up into satellites at frequencies $\Delta\omega_p = p\omega(p = 0, \pm 1, \pm 2, \ldots)$. This is a consequence of the emergence of QS of the radiator. In the formal solution of the Schrödinger equation, there is generally an infinite number of satellites. However, if the detuning $\Delta\omega$ is large enough, so that $\tau_{QS} \gg 1/\Delta\omega$, then QS did not have time to form: the relevant time in the correlation function $\tau \sim 1/\Delta\omega \ll \tau_{QS}$. Thus, the actual number of satellites is limited by some number $p_{\max}$, which can be estimated from the requirement $\tau_{QS} \leq 1/\Delta\omega = 1/(p\omega)$: $p_{\max} \sim 1/(\omega\tau_{QS})$.

The intensities of the satellites significantly depend on the relation between the lifetime τ_{life} of the excited state of the radiator and the coherence time $\tau_F \sim 1/\gamma$ of QEF. In the case where $\tau_{\text{life}} \ll \tau_F$, the radiator perceives QEF as *single-mode* field $\mathbf{E}(t) = \mathbf{E_0} \cos(\omega t + \varphi)$. Consequently, for satellites intensities $I(p, \mu)$ one obtains Blochinzew's result [1]:

$$I(p, \mu) = J^2_{|p|}(\mu), \tag{6.4}$$

where $J_p(\mu)$ are Bessel functions and $\mu = \delta_s(E_0)/\omega$ is the strength of the modulation of the "light" wave emitted by the radiator (here the "light" wave refers to any part of the spectrum of the electromagnetic radiation). In the opposite case of $\tau_{\text{life}} \gg \tau_F$, the radiator "feels" the multi-mode character of QEF. Consequently, for satellites intensities $I(p, \mu)$ one obtains Lifshits's result [2]:

$$I(p, \mu) = I_{|p|}(\mu^2/2)\exp(-\mu^2/2), \tag{6.5}$$

where $I_{|p|}(\mu^2/2)$ are modified Bessel functions.

The analytical solutions from [1, 2] were usually regarded as two mutually excluding alternatives. However, from the above, it follows that they can be considered as two limiting cases of the more general model.

We note that in the case of the weak modulation of the "light" wave by QEF ($\mu \ll 1$), there are only few satellites observed because their intensities decrease very rapidly with the number p. In the opposite case of the strong modulation of the "light" wave by QEF ($\mu \gg 1$), one encounters the multi-satellite regime. In this regime, a very useful physical quantity is the shape of the envelope of satellites intensities. For Lifshits's satellites, the envelope is very close to the shape of Gauss distribution. For Blochinzew's satellites, the envelope has a more complicated structure showing numerous oscillations. Around the most intense satellites, this shape can be well approximated by Airy function, as shown by Oks [14], what allows obtaining useful analytical formulas for halfwidths of the multi-component hydrogenic spectral lines [14]. Thus, in this case, there would practically be no need for numerical simulations of QS, which becomes computationally-expensive in the multi-satellite regime.

As the second example, we consider the interrelation between the homogeneous SB and QEF. In the case where the characteristic time of the formation of QS is much smaller than the lifetime of the excited state of the radiator, that is where

$$\tau_{\mathrm{QS}}(k, E_0, \omega) \ll \tau_{\mathrm{life}}(k, N_e, T_e, N_i, T_i, \gamma, \omega, E_0, \Delta\omega), \tag{6.6}$$

the homogeneous SB is experienced by QS, rather than by the usual states of the radiator. In particular, the contributions to the Stark widths due to electrons $\gamma_e(k, N_e, T_e, \Delta\omega)$ and/or dynamic part of ions $\gamma_i(k, N_i, T_i, N_e, \Delta\omega)$, can acquire factors $I(p, \mu)$ given by Eqs. (6.4) or (6.5), as noted in [14]. The most interesting is the case where those factors correspond to Eq. (6.4), so that $I(p, \mu) = J^2_{|p|}(\mu)$. Since, the Bessel functions $J_{|p|}(\mu)$ are oscillatory functions of their argument and they vanish at certain values of μ, then in this case QEF may dramatically diminish γ_e and/or γ_i (and thus the width of the entire line) when the amplitude and the frequency of QEF are such, that the Bessel function $J_{|p|}(\mu)$ vanishes.

We note that a certain role is also played by the relation between τ_e (and/or τ_i) and τ_{QS}. For example, this is the case where the homogeneous Stark widths due to electrons γ_e is calculated in a formalism that requires a lower cutoff for impact parameters leading to the factor $\ln(\rho_{\max}/\rho_{\min}) = \ln(\Omega_{\max}/\omega_{pe})$ in γ_e. In the subcase, where $\tau_{\mathrm{QS}} \ll \tau_e$ (and in turn, $\tau_e \ll \tau_{\mathrm{life}}$), the upper cutoff of frequencies is $\Omega_{\max} \sim 1/\tau_e$. This is the conventional subcase. However, in the opposite subcase where $\tau_e \ll \tau_{\mathrm{QS}}$ (and in turn, $\tau_{\mathrm{QS}} \ll \tau_{\mathrm{life}}$), the upper cutoff of frequencies should become $\Omega_{\max} \sim 1/\tau_{\mathrm{QS}}$. In this unconventional subcase, the homogeneous Stark width caused by electrons acquires an additional dependence on parameters of QEF — via $\ln(1/\omega_{pe}\tau_{QS})$.

Thus, in the case given by Eq. (6.6), the homogeneous Stark widths caused by electrons and the dynamic part of ions *directly* depend on the amplitude E_0 and the frequency ω of QEF: first of all, via the argument $\mu = \delta_s(E_0)/\omega$ of the Bessel functions and, in the subcase where $\tau_e \ll \tau_{QS}$ — also via $\ln(1/\omega_{pe}\tau_{QS})$. As a result, *the total homogeneous Stark width* $\Gamma = \gamma_e + \gamma_i + \gamma_F$ *depends on the QEF amplitude* E_0 *and frequency* ω *in a rather complicated way.* For example, as the QEF amplitude E_0 increases, the first two terms in the quantity Γ decrease, while the third term — increases.

In the case opposite to Eq. (6.6), where,

$$\tau_{QS}(k, E_0, \omega) \gg \tau_{life}(k, N_e, T_e, N_i, T_i, \gamma, \omega, E_0, \Delta\omega), \tag{6.7}$$

QS do not form during the lifetime of the excited state of the radiator. In this case, the homogeneous SB does not get modified by QEF. Moreover, in this case, QEF can be treated as quasistatic.

As the third example, we consider how QEF can *indirectly* affect some features of spectral lines in plasmas. Specifically, we present here how the interaction V_{FP} of QEF with the plasma modifies the electron velocity distribution and thus causes an *additional shift* of spectral lines of the subsystem R — as it was first shown in 1984 by Oks [14]. In the case where $\tau_e \ll 1/\omega$, relatively fast passages of plasma electrons by the radiator get superimposed with their relatively slow oscillations due to QEF. Under the additional condition $\tau_{life} \ll \tau_F = 1/\gamma$, the radiator perceives QEF as *single-mode* field $\mathbf{E}(t) = \mathbf{E_0}\cos(\omega t + \varphi)$ — as noted above. Then the electron velocity can be represented in the form: $\mathbf{V} = \mathbf{V}_M + \mathbf{v}$. Here, $\mathbf{V}_M$ is characterized by the isotropic Maxwell distribution and $\mathbf{v}$, being parallel to $\mathbf{E_0}$, has the distribution $\pi^{-1}(2v_0^2 - v^2)^{-1/2}$, where $v_0^2 = [eE_0/(m_e\omega)]^2/2$. *Because of the anisotropy of the distribution of the total electron velocity* $\mathbf{V}$, *the first order term in the expansion of the evolution operator does not vanish under the angular averaging — in distinction to the conventional situation.* As a result, there emerges an additional shift of energy levels of the radiator and, consequently — of the spectral lines emitted by the radiator. This shift was named in [14] the *Electron Oscillatory Shift* (*EOS*).

The dominant contribution to EOS originates from the *quadrupole* interaction of the radiator with plasma electrons. Analytical expressions for the EOS with the allowance for the quadrupole interaction and for all higher multipole interactions were derived for hydrogenic lines in [14, 19] and presented in the book [11].

In 1992, Chichkov, Shumsky, Uryupin published a paper [20], where they considered essentially the same effect of the modification of the electron velocity distribution by QEF. They made an allowance for the collisional relaxation of plasma electrons. However, in laser-produced plasmas, the frequencies of the electron–electron collisions (ν_{ee}) and of the electron-ion collisions (ν_{ei}) are usually much smaller than the inverse lifetime of the radiator $1/\tau_{life}$. Therefore, in this situation, the allowance for the collisional relaxation of plasma electrons practically does not affect shapes and shifts of spectral lines. It should be noted that the authors of [20] did not attempt to calculate shapes and shifts of spectral lines. So, their paper [20], being limited to the calculations of the electron velocity distribution, is a good work in its own right, but it has virtually no additional practical consequences for shapes and shifts of spectral lines compared to the EOS.

The general principles of the spectroscopic diagnostics of plasmas containing QEF developed as early as in 1984 [14] and outlined above, extensively employ the concept of QS, also known as "dressed" atomic states. The latter name refers to the fact that QS are not the states of a "naked" atom, but rather are the states of the combined system "atom plus QEF". For completeness, we note that later the formalism of dressed atomic states was generalized: in 1994 Ispolatov and Oks [21] introduced atomic states dressed by the *broadband* microfield of plasma electrons, rather than by QEF — see Appendices C and D.

An additional advance was presented by Sauvan and Dalimier [22], where the authors transformed a so-called Floquet–Liouville formalism (previously developed for nonlinear optics [23]) into the powerful technique for simulations of spectral line shapes in plasmas containing QEF. The advantage of this formalism is that — compared to the previous ones — it is more efficient for the lineshape calculations involving the dynamic evolution of a large number of atomic states with the allowance for their relaxation caused by the plasma environment. Details are presented in Appendix K.

We re-iterate that the interplay of the four characteristic times presented above ($\tau_e, \tau_i, \tau_{QS}, \tau_{life}$) leads to a large variety of physically-different situations and is the reason for the complexity and richness of the research area of the spectral line broadening in plasmas containing QEF. In particular, laser-produced plasmas under various experimental conditions could correspond to different scenarios outlined above.

References

[1] D.I. Blochinzew, *Phys. Z. Sow. Union* **4** 501 (1933).
[2] E.V. Lifshits, *Sov. Phys. JETP* **26** 570 (1968).
[3] M. Baranger and B. Mozer, *Phys. Rev.* **123** 25 (1961).
[4] W.S. Cooper and H. Ringler, *Phys. Rev.* **179** 226 (1969).
[5] E. Oks and V.P. Gavrilenko, *Sov. Tech. Phys. Lett.* **9** 111 (1983).
[6] V.P. Gavrilenko and E. Oks, *Sov. J. Quantum Electron.* **13** 1269 (1983).
[7] A. Cohn, P. Bakshi and G. Kalman, *Phys. Rev. Lett.* **29** 324 (1972).
[8] A.I. Zhuzhunashvili and E. Oks, *Sov. Phys. JETP* **46** 1122 (1977).
[9] V.P. Gavrilenko and E. Oks, *Sov. Phys. JETP* **53** 1122 (1981).
[10] V.P. Gavrilenko and E. Oks, *Sov. J. Plasma Phys.* **13** 22 (1987).
[11] E. Oks, *Plasma Spectroscopy. The Influence of Microwave and Laser Fields*, Springer, New York (1995).
[12] V.P. Gavrilenko, E. Oks and V.A. Rantsev-Kartinov, *JETP Lett.* **44** 207 (1985).
[13] V.P. Gavrilenko and E. Oks, *Optic. Commun.* **56** 415 (1986).
[14] E. Oks, *Sov. Phys. Doklady* **29** 224 (1984).
[15] E. Oks, *Measurement Techniques* **29** 805 (1986).
[16] Ja.B. Zel'dovich, *Sov. Phys. JETP* **24** 1006 (1967).
[17] V.I. Ritus, *Sov. Phys. JETP* **24** 1041 (1967).
[18] E. Oks and G.V. Sholin, *Sov. Phys. JETP* **41** 482 (1975).
[19] I.M. Gaisinsky and E. Oks, in: *Correlations and Relativistic Effects in Atoms and Ions*, USSR Acad. Sci. Research Council on Spectroscopy, Moscow, p. 106 (1986).

[20] B.N. Chichkov, S.A. Shumsky and S.A. Uryupin, *Phys. Rev. A* **45** 7475 (1992).

[21] Ya. Ispolatov and E. Oks, *J. Quant. Spectrosc. Rad. Transfer* **51** 129 (1994).

[22] P. Sauvan and E. Dalimier, *Phys. Rev. E* **79** 036405 (2009).

[23] T.S. Ho, K. Wang and Shih-I Chu, *Phys. Rev. A* **33** 1798 (1986).

Chapter 7

Langmuir Waves

7.1 Introductory Remarks

Langmuir waves/turbulence is the high-frequency branch of the electrostatic waves in plasmas. Its characteristic frequency is the plasma electron frequency $\omega_{\rm pe} = (4\pi e N_e/m_e)^{1/2}$. Langmuir waves are quasimonochromatic: the width of their frequency band $\delta\omega \ll \omega_{\rm pe}$.

While the low-frequency electrostatic turbulence is typically perceived by radiating atoms/ions (radiators) as quasistatic, Langmuir waves cause dynamic spectroscopic phenomena. The time evolution of the radiator may acquire a dynamic character in spite of relaxation processes: the averaged motion of the optical electron may be described in terms of precession, nutation, etc. This causes new components in the radiation spectra: for example, there may occur satellites separated from the unperturbed frequency of the spectral line ω_0 by integer multiples of $\omega_{\rm pe}$.

Here are some more details. The energy spectrum of a radiator in a plasma consists of multiplets having relatively large separations from each other of the order of ω_0. Each multiplet has a "fine structure" of the characteristic scale $\Delta \ll \omega_0$. Typically, the frequency of the Langmuir waves $\omega_{\rm pe} \ll \omega_0$, but the ratio $\omega_{\rm pe}/\Delta$ can take any value. There are three interacting subsystems: the radiator, Langmuir wave, and the plasma medium. The interaction of these three subsystems can significantly change the fine structure of the radiator multiplets. In its turn, it changes the observed spectrum of the (usually spontaneous) radiative transitions that occur at the frequency of the order ω_0.

In spectra of helium atoms and helium-like ions, as well as lithium atoms and lithium-like ions (all of which will be called here non-Coulomb radiators, for brevity), the Langmuir waves can manifest as few satellites. In relatively low-density plasmas, the nearest satellites appear at the distances $\pm\omega_{\mathrm{pe}}$ from the dipole-forbidden spectral line. (Those non-Coulomb radiators can have a dipole-forbidden spectral line in the vicinity of the dipole-allowed spectral line; the former can show up due to the intermixing of the unperturbed wave functions of the multiplet sublevels by the quasistatic electric field $\mathbf{F}$ in plasmas.) In relatively high-density plasmas, the nearest satellites can also appear at the distances $\pm\omega_{\mathrm{pe}}$ from the dipole-allowed spectral line. (This is again due to the intermixing of the unperturbed wave functions of the multiplet sublevels by the quasistatic electric field F in plasmas.) For the most frequently used spectral lines of neutral helium, the borderline values N_e^{cr}, separating the above two cases, are as follows: $N_e^{\mathrm{cr}} \sim 10^{14}\,\mathrm{cm}^{-3}$ for He I 447.2 nm and 492.2 nm, $N_e^{\mathrm{cr}} \sim 10^{16}\,\mathrm{cm}^{-3}$ for He I 501.6 nm and 667.8 nm. Detailed theories of satellites of dipole-forbidden spectral lines of non-Coulomb radiators are presented in Appendix J.

For hydrogen atoms and hydrogenlike ions (Coulomb radiators), the situation is more complicated. On the one hand, the effect of Langmuir waves in the absence of any quasistatic electric field would be satellites at the distances $\pm\omega_{\mathrm{pe}}$, $\pm 2\omega_{\mathrm{pe}}$, $\pm 3\omega_{\mathrm{pe}}, \ldots$ from the unperturbed position of the spectral line. If the Langmuir wave is strong enough, such that the instantaneous Stark splitting in the frequency scale $\Delta\omega(E_0) = (n^2 - n'^2)\hbar E_0/(Zm_e e)$ at the amplitude value E_0 of the Langmuir electric field is much greater than ω_{pe}, far satellites have a significant statistical weight: even under a relatively small additional (Doppler or instrumental) broadening with a characteristic scale $\Delta\omega_{1/2}^{\mathrm{add}} \sim \omega_{\mathrm{pe}}$, one would observe a broad satellites envelope of the characteristic width $\sim \Delta\omega(E_0)$. Here, n and n' are, respectively, the principal quantum numbers of the upper and lower energy levels involved in the radiative transition, Z is the nuclear charge of the radiator.

On the other hand, the combined action of the Langmuir electric field $\mathbf{E}_0 \cos \omega_{\mathrm{pe}} t$ and the quasistatic field $\mathbf{F}$ manifests as dips (or depressions) at the specific locations in the spectral line profile — called L-dips for brevity. The detailed theory of L-dips is presented in Appendix G for non/weakly-magnetized plasmas and in Appendix H.3 for strongly magnetized plasmas.

Both Langmuir-wave-caused satellites (L-satellites) and L-dips might appear in approximately the same range of wavelength $\Delta\lambda \sim \omega_{\mathrm{pe}}\lambda_0^2/(2\pi c)$, where λ_0 is the unperturbed wavelength of the spectral line. So, there is a legitimate question: what would show up in experimental profiles of hydrogenic spectral lines in plasmas containing Langmuir waves: L-satellites, or L-dips, or both? It was proven analytically in Sec. 7.1 of the book [1] that the primary manifestation of Langmuir waves would be L-dips, rather than L-satellites. This means that *if a sequence of peaks and troughs is seen in an experimental profile of a hydrogenic spectral line, the attention should be focused at the positions of the troughs (rather than the positions of the peaks) while attempting to interpret the observed structure as being caused by Langmuir waves.*

The maximum possible amplitude E_0 of the Langmuir field is controlled by the inequality:

$E_0 \ll (8\pi N_e T_e)^{1/2}$. Physically, this inequality means that the energy density of the Langmuir waves should be much smaller than the thermal energy density of plasma electrons. For non-Coulomb radiators, this estimate helps choosing a spectral line that would satisfy two conditions: (1) the expected L-satellites would be sufficiently intense; (2) competing broadening effects, such as Doppler, Zeeman, self-absorption, and Stark broadening by electrons and by the dynamical part of the ion microfield, would not prevent the detection of L-satellites. For Coulomb radiators, the above inequality should be used for estimating the maximum possible halfwidth $\Delta\lambda_{1/2}^{\mathrm{dip}}$ of the expected L-dips: $\Delta\lambda_{1/2}^{\mathrm{dip}} \sim [n^2\hbar E_0/(Zm_e e)]\lambda_0^2/(2\pi c)$. This estimate allows selecting hydrogenic spectral lines, for which L-dips would not be washed out by the competing broadening effects listed above.

Since, the detailed theories of the above phenomena are presented in Appendices G, H.3 and J, below we focus on examples of the experimental works, where parameters of Langmuir waves were deduced from the experimental lineshapes using the corresponding theories.

7.2 Measurements Using Satellites of Dipole-forbidden Spectral Lines of Helium and Lithium

Kunze and Griem in their paper of 1968 [2] were the first to claim the experimental observation of satellites of the lines He I 447.2 nm and 492.2 nm during the implosion phase of a low-density θ pinch in helium ($N_e \sim (3-4) \times 10^{13}\,\mathrm{cm}^{-3}$). Specifically, their experimental profile of the line He I 447.2 nm seem to show two satellites, each located on the different side compared to the would-be location of the dipole-forbidden line: the "far" satellite (the satellite located further away from the allowed line) and the "near" satellite (the satellite located closer to the allowed line), but the near satellite manifested only as a shoulder. However, first, the intensity of the near satellite was lower than the intensity of the far satellite, which, for the level of Langmuir waves presumed by the authors in [2], contradicted not only Baranger–Mozer' theory [3] (presented in Appendix J.1), but also the more accurate adiabatic theory of satellites [4] (presented in Appendix J.2), even though the latter theory is applicable not only for relatively weak Langmuir waves (to which Baranger–Mozer' theory is limited), but also for much stronger Langmuir waves. (Griem's extension of Baranger–Mozer' theory to the next order in [2] was still inferior to the adiabatic theory of satellites, as illustrated in Fig. 5.4 of the book [1]; the latter figure shows, in particular, that for the far and near satellites to be of equal intensities, as presumed in [2], it would require significantly higher level of Langmuir waves than it was deduced in [2] from the ratio of the intensity of the far satellite to the intensity of the allowed line.) Second, Kunze–Griem's experimental profile of the line

He I 492.2 nm [2] seem to show only the far satellite without any manifestation of the near satellite. Despite all these problems, the authors of [2] should be given credit for the first experiment that brought the attention of the research community to spectroscopic diagnostics of Langmuir waves in plasmas.

Yaakobi and Bekefi [5] in their paper of 1969 presented the observation of a single satellite (far satellite) of the line Li I 610.4 nm ($2P \rightarrow 3D$) in the experiment where a capacitor was discharged through a thin lithium wire, as shown in Fig. 7.1. The electron density was by five orders of magnitude higher than in experiment [2].

It is seen that the would-be position of the near satellites coincides with the position of the allowed line. In this situation, there occur resonant phenomena: they cannot be taken into account by the perturbation theory even if it would be extended to a higher order (which is what Griem did in [2]). The adequate non-perturbative theory for such resonances was developed much later in 1982: it was first presented in paper [6] and then in more detail in Sec. 5.2 of the book [1]. At the absence of the adequate

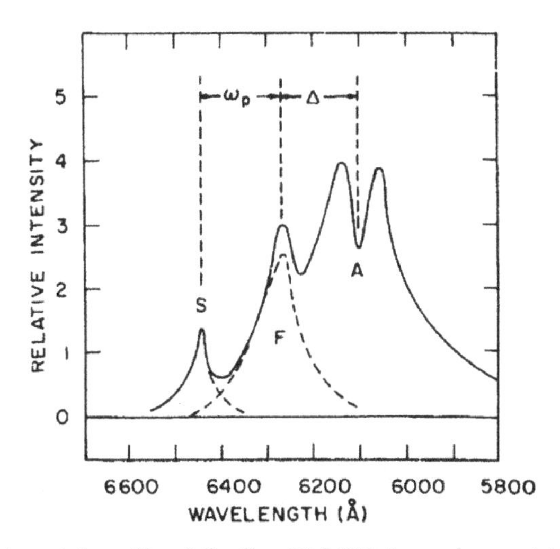

Fig. 7.1. Experimental profile of the line Li I 610.4 nm observed in the exploding lithium wire experiment [5]. The allowed line, the forbidden line, and the far satellite of the forbidden line are marked by A, F and S, respectively. The dashed curves represent the unfolding of the partially overlapping lines.

theory in 1969, Yaakobi and Bekefi could not properly interpret the observed spectrum of the allowed line. It is seen from Fig. 7.1 that the allowed line got split (became a doublet). This is exactly what follows from the adequate non-perturbative theory [6]: the resonance causes the Rabi-type splitting — the separation between the components of the doublet is equal to the generalized Rabi frequency [6], which is controlled by the amplitude of the Langmuir field. (Of course, the resonant situation is a matter of chance: it is seldom encountered in experimental spectra of helium or lithium lines in plasmas containing Langmuir waves.) Thus, there is no question that Langmuir waves were present in the lithium plasma in the experiment [5]. Interestingly enough, those were *thermal* Langmuir waves/turbulence — in distinction to supra-thermal levels of Langmuir waves found in almost all other experiments where Langmuir waves were detected spectroscopically. The amplitude of the Langmuir waves was estimated to be 200 kV/cm — consistent with a much higher electron density $N_e = 2 \times 10^{18}\,\text{cm}^{-3}$ compared to all other studies of Langmuir waves in helium or lithium plasmas.

Davis in his paper of 1972 [7] and Rutgers in his paper of 1975 [8] were among the first to observe clearly resolved a pair of satellites of dipole-forbidden lines, the near satellite having a higher intensity than the far satellite, as expected from the theories. Both experiments [7, 8] were turbulent heating experiments, but at different plasma machines: a high-voltage theta-pinch in [7] and a hollow-cathode discharge between magnetic mirrors in [8].

Figure 7.2 presents two experimental profiles of the line He I 492.2 nm obtained at two different instants of the discharge [7] with a polarizer parallel to the magnetic field. At the instant 450 ns, both the near and far satellites were well-resolved and clearly identifiable (in distinction to the instant 250 ns). Both the experimental ratio of the near satellite intensity S_{near} to the allowed line intensity I_A and the experimental ratio of the far satellite intensity S_{far} to the allowed line intensity I_A yielded practically the same amplitude of the Langmuir waves (approximately 6 kV/cm), thus confirming the interpretation of the observed features.

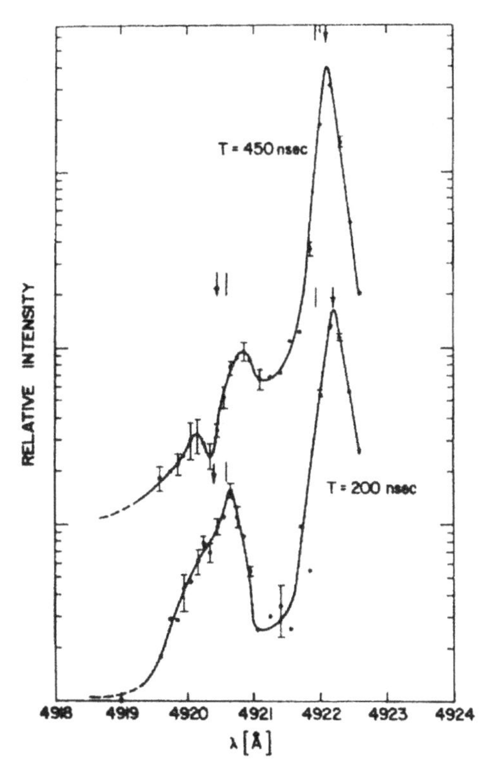

Fig. 7.2. Experimental profiles of the line He I 492.2 nm observed at two different instants of the discharge with a polarizer parallel to the magnetic field in the turbulent heating experiment [7]. The straight lines mark the unperturbed positions of the allowed and the would-be forbidden lines. The arrows mark the positions of the allowed and the would-be forbidden lines shifted due to the quadratic Stark effect.

Figure 7.3 presents the experimental profile of the line He I 402.6 nm ($2^3P \rightarrow 3^3D$) integrated over the time interval (1–1.5) μs [8]. Just as in the experimental profile from Fig. 7.1 at 450 ns, both the near and far satellites were well-resolved and clearly identifiable. Both the experimental ratio of the near satellite intensity S_{near} to the allowed line intensity I_A and the experimental ratio of the far satellite intensity S_{far} to the allowed line intensity I_A yielded practically the same amplitude of the Langmuir waves (approximately 5 kV/cm), thus confirming the interpretation of the observed features. For completeness it should be noted that the experimental profile of

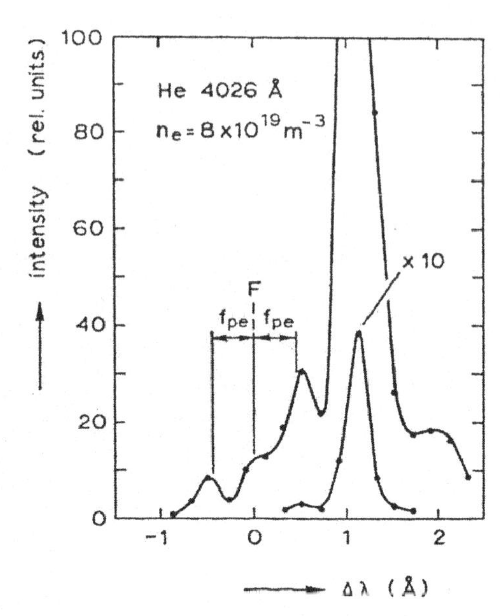

Fig. 7.3. Experimental profile of the line He I 402.6 nm observed in the turbulent heating experiment [8].

the same line He I 402.6 nm ($2^3P \rightarrow 3^3D$) integrated over the time interval (1.5–2) μs [8], exhibited several satellites of not only the forbidden line, but also the allowed line — as expected from the theory at sufficiently high amplitudes of the Langmuir waves (see, e.g., [9]), which from this experimental profile was estimated to be 12 kV/cm.

Drawin and Ramette in their paper of 1978 [10] warned that the mere fact of observing additional peaks in profiles of He lines does not necessarily mean that Langmuir-wave-caused satellites are observed. If the electron density, deduced from the separation between the experimental peaks under the assumption that they are Langmuir-wave-caused satellites, differs significantly from the electron density, determined from the experimental widths (or by other means), then the assumption is wrong. By analyzing experimental peaks in the profile of the line He I 447.2 nm in their own experiment in a linear discharge tube [10], Drawing and Ramette showed that those were most likely He_2 molecular lines. They also pointed out that such interpretation could be more appropriate for experiments by Kawasaki [11] and by Sanchez and Bengtson [12], then the interpretation via Langmuir-wave-caused satellites.

Now a word of caution from the author of this book. If in your experiment you observed a couple of satellites of a dipole-forbidden line of He/He-like ions or Li/Li-like ions and the electron density, deduced from the separation of the satellites (i.e., using the expression for the plasma electron frequency $\omega_{\mathrm{pe}} = (4\pi e N_e/m_e)^{1/2}$), is consistent with the electron density, deduced from line widths (or by other means), the best way to determine the amplitude of the Langmuir wave is from the experimental ratio of intensities of one satellite to another. This is because of the old method, based on the ratio of intensities of the satellite and the allowed line (used, e.g., in works [2, 7, 8] and in many others), could frequently lead to incorrect results. This is because in many experiments, the allowed line is emitted from a larger volume than the satellites: the satellites are emitted only from the volume where there are Langmuir waves while the allowed line is emitted from the entire volume along the direction of observation.

For determining the amplitude of Langmuir waves from the experimental ratio of intensities of one satellite to another, one can use the adiabatic theory of satellites presented in Appendix J.2. We note that in the less accurate Baranger–Mozer' theory of satellites (presented in Appendix J.1), such ratio does not depend on the amplitude of Langmuir waves and thus cannot be used. In fact, the method based on the ratio of intensities of one satellite to another effectively provides a spatial resolution even if the experimental design did not involve the spatial resolution. Indeed, despite the volume, where Langmuir waves are developed, could be significantly smaller than the total plasma volume, this method provides the information only from that small volume, which is why this method is called "quasilocal."

While the above experimental examples were limited to the employment of helium spectral lines, the above methods can be also utilized while using spectral lines of lithium, as well as of He-like and Li-like ions.

7.3 Manifestations of Langmuir Solitons in Satellites of Dipole-forbidden Spectral Lines of Helium, Lithium and of the Corresponding Ions

In paper [13], there was calculated analytically the shape of satellites of dipole-forbidden lines in a spectrum *spatially-integrated* through a Langmuir soliton (or through a sequence of Langmuir solitons separated by a distance L). The Langmuir solitons have the following form in space [14]:

$$F(x,t) = E(x)\cos\omega_p t, \quad E(x) = E_0/\mathrm{ch}(x/\lambda), \quad \lambda \ll L. \tag{7.1}$$

Here,

$$\omega_p = \omega_{\mathrm{pe}} - 3T_e/(2m_e\omega_{\mathrm{pe}}a^2), \tag{7.2}$$

where ω_{pe} is the plasma electron frequency. For diagnosing solitons, it is necessary not only to find experimentally an electric field oscillating at the frequency $\sim\omega_{\mathrm{pe}}$, but also to make sure that the spatial distribution of the amplitude corresponds to the form factor $E(x)$ from (7.1).

Under any quasimonochromatic electric field, dipole-forbidden spectral lines of helium, lithium, and of the corresponding ions can exhibit satellites — see Appendix J. In cases of a relatively large separation between the forbidden and allowed lines, or a relatively weak amplitudes, the intensities of the far (+) and near (−) satellites are $S_\pm = a_\pm E^2(x)$, where $a_\pm$ does not depend on x.

The spatially-integrated profile of the satellites, calculated with the allowance for the quadratic shift of their frequencies $bE^2(x)$, where b does not depend on x, has the form:

$$\begin{gathered} S_\pm(\Delta\omega) = (1/L)\int_{L/2}^{L/2} dx a_\pm E^2(x)\delta[f(x)], \\ f(x) = \Delta\omega \pm \omega_p - bE^2(x). \end{gathered} \tag{7.3}$$

After calculating the integral in Eq. (7.3), in paper [13], it was obtained:

$$S_\pm(\Delta\omega) = (\lambda/L)[a_\pm/(2|b|)]/[1 - (\Delta\omega \pm \omega_p)/(bE_0^2)]^{1/2}. \tag{7.4}$$

The profiles $S_{\pm}(\Delta\omega)$, formally calculated by Eq. (7.4), have singularities at $\Delta\omega \pm \omega_p = bE_0^2$. From the physical point of view, for obtaining a finite result at $\Delta\omega \pm \omega_p = bE_0^2$, it is necessary to replace the δ-function in Eq. (7.3) by a real profile, e.g., by the Lorentz profile representing the dynamical Stark broadening (SB) by electrons and by some ions:

$$\delta[f(x_0)] \rightarrow (1/\pi)\gamma/\{\gamma^2 + [f(x_0)]^2\}, \tag{7.5}$$

where x_0 is the root of the equation $f(x) = 0$. Using the Taylor expansion of $f(x)$ at $x = x_0$ and taking into account the first derivative df/dx vanishes at $x = x_0$, the right side of Eq. (7.5) can be approximated as follows:

$$(1/\pi)\gamma/\{\gamma^2 + [d^2 f(x_0)/dx^2]^2(x - x_0)^4\}. \tag{7.6}$$

Then using the integral

$$(1/\pi)\int_{-\infty}^{\infty} dz\gamma/(\gamma^2 + z^4) = 1/(2\gamma)^{1/2}, \tag{7.7}$$

we find

$$S_{\text{soliton}}^{\max} = S_{\pm}(\Delta\omega = \pm\omega_p - bE_0^2) = [\lambda a_{\pm}/(2^{1/2}\pi L b^{1/2})]E_0^2/\gamma^{1/2}, \tag{7.8}$$

where $S_{\text{soliton}}^{\max}$ is the peak intensity of the satellites in the case of solitons.

In the case of non-solitonic Langmuir waves, the peak intensity of the satellites is (see Appendix J):

$$S_{\text{non-soliton}}^{\max} = a_{\pm}E_0^2/(\pi\gamma). \tag{7.9}$$

So, for the ratio of the peak intensity of the satellite under solitons to the peak intensity of the same satellite under non-solitons, we get

$$S_{\text{soliton}}^{\max}/S_{\text{non-soliton}}^{\max} = [\lambda/(2^{1/2}L)][\gamma/(bE_0^2)]^{1/2} \gg 1 \tag{7.10}$$

for the typical situation where $\gamma \gg bE_0^2$, i.e., where the dynamical Stark width γ is much greater than the quadratic shift bE_0^2 of the satellites.

Thus, in the case of Langmuir solitons, the peak intensity of the satellites can be significantly enhanced — by orders of magnitude — compared to the case of non-solitonic Langmuir waves. This distinctive feature of satellites under Langmuir solitons allows distinguishing them from non-solitonic Langmuir waves.

We note that for a more general case, where both the dynamical SB and the Doppler broadening are taken into account, the δ-function in Eq. (7.3) should be substituted by the Voigt profile resulting in

$$S^{\max}_{\text{soliton}}/S^{\max}_{\text{non-soliton}} \sim \Delta\omega_{\text{Voigt}}/(bE_0^2)]^{1/2} \gg 1, \tag{7.11}$$

where $\Delta\omega_{\text{Voigt}}$ is the halfwidth of the Voight profile. Again, in the typical situation where $\Delta\omega_{\text{Voigt}} \gg bE_0^2$, the peak intensity of the satellites is significantly enhanced — by orders of magnitude — compared to the case of non-solitonic Langmuir waves.

Finally, we mention that Hannachi *et al.* [15] performed simulations for finding the effect of Langmuir solitons on the hydrogen Ly_α line. The effect was an additional broadening. However, even at the low electron density $N_e = 10^{14}\,\text{cm}^{-3}$, the effect was very small compared to the SB by plasma microfields. Moreover, the additional broadening rapidly diminished with the increase of N_e, so that there would be practically no additional broadening at $N_e > 10^{15}\,\text{cm}^{-3}$. Therefore, it seems that the results by Hannachi *et al.* [15] could not be useful for the experimental diagnostics of Langmuir solitons.

7.4 Measurements using Hydrogenic Spectral Lines

There are two major effects of Langmuir turbulence on profiles of hydrogenic spectral lines, as noted in Appendix H. One of the effect is an additional dynamical Stark broadening, as presented in Appendix H.1 and in the book [1]. The first experimental observation of this effect was presented by Zakatov *et al.* [16]. In this work, the excessive broadening of the H-alpha line in a beam-plasma experiment was interpreted as the effect of Langmuir turbulence. Another experimental observation of this effect was depicted by Karfidov and Lukina [17]. In their work, the excessive broadening of the H-beta line in different beam-plasma experiment was interpreted as the effect of Langmuir turbulence. We also mention that in the theoretical paper [18], the broadening of hydrogen lines by thermal Langmuir turbulence was used as the secondary effect required (together with the quasistatic broadening by thermal ion-acoustic turbulence, as the primary effect) for explaining the experimental width of the H-alpha line measured by Kielkopf and Allard in their benchmark experiment [19] performed in a laser-produced pure-hydrogen plasma at super-high (for hydrogen spectral lines) electron densities up to $1.4 \times 10^{20}\,\mathrm{cm}^{-3}$ — details are presented in Appendix I.

The broadening of hydrogenic spectral lines by Langmuir turbulence was not widely used for diagnostics compared to the second effect: the appearance of a sequence of peaks and troughs in the experimental line profile. In early experimental works, the attempts to interpret the observed structures as being caused by Langmuir waves were focused at the experimental positions of the peaks in the structure. In other words, the underlying idea of the interpretation was a sequence of satellites at the distances $\pm\omega_{\mathrm{pe}}$, $\pm 2\omega_{\mathrm{pe}}$, $\pm 3\omega_{\mathrm{pe}}, \ldots$ from the unperturbed position of the spectral line (in the frequency scale) — such as those predicted by Blochinzew [20] (see Appendix L for details) or by Lifshitz [21] while disregarding the presence of quasistatic electric fields in plasmas.

Only after it was rigorously proven analytically (as mentioned and referenced in Sec. 7.1) that, after taking into account quasistatic electric fields in plasmas (represented by the low-frequency plasma turbulence and/or the quasistatic part of the ion microfield), the primary manifestation of Langmuir waves would be L-dips — in positions close to the positions of would-be L-satellites. In other words, a local "zigzag" of intensities due to L-dips is much greater than a local "zigzag" of intensities due to a would-be L-satellite. Consequently, in a sequence of peaks and troughs seen in an experimental profile, the positions of the troughs (rather than the positions of the peaks) contain the information necessary for identifying the experimental structure, as mentioned in Sec. 7.1. The details of the theory of L-dips are presented in Appendices G and H.3.

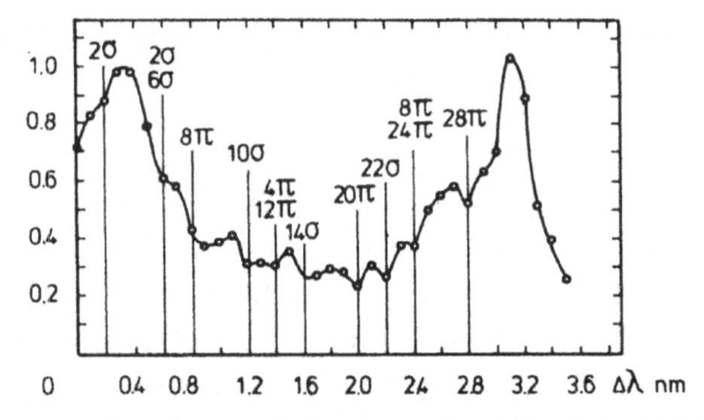

Fig. 7.4. Short-wavelength part of the observed profile of the H-delta line from Gallagher-Levine's experiment of 1973 [22]. The vertical lines are theoretically expected positions of L-dips, marked for the first time by Zhuzhunashvili and Oks in paper [24] in 1977. Above each vertical line is the label of the Stark component, in the profile of which the particular L-dip occurs.

Below are two examples of false interpretations of the experimental structures in early experiments [22, 23] performed before the theory of L-dips was developed. So, the authors of [22, 23] should not be blamed for the false interpretations.

Figure 7.4 shows the short-wavelength part of the observed profile of the H-delta line from Gallagher-Levine's experiment of 1973 [22] — the turbulent heating experiment at Tormac, the toroidal high-beta magnetic fusion machine (the long-wavelength part was not presented in [22]). Figure 7.5 shows the short-wavelength part of the observed profile of the H-gamma line from Rutgers–Kalfsbeek experiment of 1975 [23] — the turbulent heating experiment. In both figures, the vertical lines are theoretically expected positions of L-dips, marked for the first time by Zhuzhunashvili and Oks in paper [24] in 1977. Above each vertical line is the label of the Stark component, in the profile of which the particular L-dip occurs. It is seen that there is an excellent agreement between the experimental troughs and the theoretical positions of the L-dips. Thus, the consistent interpretation of the experimental profiles from [22, 23] was via L-dips, while the interpretation via L-satellites (given in [22, 23]) was not consistent. (We note in passing a misprint in Chapter 4 of Ref. [21] in book [1]: is should have been to Rutgers–Kalfsbeek experiment of 1975).

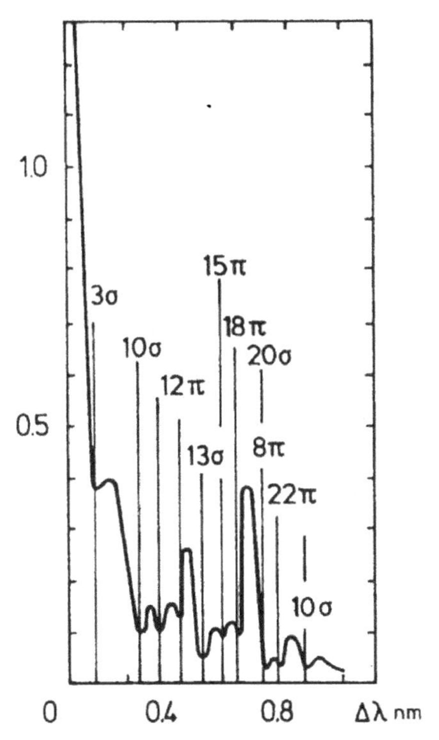

Fig. 7.5. Short-wavelength part of the observed profile of the H-gamma line from Rutgers–Kalfsbeek's experiment of 1975 [23]. The vertical lines are theoretically expected positions of L-dips, marked for the first time by Zhuzhunashvili and Oks in paper [24] in 1977. Above each vertical line is the label of the Stark component, in the profile of which the particular L-dip occurs.

A similar misinterpretation was given by Berezin *et al.* to structures in the profiles of deuterium lines D-alpha and D-beta observed in their experiment in a theta-pinch in 1983 [25]. They focused at the peaks and interpreted the observed structures as Blochinzew's type satellites presumably caused by low-hybrid waves — even though the theory (and the confirming experiment) had been already published by Zhuzhunashvili and Oks in 1977 [24]. In Sec. 7.2.2 of book [1], it is shown in detail that the interpretation by Berezin *et al.* of their experimental profiles is highly inconsistent and that in reality, the observed structures were L-dips. (So, Langmuir waves were developed in their experiment, rather than low-hybrid waves).

Paper [24] presented the first experiment, where L-dips in hydrogen lines were identified by the authors themselves (in distinction to the experiments presented in [22, 23, 25]). The experiment was performed at the magnetic mirror trap "Dimpol." The initial plasma was created by a Penning-type discharge in a DC magnetic field. Then an AC magnetic field of a high amplitude was imposed, creating the opposing magnetic fields configuration in a θ-pinch configuration.

Figure 7.6 shows the red part of the symmetric experimental profile of the H-alpha line observed along the magnetic field in [24]. The vertical lines are theoretically expected positions of L-dips. Above each vertical line is the label of the Stark component, in the profile of which the particular L-dip occurs.

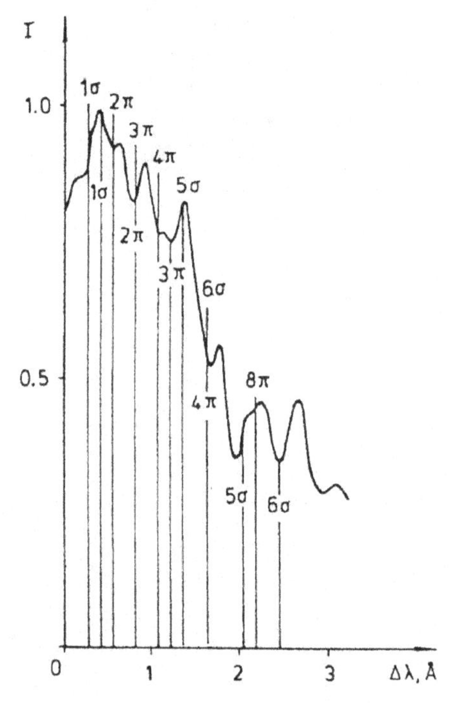

Fig. 7.6. The red part of the symmetric experimental profile of the H-alpha line observed along the magnetic field in the magnetic mirror trap "Dimpol" [24]. The vertical lines are theoretically expected positions of L-dips. Above each vertical line is the label of the Stark component, in the profile of which the particular L-dip occurs.

So, a practical question arises: if a sequence of peaks and troughs is observed in experimental profiles of hydrogenic spectra lines, how to provide a correct, consistent interpretation of such structure? Here is a practical recipe based on the theory of L-dips presented in detail in Appendix G.

Let us consider a Stark component of a hydrogenic line, corresponding to the radiative transition from the upper Stark sublevel α (characterized by the principal quantum number n_α and the electric quantum number $q_\alpha = (n_1 - n_2)_\alpha$, where n_1 and n_2 are parabolic quantum numbers of sublevel α) to the lower Stark sublevel β (characterized by the principal quantum number n_β and the electric quantum number q_β). The dips result from the resonance between the Stark splitting of the upper level or of the lower level and one or several quanta of the quasimonochromatic electric field of a frequency ω, which is equal to the plasma electron frequency $\omega_{\rm pe}$ in the case of Langmuir waves (in most experiments it was a one quantum resonance). Since, the Stark splittings of the upper and lower levels differ from each other, in the profile of the Stark component (α, β), there could appear two dips: one due to the resonance with the upper level (α-dips), another due to the resonance with the lower level (β-dips) at the following positions counted from the unperturbed wavelength:

$$\begin{aligned} \Delta\lambda_\alpha^{\rm dip} &= [k\omega_{\rm pe}\lambda_0^2/(2\pi c)]X_{\alpha\beta}q_\alpha/n_\alpha, \\ \Delta\lambda_\beta^{\rm dip} &= [k\omega_{\rm pe}\lambda_0^2/(2\pi c)]X_{\alpha\beta}q_\beta/n_\beta, \end{aligned} \tag{7.12}$$

where k is the number of quanta of the Langmuir field involved in the resonance (typically, $k = 1$), λ_0 is the unperturbed wavelength of the spectral line, and

$$X_{\alpha\beta} = n_\alpha q_\alpha - n_\beta q_{\alpha\beta}. \tag{7.13}$$

In the hydrogenic lines of the Lyman spectral series, there could be only α-dips, since the lower level is the ground level and it does not exhibit the linear Stark effect.

The expressions (7.12) for the dips position are valid in plasmas of any electron density if the quasistatic electric field is dominated by the low-frequency electrostatic turbulence. In the opposite case, where the quasistatic electric field is dominated by the ion microfield, the expressions (7.12) are still valid except for high density plasmas, for which there would be a relatively small correction term in the right sides of expressions (7.12) reflecting the spatial nonuniformity of the ion microfield over the dimensions of the radiator (see Appendix G, Eq. (G.12)).

A very important feature of the L-dips is that there positions along the spectral line profile, generally (except for the Lyman lines), are *not equidistant*, as illustrated in Fig. 7.7 for the H-alpha line for $k = 1$. While the absolute positions of the L-dips along the line profile are proportional to $\omega_{\rm pe}$ and therefore scale with the electron density as $N_e^{1/2}$, the relative positions of the L-dips with respect to one another are density independent. Therefore, after observing a sequence of peaks and troughs in an experimental profile, one can apply a "measuring comb" like the one from Fig. 7.7 and vary its absolute size (while keeping constant the relative positions of the vertical lines) trying to fit the experimental sequence of troughs. If positions of the overwhelming majority of the experimental troughs would correspond to the positions of the vertical lines in the measuring comb (at some absolute size of the measuring comb), then the hypothesis of the presence of Langmuir waves would pass the first test.

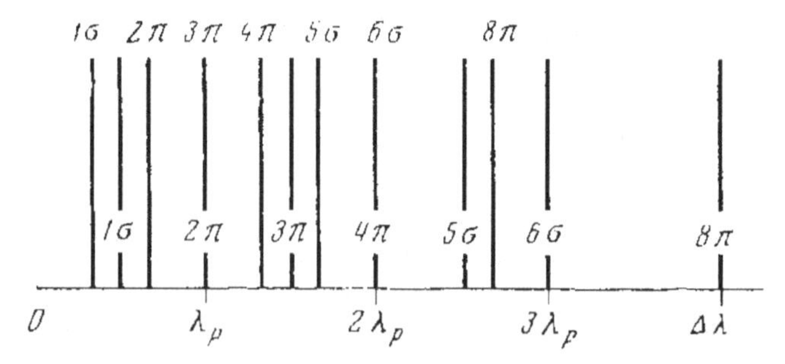

Fig. 7.7. Theoretical positions of possible L-dips in the red part of the profile of the H-alpha line at relatively low electron densities. The L-dips in the blue part should be symmetric with respect to the L-dips in the red part. Here, $\lambda_p = \omega_{\rm pe}\lambda_0^2/(2\pi c)$, where λ_0 is the unperturbed wavelength of the spectral line.

From the absolute size of the measuring comb that provided the best fit with the experimental troughs, one can easily determine $\omega_{\rm pe}$ and thus N_e. This value of N_e should be compared with N_e determined by other means. If these two values of N_e would be in a relatively good agreement with each other, then the hypothesis of the presence of Langmuir waves would pass the final test.

Observing all theoretically expected L-dips in an experimental profile requires a sufficiently high spectral resolution. If the spectral resolution is relatively low, some of the possible L-dips within the theoretical sequence for a given spectral line cannot be recorded. For this situation, it is useful to truncate the theoretical set of the L-dips by leaving only the most pronounced (the best visible) L-dips, called *reference dips*, as follows.

In paper [26], it was shown that the visibility of L-dips is proportional to the following quantity denoted as f_ν:

$$f_\nu = (I_{\alpha\beta}/X_{\alpha\beta})n_\nu(n_\nu^2 - q_\nu^2 - m_\nu^2 - 1)^{1/2}, \quad \nu = \alpha, \beta. \tag{7.14}$$

Here, $I_{\alpha\beta}$ is the relative intensity of the Stark component (in the profile of which the particular L-dip could be observed) and m_ν is the magnetic quantum number (the quantum number of the projection of the angular momentum).

Figure 7.8 presents theoretically expected positions of the *reference dips* in the red parts of the line profiles of H-alpha, H-beta and H-gamma for $k = 1$ (one quantum resonance). The α-dips are shown by vertical lines above the abscissa axis, while the β-dips — by vertical lines below the abscissa axis. Next to each vertical line is indicated the quantity f_ν from Eq. (7.14). These truncated measuring combs, showing only reference dips, are very useful for the identification of the troughs in experimental line profiles at a relatively low spectral resolution.

Each L-dip represents a structure consisting of the primary minimum of intensity, whose position is controlled by the frequency of Langmuir waves, and two surrounding bumps (Fig. 7.9). The bumps are due to a partial transfer of the intensity from the wavelength of the dip to adjacent wavelengths. One of the two bumps that is closer to the line center, being superimposed with the inclined line

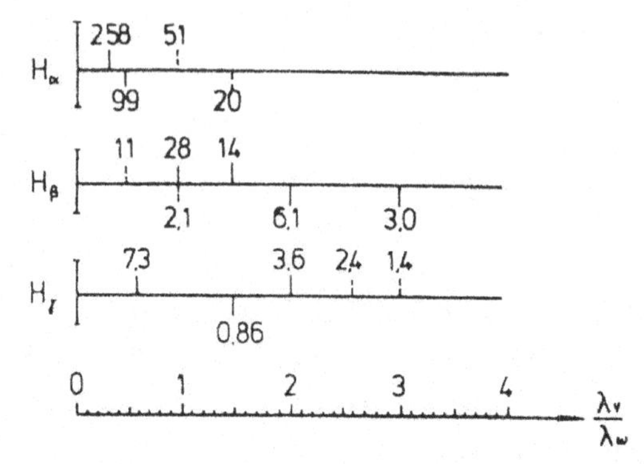

Fig. 7.8. Theoretical positions of the most pronounced L-dips (reference dips) in the red part of the profiles of the H-alpha, H-beta and H-gamma spectral lines at relatively low electron densities. The L-dips in the blue part of each spectral line should be symmetric with respect to the L-dips in the red part.

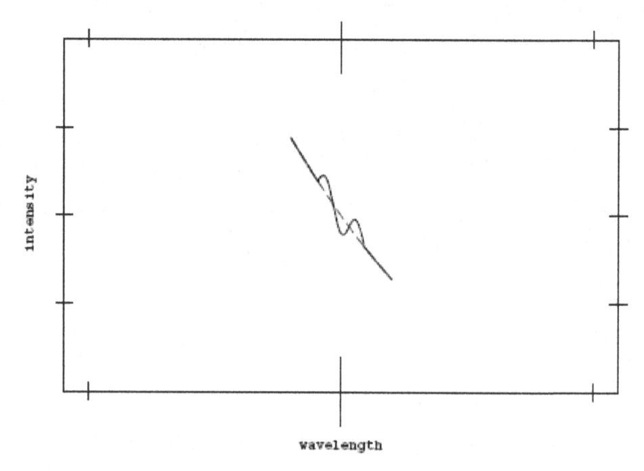

Fig. 7.9. Sketch of the detailed structure of an L-dip: the primary minimum of intensity, whose position is controlled by the frequency of Langmuir waves, and two surrounding bumps. The bumps are due to a partial transfer of the intensity from the wavelength of the dip to adjacent wavelengths. One of the two bumps that is closer to the line center, being superimposed with the inclined line profile, can lead to the appearance of a secondary minimum (or a small shoulder).

profile, can lead to the appearance of a secondary minimum (or a small shoulder). However, the position of the secondary minimum is not related to the frequency of Langmuir waves.

It is very important to emphasize that the following two conditions should be met for observing the detailed structure of the

L-dip. First, this would require a sufficiently high spectral resolution. Second, the L-dip should be well isolated from adjacent L-dips: if two L-dips structures (say, Nos. 1 and 2) are close to each other, then one of the bumps of the L-dip structure No. 1 would merge with one of the bumps of the L-dip structure No. 2: in other words, only one shared bump (rather than two bumps) would be observed between the primary minima of the two L-dip structures. An example of the observed detailed structure of the L-dip will be shown in the subsequent part of this Sec. 7.4.

The halfwidth $\delta\lambda_{1/2}$ of the L-dip (approximately equal to the separation between the primary minimum and the nearest bump) is controlled by the amplitude E_0 of the electric field of the Langmuir wave

$$\delta\lambda_{1/2} \approx (3/2)^{1/2}\lambda_0^2 n^2 E_0/(8\pi m_e ec Z_r), \tag{7.15}$$

where λ_0 is the unperturbed wavelength of the spectral line. Thus, by measuring the experimental halfwidth of L-dips, one can determine the amplitude E_0 of the Langmuir wave. For example, from the halfwidths of the dips in the H-alpha profile observed in the experiment [24] and presented here in Fig. 7.6, it was found $E_0 = (4.3 \pm 0.5)\,\text{kV/cm}$.

Using the L-dips phenomenon, it is possible to determine experimentally not only the average amplitude of Langmuir waves, but also to study whether Langmuir waves developed anisotropically and to find out the degree of the anisotropy. For this purpose, the same spectral line and the same L-dip in its profile should be observed in two orthogonal linear polarizations. The halfwidth of L-dips is sensitive to the degree of the anisotropy (the corresponding theory was developed in paper [27]).

Figure 7.10 shows, as an example, the ratio of halfwidths of an L-dip (in the profile of any π-component of a hydrogenic spectral line) at two mutually orthogonal orientations (1 and 2) of a polaroid, versus the degree of anisotropy of Langmuir waves $\mu = (\langle E_1^2\rangle/\langle E_2^2\rangle)^{1/2}$, where $\langle E_1^2\rangle$ and $\langle E_2^2\rangle$ are the mean square amplitudes of Langmuir waves in the directions 1 and 2.

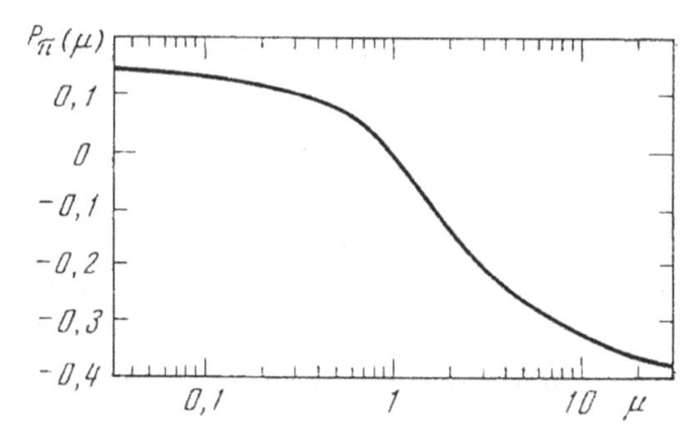

Fig. 7.10. The ratio of halfwidths of an L-dip (in the profile of any π-component of a hydrogenic spectral line) at two mutually orthogonal orientations (1 and 2) of a polaroid, versus the degree of anisotropy of Langmuir waves $\mu = (\langle E_1^2 \rangle / \langle E_2^2 \rangle)^{1/2}$, where $\langle E_1^2 \rangle$ and $\langle E_2^2 \rangle$ are the mean square amplitudes of Langmuir waves in the directions 1 and 2.

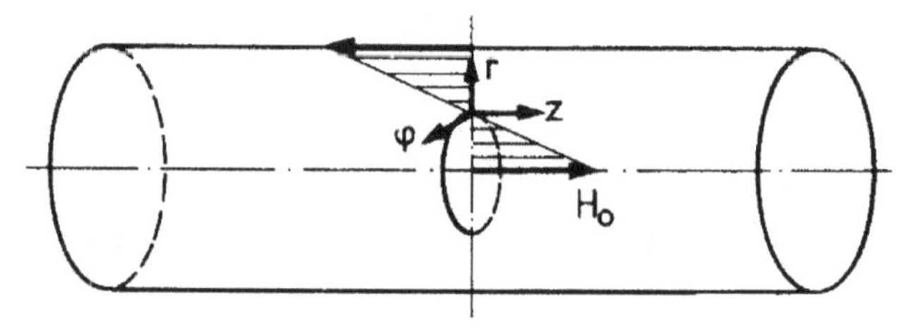

Fig. 7.11. Unit vectors of the cylindrical reference frame used for the description of the polarization analysis in the experiments at the "Dimpol" [24].

The first measurements of the degree of the anisotropy of Langmuir waves were performed by the polarization analysis in the experiments at the magnetic mirror trap "Dimpol" presented in paper [24]. In these experiments, a configuration of opposing magnetic fields was created in the theta-pinch geometry: on the initial plasma of a Penning-type discharge in a DC magnetic field there was imposed a high-amplitude AC magnetic field. For presenting the results of the experimental polarization analysis, it is convenient to introduce a reference frame attached to the plasma layer where the annihilation of the opposing magnetic fields occurs causing the development of Langmuir waves. The reference frame is shown in Fig. 7.11.

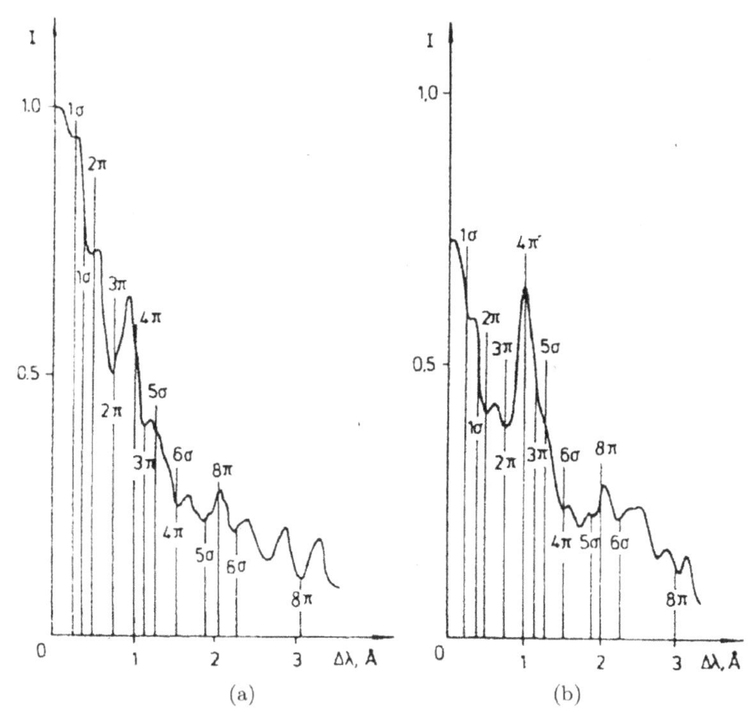

Fig. 7.12. The experimental polarization profiles of the H-alpha line at the "Dimpol" in the observation perpendicular to the magnetic field at two orientations of the polaroid: along the magnetic field (a), i.e., in the z-direction, and perpendicular to the magnetic field (b), i.e., in the r-direction [24].

Figure 7.12 shows the experimental polarization profiles of the H-alpha line in the observation perpendicular to the magnetic field at two orientations of the polaroid: along the magnetic field (a), i.e., in the z-direction, and perpendicular to the magnetic field (b), i.e., in the r-direction. Figure 7.13 shows the experimental polarization profiles of the H-alpha line in the observation along the magnetic field at two orientations of the polaroid: in the r-direction (a) and in the φ-direction (b).

While analyzing halfwidths of the L-dips in polarization profiles, it is important to keep in mind that the close proximity of two or more L-dips distorts their halfwidths. The distortion can exceed the polarization difference in the halfwidths of the L-dips. Therefore,

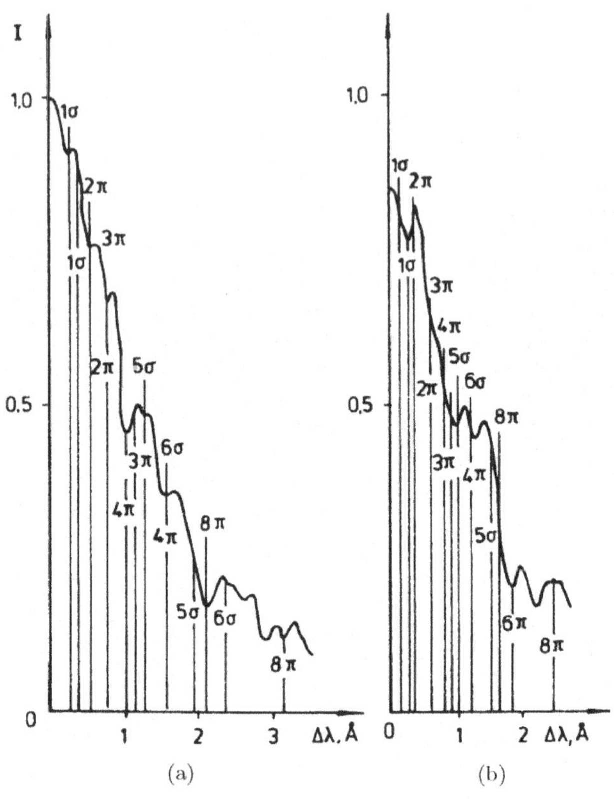

Fig. 7.13. The experimental polarization profiles of the H-alpha line at the "Dimpol" in the observation along the magnetic field at two orientations of the polaroid: in the r-direction (a) and in the φ-direction (b) [24].

the most reliable results should be obtained from the polarizations analysis of well-isolated L-dips. In the profile of the H-alpha line, the most isolated is the L-dip in the profile of the 8π-component (located at $4\lambda_p$ in Fig. 7.7). In each polarization profile in Figs. 7.12, 7.13, two bumps are observed between this L-dip and the nearest L-dip in the profile of the 6σ-component (located at $3\lambda_p$ in Fig. 7.7). There are also two other relatively well isolated L-dips in the profile of the H-alpha line. One is located at λ_p in Fig. 7.7: it is a "double-dip" — the superposition of the L-dip in the 3π-component, caused by the resonance of ω_{pe} with the Stark splitting of the upper level of $n = 3$, with the L-dip in the 2π-component, caused by the resonance of ω_{pe} with the Stark splitting of the lower level of $n = 2$. Another relatively well-isolated L-dip is located at λ_p in Fig. 7.7: it is also a "double-dip" — the superposition of the L-dip in the 6σ-component, caused by the resonance of ω_{pe} with the Stark splitting of the upper level of $n = 3$, with the L-dip in the 4π-component, caused by the resonance of ω_{pe} with the Stark splitting of the lower level of $n = 2$.

A statistical comparison of the halfwidths of these three well isolated L-dips in the experimental polarization profiles observed at the "Dimpol" revealed the following. In the observation perpendicular to the magnetic field, there was no noticeable polarization difference in the halfwidth of any of these L-dips in many pairs of the experimental polarization profiles. Consequently, in terms of the Langmuir field amplitude, this meant that $\langle E_z^2\rangle = \langle E_\varphi^2\rangle$. In the observation along the magnetic field, the analysis of the polarization difference in the halfwidth of any of these L-dips in many pairs of the experimental polarization profiles yielded. $(\Delta\lambda_{1/2}^{\mathrm{dip}})_{r\pi} > (\Delta\lambda_{1/2}^{\mathrm{dip}})_{\varphi\pi}$, $(\Delta\lambda_{1/2}^{\mathrm{dip}})_{r\sigma} > (\Delta\lambda_{1/2}^{\mathrm{dip}})_{\varphi\sigma}$. According to Fig. 7.10, this meant that $(\langle E_r^2\rangle) < (\langle E_\varphi^2\rangle)$. Thus, the distribution of the Langmuir field in the plasma layer, where the annihilation of the opposing magnetic fields occurred, had the form of an oblate spheroid with the symmetry axis in the r-direction.

Below we present more examples of confirmed observations of L-dips in the profiles of various hydrogenic spectral lines in experiments performed at various types of plasmas by different experimental groups. We limit ourselves by examples that are instructive — each in its own way, as presented below.

Figure 7.14 shows profiles of the hydrogen Lyman-alpha line (in the logarithmic scale) observed at various phases of a Z-pinch in the experiment performed in Kunze's group be Bertschinger in 1980 [28]. It was a pioneering experiment in the following two aspects. First of all, it was the first observation of L-dips in the spectral lines of the Lyman series — to the best of our knowledge. Second, in this experiment the electron density N_e was determined at each instant of time from the measurements of the continuous

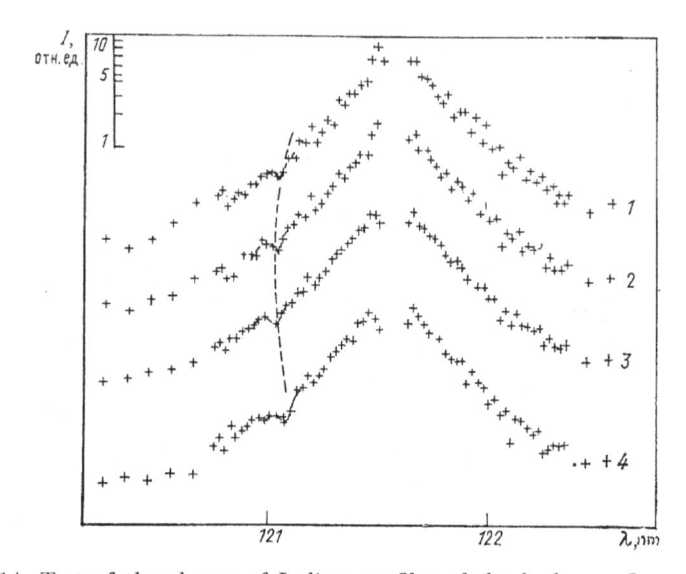

Fig. 7.14. Test of the theory of L-dips: profiles of the hydrogen Lyman-alpha line (in the logarithmic scale) observed in a Z-pinch in the experiment [28] at the following instants (counted from the instant of the maximum compression: (1) $t = -50\,\text{ns}$, (2) $t = 0$ (the instant of the maximum compression), (3) $t = 50\,\text{ns}$, (4) $t = 100\,\text{ns}$. The electron density N_e was determined at each instant of time from the measurements of the continuous spectrum (i.e., independent from the line profiles). Then using this experimental values of N_e, the author of Ref. [28] marked at the experimental Lyman-alpha profile at each instant of time by the dashed line the value of the detuning (from the line center) $\Delta\lambda = \lambda_p = \omega_{\text{pe}}\lambda_0^2/(2\pi c)$. These marks are shown only in the blue part of the profiles for not encumbering the figure.

spectrum (i.e., independent from the line profiles). Then using this experimental values of N_e (which, e.g., at the maximum compression of the Z-pinch was $1.5 \times 10^{18}\,\text{cm}^{-3}$), the author of [28] marked at the experimental Lyman-alpha profile at each instant of time the value of the detuning (from the line center) $\Delta\lambda = \lambda_p = \omega_{\text{pe}}\lambda_0^2/(2\pi c)$. It turned out that at each instant of time, i.e., at different electron densities and thus at different values of λ_p, the mark $\Delta\lambda = \lambda_p$ coincides with a trough in the experimental line profile — as shown in Fig. 7.14 by the dashed line (these marks are shown in Fig. 7.14 only in the blue part of the profiles for not encumbering the figure). Thus, it was demonstrated that the position of the experimental trough

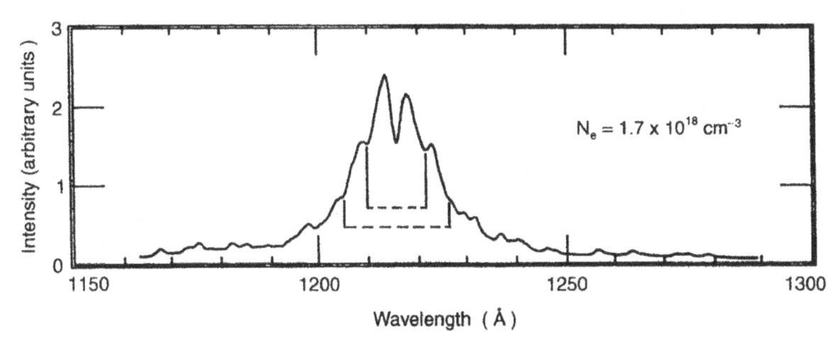

Fig. 7.15. Another test of the theory of L-dips: the comparison of the experimental and theoretical positions of the L-dips in the Ly-alpha line profile observed in Kunze's group at a gas-liner pinch [29]. The theoretical positions were calculated using the electron density that was determined experimentally from the coherent Thomson scattering (independent of the lineshapes) and then marked in the experimental profile. These marks are represented by the two pairs of vertical lines connected inside each pair by the dashed line. One pair of the vertical lines marked the theoretical positions of the L-dips, corresponding to the one quantum resonance (i.e., to $k = 1$ in Eq. (7.12)), the other pair — the theoretical positions of the L-dips, corresponding to the one quantum resonance (i.e., to $k = 2$ in Eq. (7.12)).

scales with the electron density as $N_e^{1/2}$, which was a convincing proof that the experimental troughs were L-dips.

Figure 7.15 presents the observed spectrum of the hydrogen Lyman-alpha line from yet another pioneering benchmark experiment in Kunze's group: the experiment at the gas-liner pinch [29]. The gas-liner pinch operated as a modified Z-pinch with a special gas inlet system. At each shot plasma parameters, including N_e, were measured by the coherent Thomson scattering independent of the lineshapes. So, this was another benchmark experiment for testing the theory of the L-dips. The theoretical positions were calculated using the electron density that was determined experimentally from the coherent Thomson scattering (independent of the lineshapes) and then marked in the experimental profile. In Fig. 7.15, these marks are represented by the two pairs of vertical lines connected inside each pair by the dashed line. It is seen that the positions of each theoretically expected L-dips coincide with the troughs of the experimental profile.

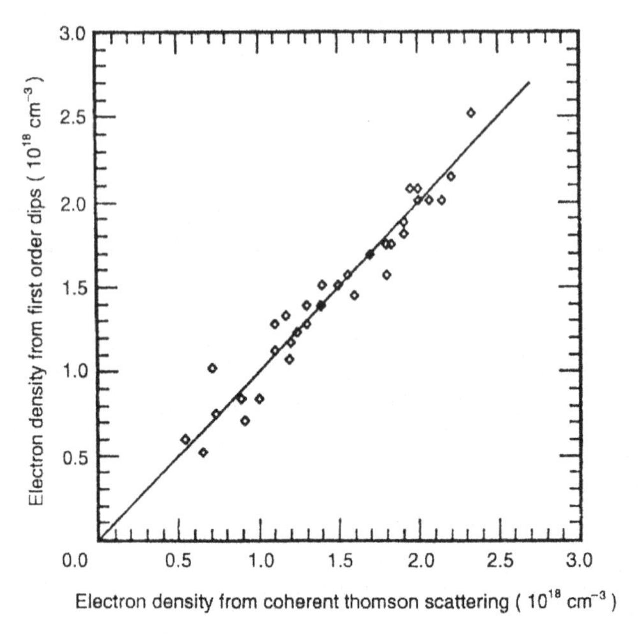

Fig. 7.16. Comparison of the electron density, determined experimentally from the coherent Thomson scattering, with the electron density, determined experimentally from the positions of the L-dips in the Lyman-alpha line in the benchmark experiment in Kunze's group at the gas-liner pinch [29]. The comparison is presented for various shots corresponding to different electron densities.

Figure 7.16 presents another very important result from the benchmark experiment [20]. Namely, at various shots corresponding to different electron densities (up to $3 \times 10^{18}\,\mathrm{cm}^{-3}$), Fig. 7.16 shows (for each shot) the comparison of the electron density, determined experimentally from the coherent Thomson scattering, with the electron density, determined experimentally from the positions of the L-dips. It is seen that the passive spectroscopic method for measuring the electron density based on the L-dips phenomenon yields the same high accuracy as the active method based on the Thomson scattering, which is much more complicated experimentally than that passive method employing the L-dips.

There was also another pioneering results obtained in the same experiment [29]. Namely, theoretically predicted shape of the L-dip structure (bump–dip–bump, see Fig. 7.9) was confirmed

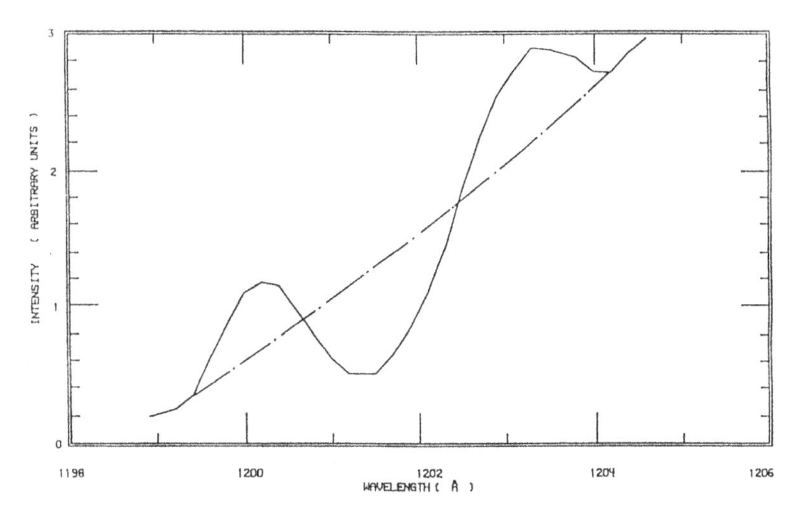

Fig. 7.17. Zoom on the observed bump–dip–bump structure of the L-dip around 1200 Å in the profile of the hydrogen Lyman-alpha line at 200 ns after the maximum compression of the gas-liner pinch [29].

experimentally for the first time — due to a sufficiently high spectral resolution, as shown in Fig. 7.17. From the halfwidth of the experimental L-dip, the amplitude of the Langmuir wave was found to be 2 MV/cm.

The experiments [28, 29] had the highest electron density out of all observations of the L-dips in hydrogen spectral lines. Later, the observations of the L-dips extended to much higher electron densities by using the X-ray range hydrogenic spectral lines emitted by multicharged ions in plasmas produced by powerful Z-pinches or in laser-produced plasmas.

For example, Jian *et al.* [30] observed the L-dips in the profiles of Al XIII Lyman-alpha and Lyman-gamma lines at the Yang accelerator Z-pinch aluminum plasmas, characterized by the electron density $N_e = 5 \times 10^{21}\,\mathrm{cm}^{-3}$, the electron temperature $T_e = 500\,\mathrm{eV}$ and the ion temperature $T_i = 10\,\mathrm{keV}$. The plasmas were produced by imploding aluminum wire-array. Figure 7.18 shows the corresponding experimental results.

In the experiment [30], the electron density was measured from the so-called plasma polarization shift of the spectral lines (the

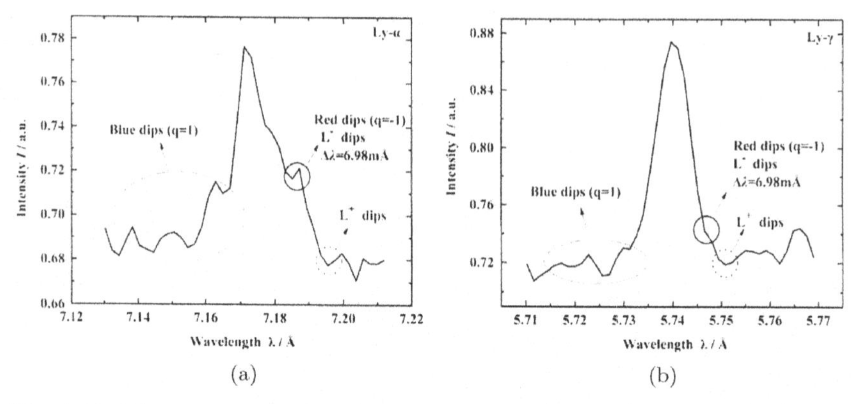

Fig. 7.18. The experimental L-dips in the red parts of the profiles of Al XIII $Ly\alpha$ and Al XIII $Ly\gamma$ lines observed at a Z-pinch facility by Jian *et al.* [30]. The would-be L-dips in the blue parts of the experimental profiles merge in the noise.

shift of the line as a whole, independent of the L-dips) using the corresponding formulas from paper [31]. Then, it was compared with the electron density determined from the positions of the experimental L-dips using the formulas from paper [29] (reproduced in this book in Appendix G, Eq. (G.10)). It was found that these two different ways of measuring the electron density yielded the same results within the accuracy of (5–6)%, which is well within the error margin of the formulas for the plasma polarization shift. (Measuring N_e from the L-dips does not require the knowledge of the temperature T — in distinction to measuring N_e from the plasma polarization shift requiring the value of T, which in the experiment [30] was estimated only within about 10% of the accuracy.) Thus, the experiment [30] can be considered as yet another test of the theory of the L-dips and of the corresponding method for measuring the electron density — at the densities by three orders of magnitude higher than the experiments [28, 29] that tested the theory of the L-dips in hydrogen plasmas.

As for observing the L-dips in laser-produced plasmas, for the first time it was done in the experiment presented in paper [32]. The experiment was performed at the nanosecond Nd:glass laser facility at LULI, France. A single laser beam with the intensity 2×10^{14} W/cm^2 was focused onto a structured target: an aluminum strip sandwiched between magnesium substrate. The plasma parameters $N_e = 5 \times 10^{22}\,\text{cm}^{-3}$ and $T_e = 300\,\text{eV}$ were estimated by hydrodynamic simulations.

The L-dips were observed in the experimental profiles of Al XIII Lyman-gamma line recorded at different distances from the target and thus, corresponding to different electron densities. The experiment confirmed the theoretically expected dependence of the L-dips positions on the electron density. The electron densities determined from the L-dips positions were in a good agreement with the electron densities obtained from the line broadening simulations.

Further observations of the L-dips in laser-produced plasmas were done at laser intensities by many orders of magnitude higher than in the experiment [32]. The first such experiment reported in paper [33] was performed at the laser intensities up to $3 \times 10^{18}\,\text{W/cm}^2$ at two laser facilities located at Kansai Photon Science Institute in Japan. That study had several very important novel features/results as follows. To begin with, this was the first study of the L-dips in spectral lines from *femtosecond* laser-driven *cluster-based* plasma. Second, in distinction to the experiment [32], the study employed a spectral line of an ion different from Al XIII, namely the Lyman-epsilon line of O VIII. Third, but perhaps most importantly, the observed L-dips turned out to be caused by Langmuir waves at the frequency $\omega_{\text{pe}}(N_e) = \omega_{\text{las}}/2$, ($\omega_{\text{las}}$ being the laser frequency), hence corresponding to the electron density equal to one quarter of the critical density. Thus, these Langmuir waves resulted from the well-known parametric instability, namely the two plasmon decay instability, and the experiment [33] constituted the first observation of the signature of this instability in spectral line profiles.

Figure 7.19 shows the profile of the O VIII Lyman-epsilon line observed in the experiment [33] at the laser intensity of $3 \times 10^{18}\,\text{W/cm}^2$, the laser irradiating the CO_2–He mixture. Two solid vertical lines correspond to the positions of two L-dips: one dip in the blue wing at $-20\,\text{mA}$ from the center of the line, another L-dip in the red wing at $37\,\text{mA}$ from the center of the line. The center of gravity of the two L-dips is shifted to the red by $9\,\text{mA}$, which is the manifestation that the multi-pole interactions higher than the

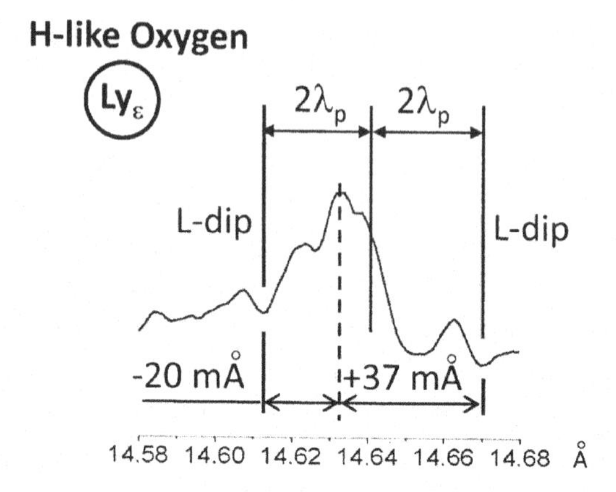

Fig. 7.19. The spectrum of the O VIII Lyman-epsilon line observed in the femtosecond laser-driven cluster-based experiment [33] at the laser intensity of 3×10^{18} W/cm^2, the laser irradiating the CO_2–He mixture. Two solid vertical lines correspond to the positions of two experimental L-dips. The bump–dip–bump structure of the L-dips is clearly visible.

dipole interaction are significant due to high electron densities (see Appendix G, Eq. (G.10)).

This was the first (and so far the best) observation of the details of the L-dip structure (bump–dip–bump) in laser-produced plasmas. The bump–dip–bump structure was most clearly visible in the blue wing. (Its superposition with a significantly inclined spectral profile created a secondary minimum at about 14.63 A of no physical significance). As for the L-dip in the red wing, its near bump was clearly visible, but the far bump was only faintly outlined because it practically merged with the noise. Here, the "near" (or "far") bump means the bump closer to (or further from) the line center with respect to the central minimum of the dip.

The two L-dips were separated from each other by $4\lambda_p$, where $\lambda_p = \omega_{\mathrm{pe}}\lambda_0^2/(2\pi c)$. They were one-quantum resonance dips in the profiles of the two most intense lateral components of the Lyman-epsilon line, originating from the Stark sublevels (311) and (131), the sublevels being labeled by the parabolic quantum numbers. The electron density deduced from the separation of the two L-dips was $N_e = 5.0 \times 10^{20}\,\mathrm{cm}^{-3}$.

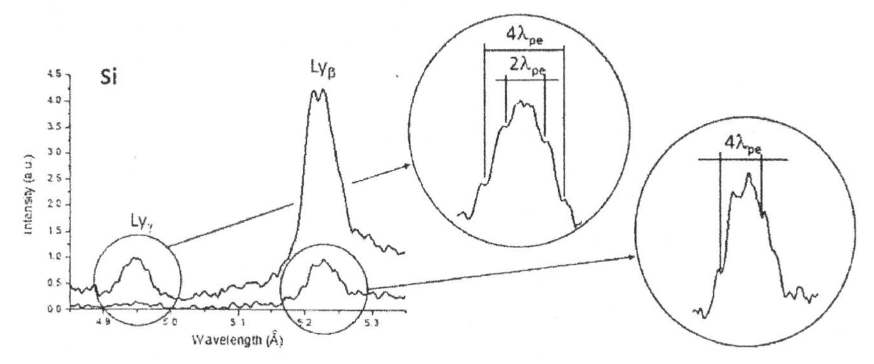

Fig. 7.20. Experimental profiles of Si XIV Lyman-beta and Lyman-gamma lines exhibiting the L-dips [34]. The upper trace was obtained at the laser intensity at the surface of the target estimated as 1.01×10^{21} W/cm^2, the lower trace — at 0.24×10^{21} W/cm^2. In the insets, positions of the dips/depressions in the profiles are marked by vertical lines separated either by $2\lambda_{pe}$ or $4\lambda_{pe}$, where $\lambda_{pe} = [\lambda_0^2/(2\pi c)]\omega_{pe}$ (λ_0 is the unperturbed wavelength of the corresponding line).

Finally, we present here the observation of the L-dips in laser-produced plasmas at the record high laser intensities up to 1.4×10^{21} W/cm^2 [34]. The experiments were performed at Vulcan Petawatt facility at the Rutherford Appleton Laboratory, United Kingdom, using Si foils as targets.

Figure 7.20 presents the experimental profiles of Si XIV Lyman-beta and Lyman-gamma lines exhibiting the L-dips. In the insets, positions of the L-dips are marked by vertical lines separated either by $2\lambda_{pe}$ or $4\lambda_{pe}$, where $\lambda_{pe} = [\lambda_0^2/(2\pi c)]\omega_{pe}$ (λ_0 is the unperturbed wavelength of the corresponding line).

There are two pairs of the L-dips in the profile of the Si XIV Ly-gamma line: in one pair the separation of the two L-dips was 28 mA, in the other pair — 56 mA. From each of the two pairs of the L-dips was deduced the same $N_e = 3.6 \times 10^{22}$ cm^{-3}. This reinforced the interpretation of the experimental troughs as the L-dips.

The electron density deduced from the experimental L-dips positions was close to the so-called relativistic critical density N_{rc} that increases with laser intensity when the latter significantly exceeds the threshold $\sim 10^{18}$ W/cm^2. (The usual critical density N_c is defined by the relation $\omega_{laser} = \omega_{pe}$, where $\omega_{pe} = (4\pi N_c e^2/m_e)^{1/2}$;

at the laser intensities of that experiment, due to the relativistic effects the electron mass m_e increased, thus leading to the increase of N_c to the value N_{rc} — to keep the relation $\omega_{laser} = \omega_{pe}$ satisfied.) Therefore, it was concluded that the Langmuir waves developed at the surface of the relativistic critical density — this conclusion was also supported by PIC simulations.

The mid-point between the two L-dips in each pair practically coincided with the unperturbed wavelength $\lambda_0 = 4.95$ Å. This was again a strong indication that the quasistatic electric field, involved in the formation of the L-dips, was dominated by the low-frequency electrostatic turbulence. If the low-frequency electrostatic turbulence would be absent, then according to Eq. (G.10) of Appendix G, the mid-point of the pair of the L-dips separated by 28 mÅ should have been shifted by 5.8 mÅ to the red with respect to λ_0 and the mid-point of the pair of the L-dips separated by 56 mÅ would have been similarly shifted by 10.7 mÅ to the red with respect to λ_0, these shifts being due to the spatial non-uniformity of the ion microfield reflected by the second term in Eq. (G.10) of Appendix G. Another strong indication of the development of the low-frequency electrostatic turbulence was obtained from the analysis of the experimental width.

In the experimental profile of the Si XIV Lyman-beta line, there was a pair of the L-dips separated from each other by 43 mÅ. The electron density deduced from the separation within the pair of these L-dips was $N_e = 1.74 \times 10^{22}\,\mathrm{cm}^{-3}$. The electron density deduced from the experimental L-dips positions was again close to the corresponding relativistic critical density. Therefore, it was concluded again that the Langmuir waves developed at the surface of the relativistic critical density.

The mid-point between the two L-dips in that Lyman-beta line again practically coincided with the unperturbed wavelength $\lambda_0 = 5.22$ Å. This was again a strong indication that the quasistatic field $\mathbf{F}$ was dominated by the low-frequency plasma turbulence. Again, another strong indication of the development of the low-frequency electrostatic turbulence was obtained from the analysis of the experimental width.

The combination of the two facts — that Langmuir waves developed at the surface of the relativistic critical density and that they were accompanied by the low-frequency electrostatic turbulence — led to the conclusion that the cause was the Parametric Decay Instability (PDI), i.e., the nonlinear process occurring at the surface of the critical (or relativistic critical) density, in which the laser wave decays into the Langmuir wave and the ion acoustic wave. Thus, the spectroscopically discovered low-frequency electrostatic turbulence was identified as the ion acoustic waves.

Finally, we note papers [35, 36] where a sequence of peaks and troughs was observed in the experimental profile of the Lyman-alpha line of F IX in a laser-produced plasma the moderate laser intensity of $\sim 10^{17}$ W/cm^2. For modeling the experimental sequence of peaks and troughs, the authors took into account the fine structure of the Lyman-alpha line of F IX while considering the combined effect of a quasistatic field and a quasimonocromatic field. It resulted in a sequence of peaks (separated by troughs) whose positions depended on E_0 and ω in a complicated way. The authors suggested that the quasimonochromatic field was represented by Bernstein modes rather than by Langmuir waves. It should be emphasized that the overwhelming majority of experimental observations of sequences of peaks and troughs in the profiles correspond to the situation where the fine structure Δ_{fs} of hydrogenic spectral lines can be neglected because $\min[\omega,\ \delta\omega_s(E_0)] \gg \Delta_{\mathrm{fs}}$ (here $\delta\omega_s(E_0)$ is the instantaneous Stark splitting in the static electric field of the strength E_0).

References

[1] E. Oks, *Plasma Spectroscopy. The Influence of Microwave and Laser Fields*, Springer, New York (1995).

[2] H.-J. Kunze and H.R. Griem, *Phys. Rev. Lett.* **21** 1048 (1968).

[3] M. Baranger and B. Mozer, *Phys. Rev.* **123** 25 (1961).

[4] E. Oks and V.P. Gavrilenko, *Sov. Tech. Phys. Lett.* **9** 111 (1983).

[5] B. Yaakobi and G. Bekefi, *Phys. Lett. A* **30** 539 (1969).

[6] V.P. Gavrilenko and E. Oks, *Proc. Int. Conf. on Plasma Phys.* (Göteborg, Sweden) p. 353 (1982).

[7] W.D. Davis, *Phys. Fluids* **15** 2383 (1972).
[8] W.H. Rutgers, *Z. Naturforsch.* **30a** 1271 (1975).
[9] W.W. Hicks, R.A. Hess and W.S. Cooper, *Phys. Rev. A* **5** 490 (1972).
[10] H.W. Drawin and J. Ramette, *Z. Naturforsch.* **33a** 1285 (1978).
[11] K. Kawasaki, *J. Phys. Soc. Japan* **43** 648 (1977).
[12] A. Sanchez and R.D. Bengtson, *Phys. Rev. Lett.* **38** 1276 (1977).
[13] E. Oks, *Sov. Phys. Doklady* **29** 224 (1984).
[14] B.B. Kadomtsev, *Collective Phenomena in Plasma*, Pergamon, Oxford (1982).
[15] I. Hannachi, R. Stamm, J. Rosato and Y. Marandet, *Europhys. Lett.* **114** 23002 (2016).
[16] L.P. Zakatov, A.G. Plakhov, V.V. Shapkin and G.V. Sholin, *Sov. Phys. Doklady* **198** 1306; **16** 451 (1971).
[17] D.M. Karfidov and N.A. Lukina, *Phys. Lett. A* **232** 443 (1997).
[18] E. Oks, *J. Phys. B: Atom. Mol. Opt. Phys.* **49** 065701 (2016).
[19] J.F. Kielkopf and N.F. Allard, *J. Phys. B: Atom. Mol. Opt. Phys.* **47** 155701 (2014).
[20] D.I. Blochinzew, *Phys. Z. Sov. Union* **4** 501 (1933).
[21] E.V. Lifshitz, *Sov. Phys. JETP* **26** 570 (1958).
[22] C.C. Gallagher and M.A. Levine, *Phys. Rev. Lett.* **30** 897 (1973).
[23] W.R. Rutgers and H.W. Kalfsbeek, *Z. Naturforsch.* **30a** 739 (1975).
[24] A.I. Zhuzhunashvili and E. Oks, *Sov. Phys. JETP* **46** 1122 (1977).
[25] A.B. Berezin,B.V. Ljublin and D.G. Jakovlev, *Sov. Phys. Tech. Phys.* **28** 407 (1983).
[26] E. Oks and V.A. Rantsev-Kartinov, *Sov. Phys. JETP* **52** 50 (1980).
[27] E. Oks and G.V. Sholin, *Opt. Spectrosc.* **42** 434 (1977).
[28] G. Bertschinger, *Messungen von VUV Linien in einem dichten Z-Pinch-Plasma.* Diss., Dokt. Der Naturwissenschaften, Ruhr University, Bochum (1980).
[29] E. Oks, St. Böddeker and H.-J. Kunze, *Phys. Rev. A* **44** 8338 (1991).
[30] L. Jian, X. Shali, Y. Qingguo, L. Lifeng and W. Yufen, *J. Quant. Spectrosc. Rad. Transfer* **116** 41 (2013).
[31] H. Nguyen, M. Koenig, D. Benredjem, M. Caby and G, Coulaud, *Phys. Rev. A* **33** 1279 (1986).
[32] O. Renner, E. Dalimier, E. Oks, F. Krasniqi, E. Dufour, R. Schott and E. Foerster, *J. Quant. Spectrosc. Rad. Transfer* **99** 439 (2006).
[33] E. Oks, E. Dalimier, A. Ya. Faenov *et al.*, Fast Track Communications, *J. Phys. B: Atom. Mol. Opt. Phys.* **47** 221001 (2014).
[34] E. Oks, E. Dalimier, A.Ya. Faenov *et al.*, *Optics Express* **25** 1958 (2017).

[35] V.P Gavrilenko, V.S. Belyaev, A.S. Kurilov *et al.*, *J. Phys. A: Math. Gen.* **39** 4353 (2006).
[36] V.S. Belyaev, V.I. Vinogradov, A.S. Kurilov *et al.*, *J. Exp. Theor. Phys.* **99** 708 (2004).

Chapter 8

Transverse Microwave-, Laser-, and/or Laser-induced Fields

In this Chapter, we present lineshape-based diagnostics of Transverse Quasimonochromatic Electric Fields (TQEFs) in plasmas, such as, electromagnetic radiation in plasmas in any part of the electromagnetic spectrum, but with the emphasis on the microwave range (the radiation of quasimonochromatic microwave sources, such as, masers) and the "extended" visible range (i.e., the range of various lasers). The laser/maser radiation cannot penetrate so-called "overdense" plasma regions, where the electron density $N_e > N_c$, where N_c is the critical density defined by the equation

$$\omega = \omega_{\text{pe}}(N_c), \tag{8.1}$$

where ω is the laser/maser frequency and $\omega_{\text{pe}}(N_c)$ is the plasma electron frequency: $\omega_{\text{pe}}(N_c) = (4\pi e^2 N_c/m_e)^{1/2}$. At super-high ("relativistic") laser intensities, such as those significantly exceeding 10^{18} W/cm^2, due to the relativistic effects (such as the relativistic increase of the electron mass) the critical density increases (even though the laser frequency is fixed) — the increased critical density depends on the laser intensity and is called the relativistic critical density N_{rc}. Below for brevity, we will use the term "critical density" in the broader sense — including the relativistic critical density.

So, the laser/maser radiation in plasmas can exist only in so-called "underdense" plasma regions (where the electron density is below the critical density) and at the surface of the critical density. The surface of the critical density usually provides a rich physics: at this surface due to various nonlinear processes the incident laser/maser radiation can get transformed in transverse electromagnetic waves of significantly higher amplitudes than the incident radiation.

The phenomenon of satellites of dipole-forbidden spectral lines of helium, lithium and of the corresponding ions can be used for diagnosing TQEFs on the same footing as for diagnosing Langmuir waves. As for the employment of hydrogenic spectral lines for diagnosing TQEFs, it brings up the following distinction compared to their use for diagnosing Langmuir waves.

The amplitude E_0 of Langmuir waves in plasmas is always smaller than the resonance value of the quasistatic electric field F_{res} involved in the formation of L-dips in hydrogenic spectral lines. (Here, F_{res} is defined by the resonance condition $\omega_{Stark}(F_{res}) = \omega_{pe}$, where $\omega_{Stark}(F_{res})$ is the Stark splitting.) The inequality $E_0 < F_{res}$ is the necessary condition for the existence of this kind of dips in profiles of hydrogenic spectral lines (see book [1], Sec. 4.2, and Appendix G). It is always fulfilled for Langmuir waves, so that they manifest by L-dips and the local "zigzags" of the intensity in the line profiles due to L-dips predominate over would-be "zigzags" of the intensity due to satellites — as pointed out in Chapter 7 (following book [1], Sec. 7.1).

In distinction to Langmuir waves, the amplitude E_0 of TQEFs can exceed (or even significantly exceed) F_{res}. In this situation, L-dips cannot exist. Instead, in the line profile, there could appear a sequence of satellites at the distances $\pm\omega_{pe}$, $\pm 2\omega_{pe}$, $\pm 3\omega_{pe}, \ldots$ from the unperturbed position of the spectral line (in the frequency scale) — see Appendix L.

8.1 Quasimonochromatic Microwave Field

The corresponding spectroscopic experiments can be first divided in two groups: those that used satellites of dipole-forbidden spectral lines and those that used hydrogen/deuterium spectral lines. Within each of these two groups, the experiments can be further divided in two groups: those that used the "passive" spectroscopy and those that used the "active" spectroscopy (i.e., used lasers to enhance the observation of microwave-caused satellites in spectral line profiles).

Below, we first describe experiments that used satellites of dipole-forbidden spectral lines (because they were historically the first), starting from "passive" experiments and then proceeding to "active" experiments. Then we will describe experiments that used satellites hydrogen/deuterium spectral lines, also starting from "passive" experiments and then proceeding to "active" experiments.

The first observation of microwave-caused satellites of a dipole-forbidden line was made by Cooper and Ringler in 1969 [2]. Namely, by applying a microwave of the frequency 31.565 GHz to a plasma discharge of a low electron density $1.2 \times 10^{12}\,\mathrm{cm}^{-3}$, they observed satellites of the line He I 438.79 nm (2^1P–5^1D). The satellite farthest from the allowed line (the far satellite) was barely visible. From the experimental separation of the satellites, they determined the microwave frequency as (32.5 ± 1.6) GHz, which was in agreement with the above frequency of the microwave source. From the intensity ratio of satellites to the allowed line, Cooper and Ringler determined the amplitude of the microwave field in the plasma to be (247 ± 35) V/cm — in agreement with the value of (228 ± 35) V/cm deduced from the Stark shift. They also measured a slight polarization of the satellite nearest to the allowed line (the near satellite). It should be noted that in the theoretical part of their paper, Cooper and Ringler used the first non-vanishing order (which turned out to be the second order) of the time-dependent perturbation theory for deriving intensities and polarization of the satellites in a linearly-polarized monochromatic electric field (as mentioned in Appendix J).

An experiment (also from the group of "passive" ones) where both satellites of a dipole-forbidden line were clearly observed was performed by Brizhinev *et al.* in 1983 [3]. The microwave source was a gyrotron producing a pulse of $200\,\mu$s duration at the wavelength $\lambda = 0.78\,\mathrm{cm}$ (the wave number $k = 1.28\,\mathrm{cm}^{-1}$). During the microwave pulse, the electron density gradually increased and reached the critical density (for the wave number $k = 1.28\,\text{cm}^{-1}$) $N_c = 1.85 \times 10^{13}\,\text{cm}^{-3}$. Profiles of the spectral lines He I 447.2 nm and 492.2 nm, exhibiting pronounced satellites, were observed at some instant t_1 when the plasma was underdense ($N_e < N_c$) and at another instant t_2 when the electron density reached the critical value and the microwave field caused nonlinear processes in the plasma.

Figures 8.1 and 8.2 show the experimental profiles of the lines He I 447.2 nm and 492.2 nm, respectively, at the instants t_2 (crosses) and t_1 (triangles) at the microwave power $P = 60\,\text{kW}$. Also shown are the experimental profiles from the region of no microwave field (circles).

One of the most instructive things about these experimental results is that along the line of site was a relatively small volume (say,

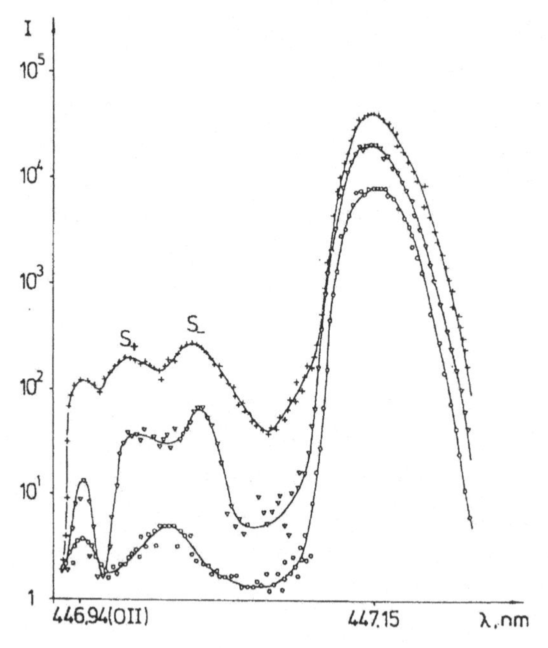

Fig. 8.1. The experimental profiles of the line He I 447.2 nm at the instants t_2 corresponding to the critical density $N_c = 1.85 \times 10^{13}\,\text{cm}^{-3}$ (crosses) and at the instant t_1 corresponding to the underdense plasma of $N_e < N_c$ (triangles) at the microwave power $P = 60\,\text{kW}$ [3]. The near and far satellites are marked as S_- and S_+, respectively. Also shown is the experimental profile from the region of no microwave field (circles). The line O II 446.9 nm is also seen.

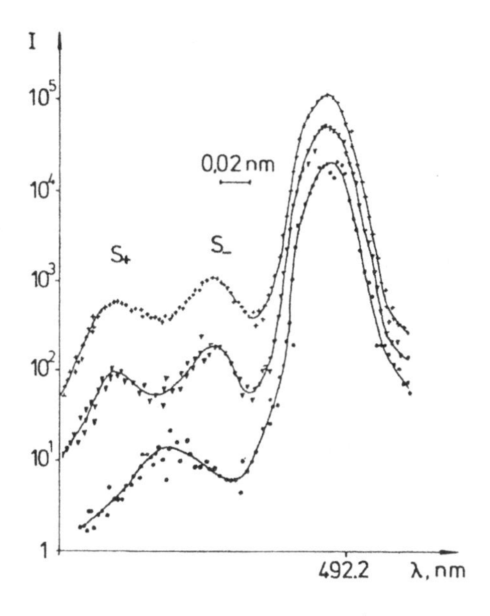

Fig. 8.2. The same as in Fig. 8.1, but for the line He I 492.2 nm [3].

of the characteristic dimension L_E) and a significantly larger volume (say, of the characteristic dimension L) containing no microwave field. Situations like this one — the presence of a spatial form-factor — was typical for many past studies of the interaction of a TQEF with plasmas and perhaps would be also relevant to future experiments. In this situation, the usual method for determining the amplitude of the TQEF by using the experimental ratio of intensities such as S_+/I_A (the ratio of intensities of the far satellite to the intensity of the allowed line) or S_-/I_A (the ratio of intensities of the near satellite to the intensity of the allowed line) obviously fails. One has to use the advanced method based on the adiabatic theory of satellites [4] (see Appendix J.2). This method is based on measuring the experimental ratio S_-/S_+. In Baranger–Mozer's theory [5] and in Cooper–Ringler's theory [6], this ratio did not depend on the TQEF amplitude E_0 (because those theories used only the first nonvanishing order of the standard perturbation theory). In distinction, in the adiabatic perturbation theory, which is

Table 8.1. The amplitude E_0 of the incident microwave radiation, the amplitude E_{0r} of the enhanced field at the critical electron density of the plasma (determined from the experimental ratio of the satellites intensities S_-/S_+ of the lines He I 492.2 nm and 447.2 nm), and the field gain $K = E_{0r}/E_0$ at different powers P of the incident microwave radiation [3].

P (kW)	E_0 (kV/cm)	E_{0r} (kV/cm) from 492.2 nm	E_{0r} (kV/cm) from 447.2 nm	$K = E_{0r}/E_0$
7.5	1.9 ± 0.2	6.4 ± 0.4	$6.5^{+0.5}_{-1}$	3.4 ± 0.6
15.0	2.7 ± 0.3	6.1 ± 0.4	$7.0^{+0.5}_{-1}$	1.9 ± 0.2
60.0	5.4 ± 0.5	6.2 ± 0.4	$8.0^{+0.5}_{-1}$	1.9 ± 0.2

more accurate than the standard perturbation theory in any order, the ratio S_-/S_+ significantly depends on E_0 and thus allows the experimental determination of this quantity.

Table 8.1 presents the TQEF amplitude E_{0r}, determined from the experimental ratio S_-/S_+, at the critical electron density for different powers P of the incident microwave radiation. Table 8.1 also presents, the corresponding amplitude E_0 calculated by the formula

$$E_0 \text{ (V/cm)} = 22[P(W)]^{1/2}, \tag{8.2}$$

and the field gain $K = E_{0r}/E_0$.

For estimating the spatial form-factor, the TQEF "amplitude" was also determined from the experimental ratio S_-/I_A (also using the adiabatic perturbation theory yielding more accurate values of this ratio than the standard perturbation theory). The word amplitude is in quotation marks because it is not the actual, real amplitude E_{0r}, which had been determined from the experimental ratio S_-/S_+; the "amplitude" determined from the ratio S_-/I_A is denoted E_0^{exp}. Figure 8.3 presents the relationship between E_0^{exp} and E_{0r} at the critical electron density (E_{0r} is denoted as E_0^{real} in Fig. 8.3). The results are averaged over the two lines 492.2 nm and 447.2 nm.

The straight line in Fig. 8.3 corresponds approximately to the case of the underdense plasma, where there was no enhancement of the incident microwave radiation. It was drawn through the vertical error bar of the value E_{0r} determined from the experimental ratio

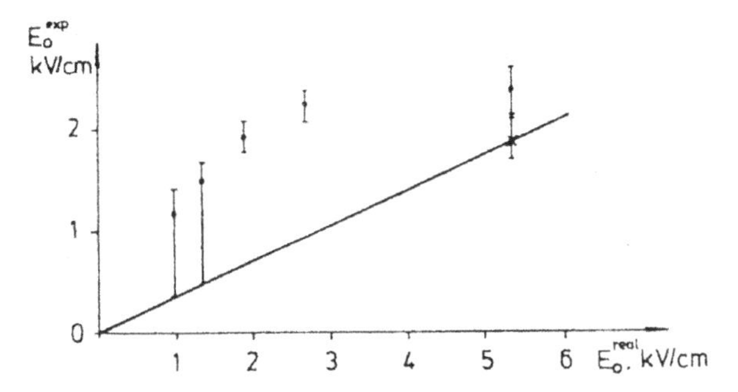

Fig. 8.3. The relationship, caused by the presence of a spatial form-factor, between the amplitude E_0^{exp} determined from the experimental ratio S_-/I_A (averaged over the two lines He I 492.2 nm and 447.2 nm) and the actual, real amplitude E_0^{real} [3].

S_-/I_A in the underdense plasma at $P = 60\,\text{kW}$. From the slope of this straight line, it follows that the amplitude E_0^{real} is approximately 2.8 times greater than E_0^{exp}. This allows estimating the spatial form-factor as $L/L_E \sim (E_0^{\text{real}}/E_0^{\text{exp}})^2 \sim (2.8)^2 \sim 8$. This value of $L \sim 8L_E \sim 8\,\text{cm}$ agreed with the halfwidth of the spatial distribution of the electron density over the chamber diameter.

For a rough estimate of the actual, real amplitude E_0^{real} at the critical electron density, each value of E_0^{exp} at the critical electron density in Fig. 8.3 should be also multiplied by 2.8. At the incident microwave powers of 7.5, 15 and 60 kW, these estimates of E_0^{real} were compared to the more rigorously obtained results from Table 8.1 and the agreement between them was verified. The results demonstrated that the enhancement of the incident microwave radiation at the critical electron density has the maximum at $E_{0r} \sim 2\,\text{kV/cm}$.

It should be emphasized that the passive spectroscopic method for measuring the microwave amplitude E_0, based on the ratio of the satellites intensities to each other S_-/S_+, effectively provides a "spatial resolution": it provides the information about E_0 from a small volume $v \ll V$, where v is the plasma volume containing the microwaves and V is the total plasma volume, from which the light is collected. This is why this method is called quasilocal. Otherwise, for achieving the spatial resolution, it would require an active method using the laser irradiating the small plasma volume containing microwaves — such active method is much more complicated from the experimental point of view.

Another instructive thing about the experiment [3] is the polarization measurements performed for determining the predominant direction of the enhanced TQEF at the critical electron density. The intensity of the near satellite of the line He I 449.2 nm was recorded at the two orthogonal orientation of the polarizer: the intensity $S_-^{(z)}$ with the polarizer parallel to the direction of the electric field in the incident microwave radiation (coinciding with the direction of the magnetic field **B** in the plasma) and the intensity $S_-^{(x)}$ with the polarizer along the wave vector **k** of the incident microwave radiation. The experimental results are presented in Table 8.2.

The experimental information from Table 8.2 served as the input data for the equations of the theory of the polarization measurements of the satellites from [1,4] presented in Appendix J.4. The first step was to find the value of the function $f_-(E_{0r})$, defined in the second line of Eq. (J.15), whose dependence on E_{0r} is presented in Fig. J.7. This step involved using the experimental value of E_{0r} determined from non-polarization measurements described above. Then by substituting into the first line of Eq. (J.15), the obtained experimental value of $f_-(E_{0r})$ and the experimental value of $S_-^{(z)}/S_-^{(x)}$, the value of $\cot^2\gamma = E_z^2/E_x^2$ was determined.

Figure 8.4 presents the obtained experimental dependence of E_z^2/E_x^2 at the critical electron density on the amplitude E_0 of the incident microwave field. It is seen that the enhancement of the TQEF in the region of the critical density is mostly due to the appearance of the E_x component, i.e., the component along the wave vector **k** of the incident wave (this component was practically

Table 8.2. The experimental ratio of the intensities $S_-^{(z)}/S_-^{(x)}$ of the near satellite of the line He I 492.2 nm in two orthogonal linear polarizations at the critical electron density versus the power of the incident microwave radiation [3].

P(kW)	1.3	4	13	40
$S_-^{(z)}/S_-^{(x)}$	1.008 ± 0.026	0.956 ± 0.014	0.942 ± 0.017	0.994 ± 0.014

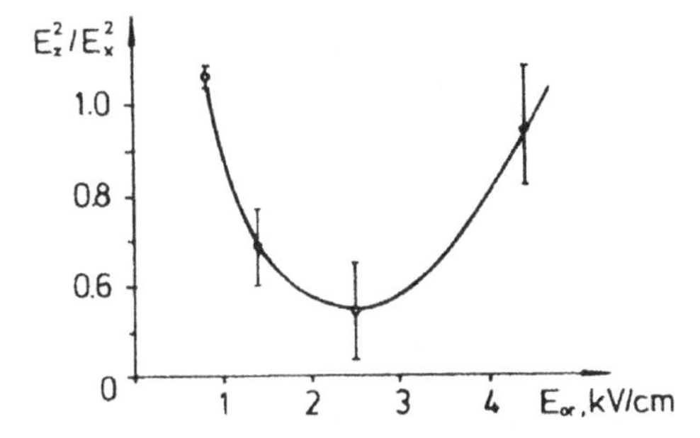

Fig. 8.4. The experimental dependence of E_z^2/E_x^2 at the critical electron density on the amplitude E_0 of the incident microwave field [3].

non-existent in the incident wave). This effect is the greatest at $E_0 =$ 2.5 kV/cm. As the amplitude E_0 of the incident wave increases from 2.5 kV/cm, both components of the enhanced TQEF became of the same order.

Now we proceed to those experiments, based on the satellites of dipole-forbidden spectral lines, that used active methods. The first such experiment was performed by Burrell and Kunze in 1972 [7]. In this experiment, where a discharge plasma of the electron density $\sim 10^{13}\,\text{cm}^{-3}$ was subjected to a microwave at the frequency 74 GHz (generated by a magnetron at the power 5 kW), the radiation of a tunable dye laser was used for enhancing the observed intensities of the satellites of the He I line 447.2 nm $(2^3P - 4^3D)$. In one of their experiments, the laser was tuned to the frequency ω_A of the allowed line. Due to the collisional coupling of the levels 4^3D and 4^3F, the intensity of satellites in the *emission* spectrum was enhanced and the near satellite become clearly visible (while the satellites were not observed at the absence of the laser pumping). This was an application of the *laser fluorescence* technique.

In another experiment reported in paper [7], the laser was tuned to the frequency $\omega_F - \omega_M$, where ω_F is the frequency of the dipole-forbidden transition 2^3P–4^3F and ω_M is the microwave frequency, the simultaneous absorption of the laser photon and the microwave photon increased the population of the level 4^3F, resulting in the enhanced intensity of the forbidden line and its near satellite in *absorption*. When the laser was tuned to the frequency $\omega_F + \omega_M$, the simultaneous absorption of the laser photon and the stimulated emission of the microwave photon also increased the population of the level 4^3F, resulting in the enhanced intensity of the forbidden line and its far satellite in *absorption*. Those enhancements were stronger than in the laser fluorescence experiment.

A later experiment in Kunze' group [8] was a variation of the laser fluorescence technique as it was employed in a non-plasma experiment as follows. A low-energy lithium beam was subjected to microwaves at the frequency 9.55 GHZ produced by a magnetron. Two dye lasers were employed. The first laser at the wavelength 670.8 nm populated one of the sublevels of the level 2^2P (in a plasma this sublevel would be excited by collisions, so that the first laser would not be necessary). The second laser was tuned in two different manners. In the first manner, it was tuned to the wavelength 460.3 nm, thus populating the lithium level 4^2D; the fluorescence intensity was recorded at the same transition 4^2D–2^2P. In the second manner, the second laser was tuned to a near or far satellite of the dipole-forbidden transition 2^2P–4^2F and together with microwave quanta there was achieved a population of the level 4^2F. Then a fluorescence intensity was recorded at the transition 3^2D–2^2P corresponding to the wavelength 610.4 nm.

In plasmas, due to the collisional mixing of the states 4^2D and 4^2F, it would be then possible to measure the satellites profiles by observing the 4^2D–2^2P transition at 460.3 nm, but this cannot be done in the collisionless beam experiment [8]. Therefore, for measuring the population of the 4^2F states, the authors of [8] observed the fluorescence intensity of the 3^2D–2^2P transition corresponding to the wavelength 610.4 nm. (Obviously, the fluorescence at the 3^2D–2^2P transition was preceded by the fluorescence at the 4^2F–3^2D transition, but the direct observation of the latter was difficult because of its wavelength of 1.87 μm.)

Figure 8.5 shows the profiles of the allowed line Li I 460.3 nm and of the forbidden line satellites measured in the way described above at the microwave power $P = 18$ kW and $P = 50$ kW [8]. At $P = 16$ kW, the experimental satellites intensity ratio S_-/S_+ was equal to 4/3, thus corresponding the theoretical results by Baranger–Mozer [5] and

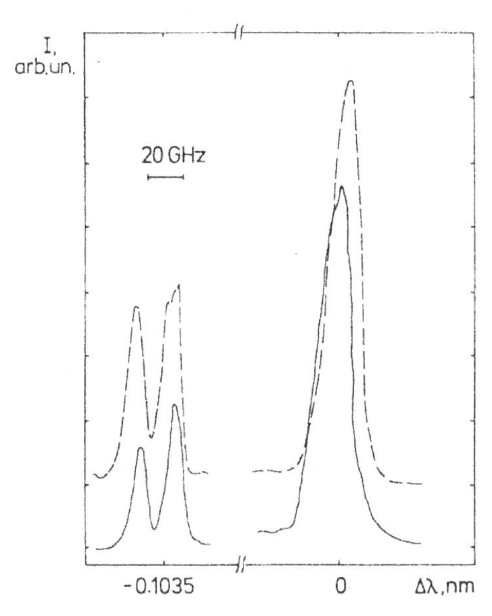

Fig. 8.5. Laser-induced fluorescence spectra from [8] near the line Li I 460.3 nm at the microwave power 18 kW (solid line) and 50 kW (dashed line). *Left*: the emission intensity at 610.4 nm; *right*: the intensity at 460.3 nm (the allowed line was saturated).

by Cooper–Ringler [2] valid for weak fields only. At $P = 50\,\mathrm{kW}$, the experimental ratio S_-/S_+ significantly decreased and this could not be explained within the theories from papers [2, 5], as noted by the authors of paper [8], who also emphasized that the decrease of the experimental ratio S_-/S_+ was not due to any saturation effects. Later by the application of the adiabatic perturbation theory [4], from experimental ratio S_-/S_+ at $P = 50\,\mathrm{kW}$, the microwave amplitude was easily determined to be $E_0 = 2.6\,\mathrm{kV/cm}$ (book [1], Sec. 7.6).

Further experiments at the same experimental setup as in [8], but with higher microwave amplitudes, were described in the paper [9]. The results are presented in Fig. 8.6. It is seen that as the microwave amplitude gets higher, the satellites structure becomes more and more complicated. Nevertheless, the experimental profiles, corresponding to the transitions (2^2P–4^2D, 4^2F) in lithium, can be simulated using the adiabatic perturbation theory [4] in the entire

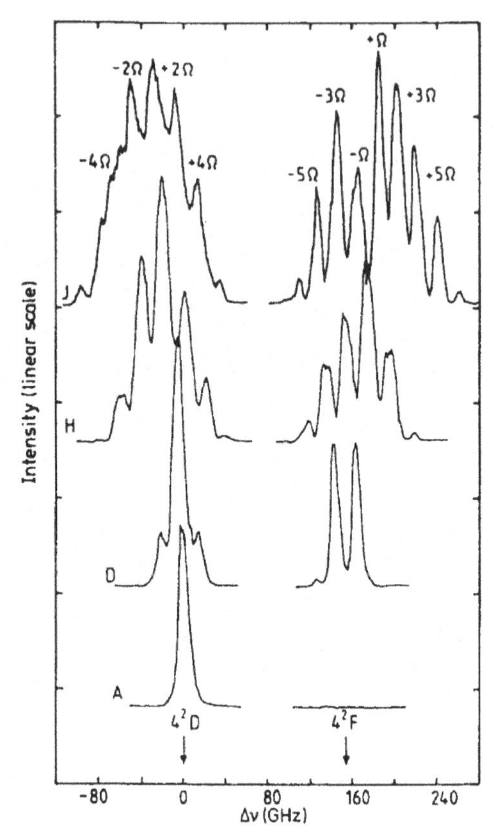

Fig. 8.6. Satellites of 2^2P–4^2D and 2^2P–4^2F transitions in Li I at the microwave frequency 9.5 GHz and different microwave amplitudes observed in the experiment [9]. The spectra are presented at the following microwave amplitudes: (A) 0 kV/cm, (D) 2.5 kV/cm, (H) 8.2 kV/cm, (I) 11.5 kV/cm.

range of the microwave amplitudes (from 0 to 11.5 kV/cm) from the experiment [9].

Next, we proceed to the experiments that were based on hydrogen/deuterium spectral lines and used passive spectroscopic methods. It should be emphasized upfront that the only thing observed in most of these experiments under the action of the microwave was some additional broadening of hydrogen/deuterium spectral lines mostly in the wings. For attempting to determine the microwave amplitude in these plasmas, it was necessary to subtract from the experimental profile at the "microwave on" (or to benchmark it to) the experimental profile at the "microwave off." This procedure is ambiguous: the experimental profile at the "microwave on" could correspond to plasma parameters (such as the temperature and the electron density) different from the experimental profile at the "microwave off." Therefore, the increase of the broadening of the experimental profile at the "microwave on" compared to the experimental profile at the "microwave off", could be partially due to the microwave-cause change of the plasma parameters affecting the usual (non-microwave) mechanisms of the broadening in plasmas (such as, Doppler and Stark broadenings). For this reason, an unambiguous determination of the microwave amplitude in the passive-type experiments employing hydrogen/deuterium spectral lines seems unlikely. Therefore, below we list these kind of experiments only briefly (and then pay more attention to the active-type experiments using hydrogen/deuterium spectral lines).

In 1979, Shefer and Bekefi [10] observed an additional broadening of the H-beta line under the microwave of the frequency 4.6 GHz produced by a relativistic electron beam magnetron, as shown in Fig. 8.7. By using some very rough theoretical estimates, they claimed the microwave amplitude to be ~100 kV/cm.

In 1981, Brizhinev *et al.* [11] observed an additional broadening in the wings of the H-beta profile under a microwave of the wavelength 0.78 cm at the incident microwave amplitude of 4.5 kV/cm, as shown in Fig. 8.8. The experimental profiles were obtained from the underdense plasma (the top frame) and from the plasma of the critical electron density (the bottom frame). The experimental profiles are presented as lg (Intensity) versus $\Delta\lambda^2$, so that the Gaussian profile would be a straight line. It is seen that the bulk of the profile has the Gaussian shape, but the wings are broader than that Gaussian. The insets (both in the top and bottom frames) show the difference profile obtained by subtracting from the experimental profile the corresponding Gaussian profile. From the analysis of the difference profile, by using the corresponding theoretical results from the paper [12] (presented in Appendix L of this book) it was determined that the microwave amplitude in the plasma at the

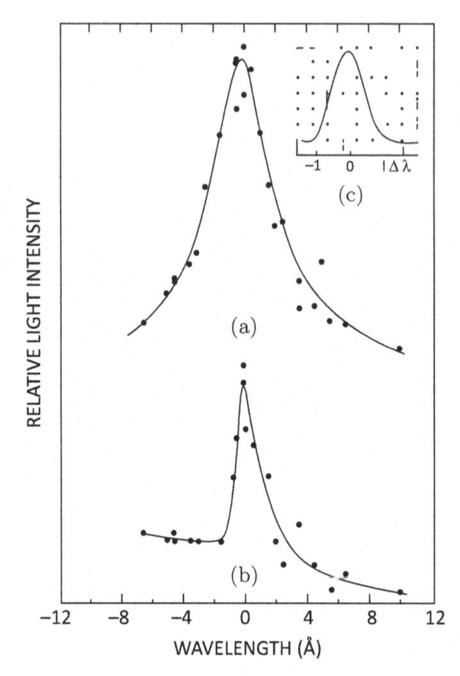

Fig. 8.7. Profiles of the H-beta line observed in the experiment [10]: (a) during the microwave pulse; (b) in the afterglow; (c) the instrumental profile.

critical electron density got enhanced from 4.5 kV/cm to 6.1 kV/cm. The results of the enhancement at the critical electron density, obtained in the same way at two other amplitudes of the incident microwave radiation, were as follows: the incident microwave field of 2.2 kV/cm or 3.2 kV/cm got enhanced to 5.1 kV/cm or 5.5 kV/cm, respectively.

In 1988, Kamp and Himmel [13] observed profiles of deuterium lines *D*-beta, *D*-gamma, and *D*-delta under a microwave of the frequency 2.45 GHz emitted by a deuterium plasma of the electron density $N_e < 5 \times 10^{12}\,\mathrm{cm}^{-3}$. These profiles were compared with the profiles observed immediately after the microwave pulse had passed. Then the difference profile for each line (i.e., the profile at "microwave on" – the profile at "microwave off") was analyzed. Figure 8.9 shows two such experimental profiles of the *D*-gamma line. Also a polarization analysis of the *D*-beta line was performed. From all the

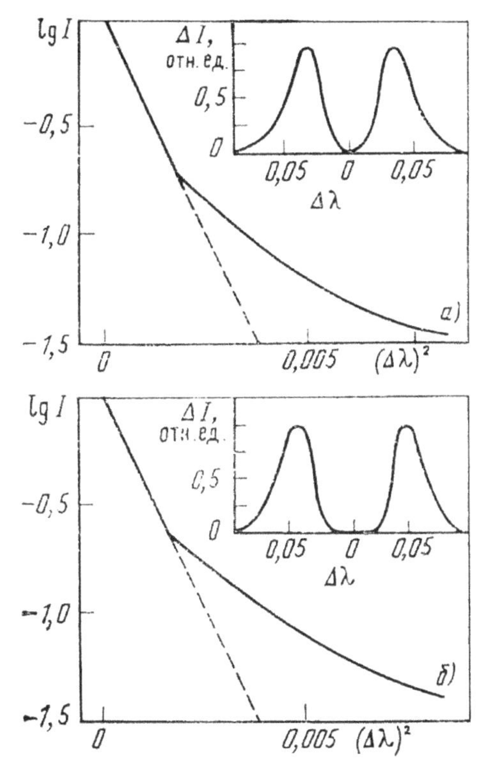

Fig. 8.8. Profiles of the H-beta line observed in the experiment [11] in the underdense plasma (the top frame) and in the plasma of the critical electron density (the bottom frame). The distance $\Delta\lambda$ from the line center is in nm. The insets (both in the top and bottom frames) show the difference profile obtained by subtracting from the experimental profile the corresponding Gaussian profile.

observed profiles, the microwave amplitude was estimated to be in the range 1.00–1.35 kV/cm.

For completeness we also note that in 1988, Kulikov and Mitsuk [14] reported "unresolved structure" (as they wrote it) of the H-beta line under a microwave of the wavelength 2.2 cm. However, the experimental profile was not presented in [14], so that their claim of determining the microwave amplitude up to 3 kV/cm from the experimental profile could not be verified.

Finally, we proceed to the active-type experiments using hydrogen/deuterium spectral lines. In 1985, Polushkin *et al.* [15] used the

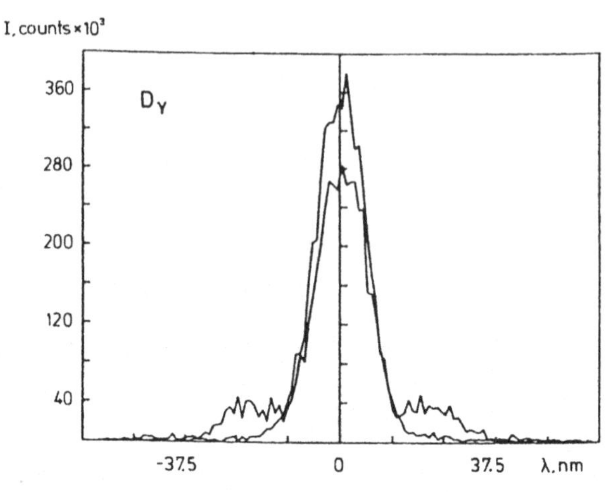

Fig. 8.9. The experimental profiles of the D-gamma line from the experiment [13]: the top profile — at "microwave on", the bottom profile — at "microwave off".

advanced principle of laser-induced fluorescence measurements of the microwave field in plasmas developed in paper [16] and presented in Appendix M of this book. Polushkin *et al.* [15] used the same microwave source at the wavelength $\lambda = 0.78\,\text{cm}$ (the wave number $k = 1.28\,\text{cm}^{-1}$), as in the experiments described in papers [3, 11]. The microwave beam was focused at a discharge tube filled with hydrogen. The electron density was $6 \times 10^{11}\,\text{cm}^{-3}$. A dye laser was tuned to the wavelength of the H-alpha line (656.3 nm). The electric fields of the laser and microwave radiations were parallel to each other, while their wave vectors were orthogonal.

Figure 8.10 shows the inverse fluorescence signal $1/B$ versus the inverse intensity of the laser radiation I_l for the following three values of the microwave power P: (1) $P = 0$, (2) $P = 45\,\text{kW}$, (3) $P = 60\,\text{kW}$. According to the theory from paper [16] (presented here in Appendix M), the dependence of $1/B$ versus $1/I_l$ should be a straight line, whose slope can be used for measuring the microwave field.

Further, out of the two methods for measuring the microwave amplitude in plasmas, developed in paper [16], Polushkin *et al.* [15] used the method based on the experimental ratio of the slope of the dependence of $1/B$ versus $1/I_l$ at some non-zero P to the slope

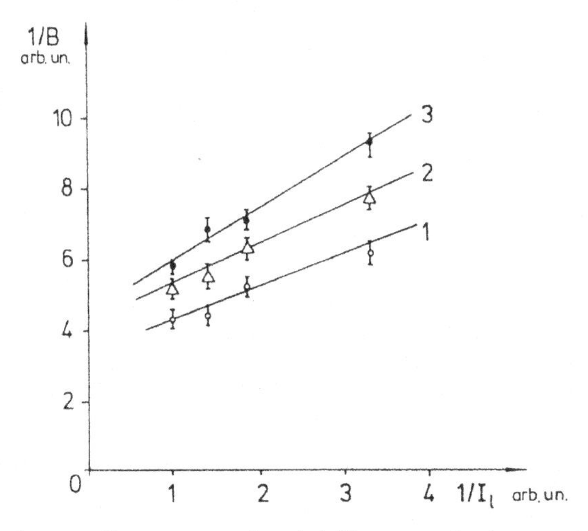

Fig. 8.10. The inverse fluorescence signal $1/B$ versus the inverse intensity of the laser radiation I_l for the following three values of the microwave power P in the experiment [15]: (1) $P = 0$, (2) $P = 45\,\text{kW}$, (3) $P = 60\,\text{kW}$.

Table 8.3. The microwave amplitude E_0 at different values of the microwave power P measured by the laser induced fluorescence method in experiment [16].

P (kW)	45	60
E_0 (kV/cm) in plasma (measured)	4.1 ± 0.7	6.0 ± 0.9
E_0 (kV/cm) in vacuum (calculated)	4.7 ± 0.2	5.4 ± 0.2

of the dependence of $1/B$ versus $1/I_l$ at $P = 0$. Then by using the theoretical dependence of this ratio on the microwave amplitude from [16] (see Appendix M, Fig. M.1), they determined the microwave amplitude E_0 in the plasma for each value of the incident microwave power P. The results are presented in Table 8.3.

The last row in Table 8.3 shows the value of the microwave amplitude E_0 in vacuum, calculated by formula (8.2). These values are in a good agreement with the corresponding values measured in the plasma, thus confirming the effectiveness of this laser induced fluorescence method.

It is very important to emphasize that by using the theoretical basis of the above laser-induced fluorescence method from [16] and presented in Appendix M of this book, the method can be extended in the following two ways. In low-temperature plasmas, it can be used for local measurements of Langmuir waves or Bernstein modes (rather than only microwaves penetrating into a plasma from an external source). In high-temperature plasmas, it is possible to map the distribution of a powerful infra-red laser field in the plasma by stimulating — with a near-UV laser — the corresponding spectral lines of hydrogenlike ions (e.g., the line Li III 208.2 nm) or helium-like ions (e.g., the line Be III 208.0 nm or 212.2 nm).

In a later experiment [17] performed in the same group, the authors used the intracavity laser spectroscopy for measuring microwave fields in a plasma of a hydrogen-deuterium mixture at the ratio of hydrogen and deuterium partial pressures between 5 and 50. The plasma was subjected to microwaves generated by a cyclotron resonance maser (gyrotron) of the frequency 38.5 GHz and of the power 200 kW. The gas discharge tube, having optical windows at both ends, was placed inside the laser cavity bounded by mirrors. As any method based on absorption, the intracavity laser spectroscopy has a limited dynamic range because the measurement errors increase for both small (>0.05) and large (>3) values of the effective optical thickness.

Figure 8.11 shows the experimental absorption profiles in a range of the frequency $\nu(\mathrm{cm}^{-1})$ near the Balmer-alpha lines of hydrogen and deuterium for two different concentrations of the absorbing atoms. In Fig. 8.11(a), where the concentration of absorbing atoms is smaller than in Fig. 8.11(b), the $H_\alpha^{\pm 1}$ satellites (i.e., the satellites separated from the H-alpha line by plus or minus the microwave frequency) are clearly visible — in distinction to the $D_\alpha^{\pm 1}$ satellites. At the higher concentration of absorbing atoms (Fig. 8.11(b)), the $D_\alpha^{\pm 1}$ satellites become visible, but the absorption coefficients for the H-alpha and D-alpha lines become so large that they exceed the dynamic range of this experimental method.

The authors of the paper [17] did not attempt measuring the absolute values of the absorption coefficients. We should point out that the observed satellites seemed to be of Blochinzew's type [18]. The underlying theory for real (i.e., multi-component) hydrogenic

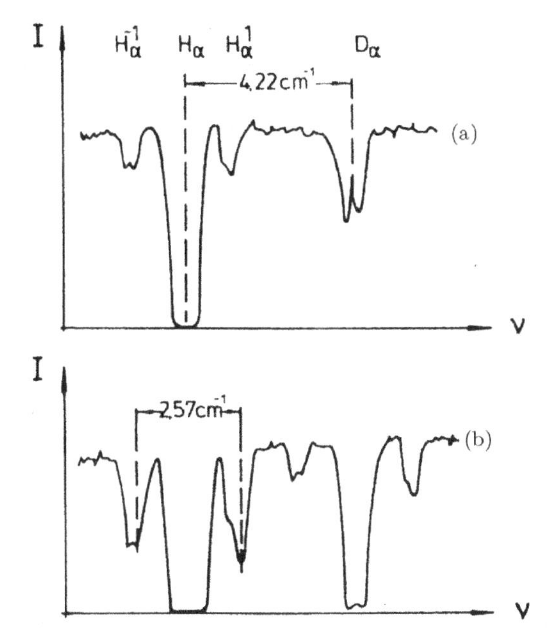

Fig. 8.11. The experimental absorption profiles in a range of the frequency $\nu(\mathrm{cm}^{-1})$ near the Balmer-alpha lines of hydrogen and deuterium, observed in the intracavily laser spectroscopy experiment [17] in the H–D mixture for two different concentrations of the absorbing atoms: in (b) the concentration is higher than in (a).

lines was developed in paper [12] and presented also in book [1], Sec. 3.1 (as well as in Appendix L of the present book).

8.2 Laser- and/or Laser-induced Field

In 1997, Elton *et al.* in the experiment with a laser produced plasma of the electron density $\sim 10^{20}\,\mathrm{cm}^{-3}$ [19], observed satellites of the dipole forbidden components of the He-like line Mg XI 5.264 nm and of the Li-like line Mg X 6.32 nm — the satellites caused by the laser radiation in the underdense plasma (for brevity, laser satellites). The authors used a frequency tripled 1.06 μm laser OMEGA-upgrade, located at the University of Rochester Laboratory for Laser Energetics (USA), at the laser intensities $(10^{14}$–$10^{16})$ W/cm^2. The laser wavelength was 0.35 μm and the nominal pulse duration was 1 ns. Only S_+ satellite (the far satellite) was observed at the He-like line Mg XI 5.264 nm and only S_- satellite (the near satellite) was observed at the He-like line Mg X 6.32 nm.

In 2009, Sauvan *et al.* [20] reported the observation of laser satellites of the He-like line Al XII 0.6635 nm (hereafter, Al He-beta). Two laser beams were simultaneously used in this experiment performed at the Jena (Germany) multi-terawatt Ti:sapphire laser system JETI. The plasma producing laser beam (0.2 J, 5×10^{15} W/cm^2) was introduced to the chamber via a delay line and focused to a flat tip of the tapered Al target. The second beam (0.45 J, 1.2×10^{16} W/cm^2) with the axis parallel to the target surface was hitting the plasma plume transversally. The axis of the plasma expansion, the electric field of the second beam and the direction of the X-ray spectra observation were mutually perpendicular. The electric field vector of the second laser beam was approximately perpendicular to the line of sight of the spectra, i.e., the transverse-field-affected line profiles were observed. Both focused laser beams were spatially overlapped and temporally synchronized with the precision better than 1 ps. The laser wavelength of 0.8 μm corresponded to the critical electron density $N_c = 1.7 \times 10^{21}$ cm^{-3}.

The reproducibility of the satellites in the experimental profiles of the Al He-beta line obtained in different shots is demonstrated in Fig. 8.12, where the measured points were smoothed using the 5 point fast Fourier transform. The positions of the satellites are shown by dashed-dotted vertical lines.

Figure 8.13 presents the comparison of one of the experimental profiles of the Al He-beta line with simulations [20]. The simulations were based on the Floquet–Liouville formalism described in [20] (presented in Appendix K of this book).

Sauvan *et al.* [20] noted that in the red part of the experimental profile, between the peaks identified as the first and the second red satellites, there is an additional local maximum denoted by an arrow in Fig. 8.12. Its position agrees with the dipole forbidden transition 1s3s $^1S_0$–1 s^2 $^1S_0$. Sauvan *et al.* [20] brought up a hypothesis that this was due the coupling of singlet $^1S_0$ and $^1D_2$ states via the quadrupole interaction with the ion microfield. (This interaction, being due to the intrinsic inhomogeneity of the ion microfield, can

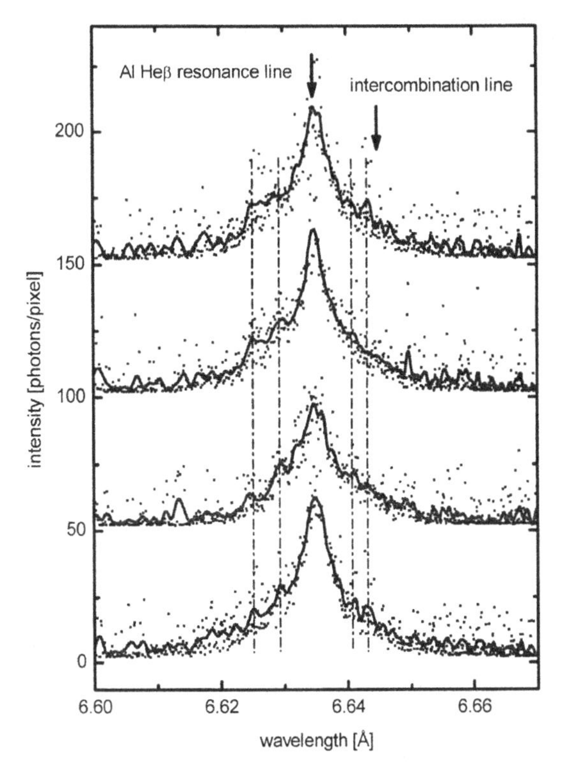

Fig. 8.12. The reproducibility of the satellites in the experimental profiles of the Al He-beta line obtained in different shots [20]. The measured points were smoothed using the 5 point fast Fourier transform. The positions of the satellites are shown by dashed-dotted vertical lines.

significantly enhance some dipole forbidden spectral lines, as it was first shown in paper [21].) However, in the subsequent work [22], Sauvan *et al.* showed that this effect was not strong enough to explain the experimental feature. It was shown in [22] that the quadratic Stark effect introduced via the enlarged atomic basis that turned out to be important for the consistent interpretation.

The TQEF in the plasma, determined in [20, 22] from the simulations best fitting the experimental profiles including the satellites, turned out to be stronger than in the incident laser radiation (~3 GV/cm). The kinetic Particle-in-Cell (PIC) simulations, presented in [20,22], showed that the dominant process, responsible

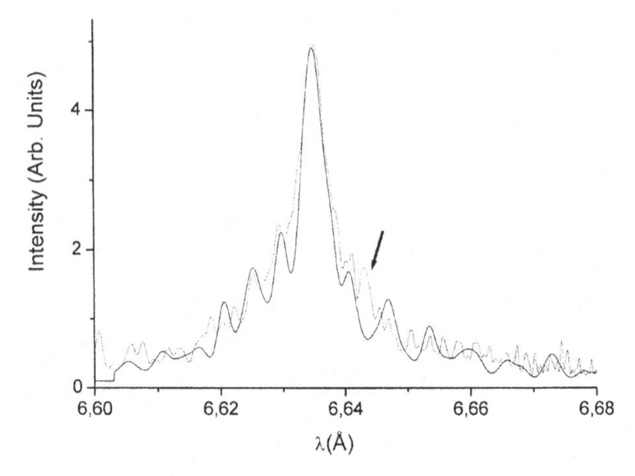

Fig. 8.13. The comparison of one of the experimental profiles of the Al He-beta line with simulations [20]. The simulated profile is shown by the bolder line than the experimental profile. The arrow is explained in the text below.

for the enhancement of the TQEF in the plasma, was the stimulated Brillouin backscattering in the so-called strong coupling regime leading to the formation of transient phenomena, such as plasma cavities and transverse electromagnetic solitons.

In 2015, Oks *et al.* [23] presented the experiment dealing with a femtosecond laser-driven cluster-based plasma, where by analyzing the nonlinear phenomenon of (non-Blochinzew) satellites of spectral lines of Ar XVII, it was revealed the nonlinear phenomenon of the generation of the second harmonic of the laser frequency in the plasma produced from clusters — see also the subsequent paper by Faenov *et al.* [24]. The incident laser intensity was 3×10^{18} W/cm^2, the pulse duration — 40 fs. The experiments were performed at Kansai Photon Science Institute (Japan).

Experimental spatially-integrated spectra of the lines Ar XVII Ly-delta (Ly-5) and especially Ly-epsilon (Ly-6) exhibited satellites separated from the line center by $2\omega[\lambda_n^2/(2\pi c)]$, where ω is the laser frequency and λ_n is the unperturbed wavelength of the Ly-n line of Ar XVII. In other words, the experimental spectra showed satellites at the frequencies $\pm 2\omega$ (counted from the line center in the frequency scale), but did not show satellites at the frequencies $\pm\omega$. If the emission would be only from the plasma region where only the wave at the frequency 2ω (wave t_2) existed, it would be fairly easy to interpret the experimental spectra. But this scenario could occur only if the emission originates from the overdense plasma at the density N_e such that

$$N_{ec} < N_e < 4N_{ec}, \tag{8.3}$$

which for the laser frequency $\omega = 2.4 \times 10^{15}\,\mathrm{s}^{-1}$ translates into the interval $1.8 \times 10^{21}\,\mathrm{cm}^{-3} < N_e < 7.2 \times 10^{21}\,\mathrm{cm}^{-3}$. For clarity: in large clusters the laser radiation pump wave at the frequency ω (wave t_1) cannot penetrate the plasma of the electron density higher than the critical density N_{ec}, while its second harmonic t_2 can — but only as long as the density does not exceed $4N_{ec}$. (It should also be noted that at the laser intensity $I = 3 \times 10^{18}\,\mathrm{W/cm}^2$, the threshold for concept of the relativistic critical density to become effective has not been reached.)

However, the broadening of the experimental Ar XVII Ly-lines delta, epsilon and zeta lines corresponded to noticeably lower densities and thus did not support such scenario. Therefore, it had to be considered the situation where the observed spectra were formed under the joint action of a relatively strong wave t_1 at the frequency ω and a weaker wave t_2 at the frequency 2ω. (The second harmonic generation produces a wave weaker than the pump wave.) This situation presented quite a challenge for interpreting the experimental spectra. Indeed, it would seem that the stronger wave t_1 at the frequency ω should produce relatively strong satellites at the frequencies $\pm\omega$; however, these satellites were not observed.

At the electron densities $\sim 10^{21}\,\mathrm{cm}^{-3}$, Stark broadening (SB) of the highly-excited lines Ar XVII Ly-7, Ly-6, and even Ly-5 is practically the same as for hydrogenic systems. This is because the SB scales as $\sim n^2$ while the fine structure splitting scales as $\sim 1/n^3$. So, the ratio of the SB to the fine structure scales as $\sim n^5$. Therefore, there exists a threshold value of n, starting from which the spectral lines become essentially hydrogenic. In the conditions of the experiments presented in [23, 24], this threshold value was $n = 5$: the Ar XVII lines from $n = 5, 6, 7$ (and higher) were essentially hydrogenic, while the Ar XVII lines from $n = 3$ and 4 were not.

While the theory of satellites under the action of a monochromatic field at some frequency ω was developed for an isolated lateral Stark component of a hydrogenic line by Blochinzew [18] and for real, multicomponent hydrogenic lines in paper [12] (and presented in the book [1], Sec. 3.1, and in Appendix L of this book), there was no theory of satellites under a bi-chromatic field, such as a relatively strong wave t_1 at the frequency ω and a weaker wave t_2 at the frequency 2ω. Thus, first of all, such theory was developed in [23], as follows.

Under a *monochromatic* field, the intensity of each satellite is expressed through just one Bessel function, as follows. The spontaneous emission spectrum of a hydrogenic system under a monochromatic field $E(t) = E_1 \cos(\omega t)$ is controlled by the Fourier expansion of the product of the reduced wave functions

$$\Phi^*_{n\alpha\alpha}\Phi_{n\beta\beta} = \exp(ix\varepsilon_1 \sin\omega t) = \sum_{p=-\infty}^{\infty} J_p(x\varepsilon_1)\exp(ip\omega t), \tag{8.4}$$

as shown by Blokhinzew [18]. Here, x is the constant of the linear Stark effect, characterizing this Stark component:

$$x = n_\alpha(n_1 - n_2)_\alpha - n_\beta(n_1 - n_2)_\beta, \tag{8.5}$$

n_1, n_2 being the parabolic quantum numbers, and n being the principal quantum numbers of the upper (subscript α) and lower (subscript β) Stark sublevels involved in the radiative transition; ε_1 is the following dimensionless parameter

$$\varepsilon_1 = 3\hbar E_1/(2Z_{\text{eff}} m_e e\omega), \tag{8.6}$$

where Z_{eff} is the effective charge perceived by the outer electron ($Z_{\text{eff}} = 17$ for Ar XVII). Therefore, the intensity I_p of the satellite at the frequency $p\omega$ counted from the unperturbed frequency ω_0 of the spectral line ($p == 0, \pm 1, \pm 2, \ldots$), is

$$I_p = \sum_x j_x[J_{p-2s}(x\varepsilon_1)]^2, \tag{8.7}$$

where j_x are the relative intensities of the Stark components.

In distinction, in a *bi-chromatic* field the intensity I_p of the satellite at the frequency $p\omega$ (where p is a positive or negative integer) counted from the unperturbed frequency ω_0 of the spectral line is expressed through a sum of products of two different Bessel functions and the number of terms in the sum is generally infinite, as follows. In a bi-chromatic field $E(t) = E_1 \cos(\omega t) + E_2 \cos(2\omega t)$, the right side of Eq. (8.4) becomes

$$\Phi^*_{n_\alpha \alpha} \Phi_{n_\beta \beta} = \exp[ix(\varepsilon_1 \sin \omega t + \varepsilon_2 \sin 2\omega t)], \tag{8.8}$$

where

$$\varepsilon_2 = \frac{3\hbar E_2}{2Z_{\text{eff}} m_e e(2\omega)}, \tag{8.9}$$

This leads to the following expression for the satellite intensity I_p:

$$I_p = \sum_x j_x \left[\sum_{s=-\infty}^{\infty} J_s(x\varepsilon_2) J_{p-2s}(x\varepsilon_1) \right]^2, \tag{8.10}$$

which is controlled by the two (rather than one) dimensionless parameters: ε_1 defined by Eq. (8.6) and ε_2 defined by Eq. (8.9).

There was also no theory of satellites for *central* Stark components (the components for which $x = 0$ in Eq. (8.5)) — whether in a monochromatic or in a bi-chromatic field. The previous theories in a monochromatic field, as well as the newly developed in [23] theory in a bi-chromatic field, relate to *lateral* Stark components. The Ly-n lines with even values of n have central Stark components. Since, at the fairly high laser intensity in the experiment [23], the satellites of the central Stark component of the Ly-6 line could not be neglected, a theory of satellites for central Stark components was also developed in [23], as follows.

For *central* Stark components ($x = 0$), the instantaneous Stark shift, being quadratic with respect to the instantaneous electric field, is proportional to $[E_1 \cos(\omega t)]^2 = E_1^2(1 + \cos^2 \omega t)/2$. Therefore, the right side of Eq. (8.4) becomes:

$$\Phi^*_{n_\alpha \alpha} \Phi_{n_\beta \beta} = \exp[i\varepsilon_{\text{quadr}} \sin 2\omega t], \tag{8.11}$$

where,

$$\varepsilon_{\text{quadr}} = \frac{n^4(17n^2 - 9m^2 + 19)a_0^3 E_1^2}{64\hbar Z_{\text{eff}}^4 \omega}. \tag{8.12}$$

Here, m is the magnetic quantum number of the Stark sublevels, involved in the emission of the central component, and a_0 is the Bohr radius. The Fourier expansion of the right side of Eq. (8.11) leads to the following result for the central component contribution $I_{c,2p}$ to the intensity of the satellite at the frequency $2p\omega$ counted from the unperturbed frequency ω_0 of the spectral line ($p = 0, \pm 1, \pm 2, \ldots$)

$$I_{c,2p} = j_c[J_p(\varepsilon_{\text{quadr}})]^2, \tag{8.13}$$

where j_c is the relative intensity of the central Stark component.

The primary task in interpreting the experimental spectra, based on these newly developed theories, was to find a pair (or pairs) of ε_1 and ε_2 satisfying the following criterion: the intensity I_2 of the 2nd satellite significantly exceeds the intensity I_1 of the 1st satellite. It turned out that if in a particular laser shot the satellites was observed only in one spectral line, then there were several pairs of ε_1 and ε_2 satisfying the criterion, thus making it impossible to select a single pair of ε_1 and ε_2.

However, if in a particular laser shot the satellites were observed in at least two spectral lines (such as, Ly-5 and Ly-6 lines), then it was possible to find a single pair of ε_1 and ε_2 satisfying the above criterion. Such experimental spectra are presented in the bottom part of Fig. 8.14 by dotted lines. In particular, for this case the theoretical analysis, through variations of different plasma and laser parameters, led to the following pair: $\varepsilon_1 = 4$, $\varepsilon_2 = 1$. According to Eq. (8.6), this translated into the electric field amplitudes $E_1 = 14\,\text{GV/cm}$ and $E_2 = 7\,\text{GV/cm}$ for the first and second harmonics, respectively. In these shots the laser intensity was $3 \times 10^{18}\,\text{W/cm}^2$, corresponding to the electric field amplitude in vacuum $E_0 = 48\,\text{GV/cm}$. Thus, the conversion efficiency, defined as the ratio of the energy density in the generated second harmonic in the plasma to the energy density of the incident laser field in vacuum, was 2%.

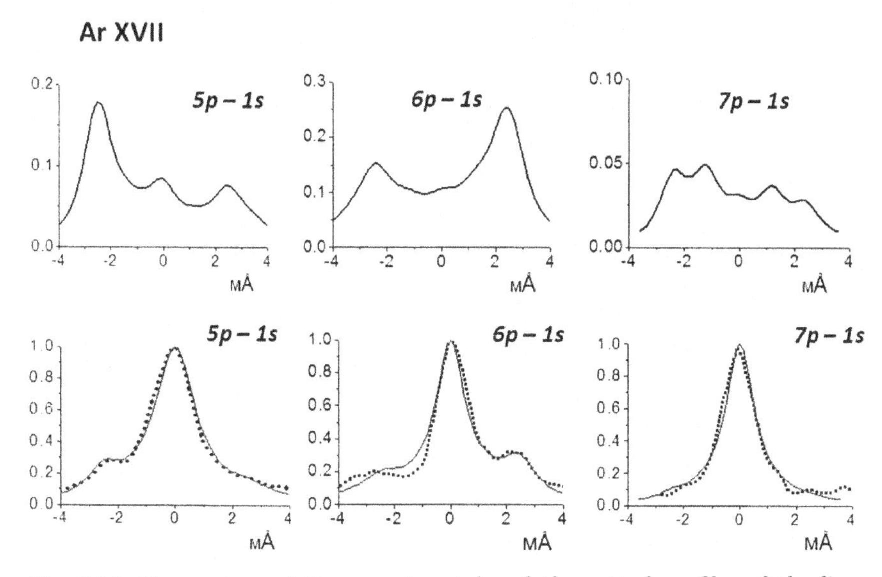

Fig. 8.14. Comparison of the experimental and theoretical profiles of the lines Ar XVII Ly-5, Ly-6, and Ly-7 from [24] at the laser intensity 3×10^{18} W/cm^2. The experimental profiles are shown by dotted lines in the bottom part of the figure. The top part of the figure shows the theoretical profiles from the region containing both the first and the second harmonics of the incident laser radiation. The complete theoretical profiles, including the emission from both the region of the bi-chromatic field and the no-periodic-field region, are shown in the bottom part of the figure by solid lines.

For fitting the experimental profiles of the Ly-delta and Ly-epsilon lines, it was necessary to take into account that the observed emission was a superposition of the emission from a more dense plasma region, where there were waves t_1 and t_2, with the emission from a less dense, but a larger plasma region, where there were no periodic electric fields. In this situation, the theoretical spectrum of a line consists of a relatively broad pedestal, emitted from the periodic-field-region, and a relatively narrow, but more intense component, emitted from the no-periodic-field-region.

Figure 8.14 shows also the theoretical profiles, calculated at $E_1 = 14\,\mathrm{GV/cm}$ and $E_2 = 7\,\mathrm{GV/cm}$, providing the best fit to the corresponding experimental profiles under laser intensity $3 \times 10^{18}\,\mathrm{W/cm^2}$ [24]. The top part of Fig. 8.14 presents the theoretical profiles from the region of the bi-chromatic field. The total theoretical profiles are shown in the bottom part of Fig. 8.14 by solid lines. It is seen that the bi-chromatic field, taken into account via the newly developed theory from [23], yielded a reliable interpretation of the experimental data. For the experimental Ar XVII Ly-7 line, at the possible location of the 2ω satellites, the intensity of this spectral line declined (compared to the center of the line) to the level close to the level of noise, so that there was zero or little visibility of the 2ω satellites. The average root-mean-square deviation of the simulated profiles from the corresponding experimental profiles was less than 5%.

From the lineshape analysis in [23], [24] it was found, in particular, that the efficiency of converting the short (40 fs) intense (3×10^{18} W/cm^2) incident laser light into the second harmonic was 2%. This result was in the excellent agreement with the 2D PIC simulation that were also performed in [23, 24].

It should be emphasized that the dominance of the 2nd satellites in the spectrum of just one line could be explained by a monochromatic field. However, the dominance of the 2ω satellites in the spectra of two lines (Ly-5 and Ly-6) could be explained only by a bi-chromatic field.

At the laser intensity 4×10^{17} W/cm^2, which was an order of magnitude lower than the laser intensity, at which the spectra presented in Fig. 8.14 were obtained, the experimental profiles did not show a signature of the second harmonic generation. These profiles could be fitted by using just the *monochromatic* (not bi-chromatic) field at the laser frequency.

Finally, we note that the distribution of a laser field in high-temperature plasmas can be mapped by the laser-induced fluorescence method using the theoretical basis from paper [16] and presented in Appendix M of this book. For example, the distribution of a powerful infra-red laser field in the plasma can be mapped by using a near-UV laser tuned to the corresponding spectral lines of hydrogen-like or helium-like ions.

References

[1] E. Oks, *Plasma Spectroscopy. The Influence of Microwave and Laser Fields*, Springer, New York, (1995).
[2] W.S. Cooper and H. Ringler, *Phys. Rev.* **179** 226 (1969).
[3] M.P. Brizhinev, V.P. Gavrilenko, S.V. Egorov *et al.*, *Sov. Phys. JETP* **58** 517 (1983).
[4] E. Oks and V.P. Gavrilenko, *Sov. Phys. Tech. Lett.* **9** 111 (1983).
[5] M. Baranger and B. Mozer, *Phys. Rev.* **123** 25 (1961).
[6] W.S. Cooper and H. Ringler, *Phys. Rev.* **179** 226 (1969).
[7] C.F. Burrell and H.-J. Kunze, *Phys. Rev. Lett.* **29** 1445 (1972).
[8] U. Rebhan, N.J. Wiegart and H.-J. Kunze, *Phys. Lett. A* **85** 228 (1981).
[9] U. Rebhan, *J. Phys. B: Atom. Mol. Opt. Phys.* **19** 3487 (1986).
[10] R.E. Shefer and G. Bekefi, *Phys. Fluids* **22** 1584 (1979).
[11] M.P. Brizhinev *et al.*, *Proceedings of 15th International Conference on Phenomenon in Ionized Gases*, Minsk, USSR, p. 971 (1981).
[12] E. Oks, *Sov. Phys. Doklady* **29** 224 (1984).
[13] A. Kamp and G. Himmel, *Appl. Phys. B* **47** 177 (1988).
[14] V.N. Kulikov and V.E. Mitsuk, *Sov. Phys. Tech. Phys. Lett.* **14** 104 (1988).
[15] I.N. Polushkin, M.Yu. Ryabikin, Yu.M. Shagiev and V.V. Yazenkov, *Sov. Phys. JETP* **62** 953 (1985).
[16] V.P. Gavrilenko and E. Oks, *Sov. Phys. Tech Phys. Lett.* **10** 609 (1984).
[17] R.A. Akhmedzhanov, I.N. Polushkin, Yu.V. Rostovtsev *et al.*, *Sov. Phys. JETP* **63** 30 (1986).
[18] D.I. Blochinzew, *Phys. Z. Sov. Union* **4** 501 (1933).
[19] R.C. Elton *et al.*, *J. Quant. Spectrosc. Rad. Transfer* **58** 559 (1997).
[20] P. Sauvan, E. Dalimier, E. Oks, O. Renner, S. Weber and C. Riconda, *J. Phys. B: Atom. Mol. Opt. Phys.* **42** 195001 (2009).
[21] I.M. Gaysinsky and E. Oks, *J. Quant. Spectrosc. Rad. Trasnfer* **41** 235 (1989).
[22] P. Sauvan, E. Dalimier, E. Oks, O. Renner, S. Weber and C. Riconda, *Intern. Review Atom. Mol. Phys.* **1** 123 (2010).
[23] E. Oks, E. Dalimier, A.Ya. Faenov *et al.*, *Optic. Express* **23** 31991 (2015).
[24] A.Ya. Faenov, E. Oks, E. Dalimier *et al.*, *Quant. Electron.* **46** 338 (2016).

Appendix A

Brief Overview of Stark Broadening Theories

We start this overview with *analytical theories* of the Stark Broadening (SB). They offer a much better physical insight than *simulation models*.

A *full-quantal* formulation of the SB was first discussed by Jablonski [1]. It was based on calculating how the field of the radiating atom/ion (hereafter, radiator) affects the motion of the "broadening" charged particle. In this theory, the key role is played by the "back reaction" of the radiator on the broadening particle: namely, by the difference between the energies of the broadening particle, interacting with the radiator in the upper and lower states. The fully-quantal theory, yields a complicated, "user-unfriendly" formalism that lacks the physical insight.

Most of the theoretical works in the SB was made in the "*no back reaction*" approximation resulting in the following two types of theories. One type is the limited-quantal theories where both the radiator and the perturbing charges are described quantally. The other type is the semiclassical theories where only the radiator is described quantally, while the perturbing charges are described classically. In both types of theories, the motion of the broadening particles is assumed given and one calculates the change in the wave function of the radiator.

The first semiclassical theory was a so-called *Conventional Theory* (CT) — also known as the "standard theory". In this theory, it was assumed that the ion microfield (i.e., the electric field due to ionic perturbers) is quasistatic from the point of view of the radiator, while the electron microfield was treated dynamically in a so-called impact approximation [2, 3]. The impact approximation considers a sequence of binary collisions of the perturbing electrons with the radiator and these collisions are considered to be completed. The CT engages the impact approximation for obtaining an approximate solution of the time-dependent Schrödinger equation (for the radiating bound electron) in the second order of the Dirac's time-dependent perturbation theory (the first order vanishes). More details are presented in Appendix B by the example of hydrogen lines.

A modified version of the collisional theory of the SB was called the *unified theory* [4, 5]. Its primary distinction from the CT — from the physical point of view — was the allowance for incomplete collisions.

An alternative to the collisional theories was called the *relaxation theory* [6, 7]. It combined techniques of Zwanzig's projection operator and of Green's function for deriving an integro-differential equation for the averaged time evolution operator of the radiating bound electron. However, for solving analytically the obtained master equation of the relaxation theory, one should make the same set of simplifying assumptions as in the CT (except for the assumption of completed collisions), thus reducing the relaxation theory essentially to the unified theory.

Speaking of simulation models, we begin with the *Model Microfield Method* (MMM) [8, 9]. In this method, there was chosen a simple approximate form of the field correlation function in such a way as to correctly fit the long and short time limits of the actual dynamics of the electric microfield. Some improvements to the MMM — both analytical and numerical — were made by various authors, especially, by Boercker *et al.* [10], Seidel [11], Stehle *et al.* [12, 13], as well as Talin *et al.* [14].

In some applications (such as, studies of opacities, detailed radiative transfer, etc.) it is necessary to be able to calculate hundreds of spectral line profiles at one time, which might consume too much time using rigorous theories. For this purpose, the *Frequency Fluctuation Model* (FFM) was developed by Calisti *et al.* [15]. In this approach, the fluctuation of the ion microfield during the radiative transition was modeled by the process where the emission switches from one Stark component to another, the result being a mixture of these components of the spectral line.

More advanced simulation models use the approaches called the *Molecular Dynamics* (MD) [16, 17] and *independent quasiparticle* technique [18–23]. These models were based on numerical simulations of the motion of charged particles surrounding the radiator: plasma electrons and, separately, of plasma ions. Then numerical solutions of the corresponding Schrödinger equations were used to yield the simulated electron and ion microfields, and finally to calculate numerically their effects on the radiator.

It should be emphasized that all of the above simulation methods have the following deficiencies. First, they do not allow for one of the major couplings between the electron and ion microfields. Second, they have problems in calculating Stark shifts, especially in dense plasmas of relatively low temperatures. Third, they lack the physical insight.

Further analytical advances were made for the SB of hydrogen lines (H-lines) and hydrogenlike lines (HL-lines). Some of them were achieved within the frames of the no-coupling approximation, where the electron microfield and the ion microfield are treated as being statistically independent of each other and so are their effects on the radiator.

One of the achievements within the no-coupling approximation was a simple binarization scheme. The binarization, i.e., the reduction of the corresponding dynamic multi-particle problem to a sequence of binary collisions, was required for both the CT and the unified theory. A rigorous proof of the binarization was provided by Derevianko and Oks [24].

Another substantial analytical achievement in the no-coupling approximation was due to Stehle *et al.* [12, 25, 26]. Previously, it was considered that within the CT or the unified theory, the Stark profile of the H-lines or HL-lines, caused by the electrons in the absence of the ion microfield, cannot be reduced to a superposition of Lorentzians, corresponding to Stark components of the line. However, *et al.* showed that the Stark profile in this case actually reduces to a single Lorentzian (which was counterintuitive) and derived a simple analytical result for the Stark width of arbitrary H-line or HL-lines in this case.

The allowance for the quasistatic ion microfield **F** makes the line profile broadened by electrons much more complicated. For the H-lines L_α and L_β, this was calculated analytically by Strekalov and Burshtein [27] assuming that the field **F** does not affect the electron broadening operator Φ (no-coupling approximation). They demonstrated that when the Stark splitting by the field **F** significantly exceeds the non-diagonal matrix elements of Φ, the Stark sublevels are broadened by electrons, but remain distinct. However, as the field F diminishes below some critical value F_{cr}, there occurs a "collapse" of the Stark sublevels: for $F < F_{\mathrm{cr}}$ the Stark sublevels become degenerate.

If all "perturbing" charges are of the same sort, further advances, exact (non-perturbative) analytical solutions of the problem are possible. The first analytical solution was found by Lisitsa and Sholin within a further approximation — the binary approximation [28]. Later Derevianko and Oks eliminated the binary assumption: they obtained the exact analytical solution for the most general, multi-particle description of the interaction of the electron or ion microfield with the radiator [24].

As for going beyond the no-coupling approximation, the first analytical attempt was made by Sholin *et al.* [29]. However, they found only a very weak, logarithmic coupling between the electron and ion microfields.

In reality, the coupling between the electron and ion microfields can be strong. This was shown with the development of the most advanced analytical theory of the SB called the *Generalized Theory* (GT) [30]. The coupling, facilitated by the radiator, becomes stronger and stronger with the increase of the electron density N_e and/or the principal quantum number n, as well as with the decrease of the temperature T [30,31]. The GT achieved this result by going beyond the fully-perturbative description of the electron microfield used in the CT.

The GT was developed analytically to the same level as the CT: despite more complicated starting formulas, it turned out to be possible expressing the so-called "broadening function" through elementary functions, like in the CT. For this purpose, those results of the GT were derived in the same impact approximation used by the CT. Later the GT was extended to allow for incomplete collisions, thus creating the "unified" version of the GT [32].

References

[1] A. Jablonski, *Phys. Rev.* **68** 78 (1945).
[2] M. Baranger, *Phys. Rev.* **111** 481; 494 (1958).
[3] A. Kolb and H.R. Griem, *Phys. Rev.* **111** 514 (1958).
[4] E. Smith, J. Cooper and C. Vidal, *Phys. Rev.* **185** 140 (1969).
[5] C. Vidal, J. Cooper and E. Smith, *J. Quant. Specrosc. Rad. Transfer* **10** 1011 (1970); **11** 263 (1971).
[6] E. Smith and C. Hooper, *Phys. Rev.* **157** 126 (1967).
[7] E. Smith, *Phys. Rev.* **166** 102 (1968).
[8] U. Frisch and A. Brissaud, *J. Quant. Specrosc. Rad. Transfer* **11** 1753 (1971); 1767.
[9] A. Brissaud, C. Goldbach, J. Leorat, A. Mazure and G. Nollez, *J. Phys. B* **17** 1477 (1984).
[10] D. Boercker, C. Iglesias and J. Dufty, *Phys. Rev. A* **36** 2254 (1987).
[11] J. Seidel, *Z. f. Naturforsch.* **32a** 1195 (1977).
[12] C. Stehle, A. Mazure, G. Nollez and N. Feautrier, *Astron. & Astrophys.* **127** 263 (1983).
[13] C. Stehle, *Astron. & Astrophys.* **292** 699 (1994); *Physica Scripta* **65** 183 (1996).
[14] B. Talin, R. Stamm, V.P. Kaftanjian and L. Klein, *Astrophys. J.* **322** 804 (1987).
[15] A. Calisti, F. Khelfaoui, R. Stamm, B. Talin and R.W. *Lee, Phys. Rev. A* **42** 5433 (1990).
[16] J.-P. Hansen, I.R. McDonald and E.L. Pollock, *Phys. Rev. A* **11** 1025 (1975).
[17] M.A. Berkovsky, J.W. Duffy, A. Calisti, R. Stamm and B. Talin, *Phys. Rev. E* **54** 4087 (1996).

[18] R. Stamm, Y. Botzanowski, V.P. Kaftandjian, B. Talin and E.M. Smith, *Phys. Rev. Lett.* **52** 2217 (1984); **54** 2170 (1985).
[19] R. Stamm and E.W. Smith, *Phys. Rev. A* **30** 450 (1984).
[20] R. Stamm, E.W. Smith and B. Talin, *Phys. Rev. A* **30** 2039 (1984).
[21] M.A. Gigosos, J. Fraile and F. Torres, *Phys. Rev. A* **31** 3509 (1985).
[22] M.A. Gigosos and V. Cardenoso, *Phys. Rev. B* **20** 6005 (1987).
[23] M.A. Gigosos and V. Cardenoso, *J. Phys. B* **22** 1743 (1989); **29** 4795 (1996).
[24] A. Derevianko and E. Oks, in *Physics of Strongly Coupled Plasmas*, Eds. W.D. Kraeft and M. Schlanges World Scientific, Singapore 199, p. 286; p. 291.
[25] C. Stehle and N. Feautrier, *J. Phys. B* **17** 1477 (1984).
[26] C. Stehle, *Astron. & Astrophys.* **305** 677 (1996).
[27] M.L. Strekalov and A.I. Burshtein, *Sov. Phys. JETP* **34** 53 (1972).
[28] V.S. Lisitsa and G.V. Sholin, *Sov. Phys. JETP* **34** 484 (1972).
[29] G.V. Sholin, A.V. Demura and V.S. Lisitsa, *Sov. Phys. JETP* **37** 1057 (1973).
[30] Ya. Ispolatov and E. Oks, *J. Quant. Spccrosc. Rad. Transfer* **51** 129 (1994).
[31] E. Oks, A. Derevianko and Ya. Ispolatov, *J. Quant. Specrosc. Rad. Transfer* **54** 307 (1995).
[32] E. Oks, *Proceedings of 18th International Conference "Spectral Line Shapes"*, Auburn, Alabama, USA, 2006, *AIP Conference Proceedings* 874, p. 19 (2006).

Appendix B

Versions of the Conventional Theory of the Stark Broadening of Hydrogen Lines in Non/Weakly-Magnetized Plasmas

B.1 Overview of the Conventional Theory (CT) where Perturbing Electrons Move as Free Particles

In the entire Appendix B, we use the term "CT" for any SB theory that: (A) employs the impact approximation and the perturbation theory for all components of the electron microfield; the impact approximation considers a sequence of binary collisions of the perturbing electrons with the radiator and these collisions are considered to be completed; (B) neglects the ion dynamics, i.e., treats the ion microfield in the quasistatic approximation; (C) neglects any coupling between the electron and ion microfields; (D) neglects the acceleration of the perturbing electron by the ion field in the vicinity of the radiating atom.

In the general case, including overlapping lines corresponding to radiative transitions between degenerate or quasidegenerate energy levels a and b, the electron impact broadening operator is defined as follows

$$\Phi_{ab} = -\int dv f(v) N_e v \sigma(v), \tag{B.1}$$

where the operator $\sigma(v)$ has the form:

$$\sigma(v) = \int d\rho 2\pi\rho[1 - S_a(\rho, v)S_b^*(\rho, v)]_{\text{ang.av}}. \tag{B.2}$$

Here, v is the velocity of the perturbing electron, $f(v)$ is the distribution of the velocities (usually assumed to be Maxwellian), ρ is the impact parameter of the perturbing electron, S_a and S_b are the corresponding scattering matrices, the symbols * and $[\cdots]_{\text{ang.av}}$ stand for the complex conjugation and the angular average, respectively. For the particular case where non-diagonal matrix elements of the Φ_{ab} are relatively small, the corresponding lineshape is a sum of Lorentzians, whose width $\gamma_{\alpha\beta}$ and shift $\Delta_{\alpha\beta}$ are equal to the negative of the real and imaginary parts of diagonal matrix elements $\langle\alpha|\langle\beta|\Phi_{ab}|\beta\rangle|\alpha\rangle$, respectively:

$$\gamma_{\alpha\beta} = -\mathrm{Re}[_{\alpha\beta}(\Phi_{ab})_{\beta\alpha}], \quad \Delta_{\alpha\beta} - -\mathrm{Im}[_{\alpha\beta}(\Phi_{ab})_{\beta\alpha}]. \tag{B.3}$$

Here, α and β correspond to upper and lower sublevels of the levels a and b involved in the radiative transition, respectively. Here and below, for any operator f, for brevity, we denote its matrix elements $\langle\alpha|\langle\beta|f|\beta\rangle|\alpha\rangle$ as $_{\alpha\beta}f_{\beta\alpha}$.

Griem's CT, as presented in Kepple–Griem paper [1] and in Griem's book [2], assumes rectilinear trajectories of the perturbing electron

$$\mathbf{r}(t) = \boldsymbol{\rho} + \mathbf{v}t, \tag{B.4}$$

where vector $\boldsymbol{\rho}$ is directed from the radiating atom to the point of the closest approach of the perturbing electron and is perpendicular to vector $\mathbf{v}$, so that

$$r(t) = (\rho^2 + v^2t^2)^{1/2}. \tag{B.5}$$

Under this assumption, after calculating the scattering matrices by the standard time-dependent perturbation theory, the operator σ takes the form:

$$\sigma = \int d\rho 2\pi\rho(W^2/\rho^2), \tag{B.6}$$

where,

$$W^2 = [2\hbar^2/(3m_e^2v^2)][(\mathbf{R}_a^2 - 2\mathbf{R}_a\mathbf{R}_b^* + \mathbf{R}_b^{*2})/a_B{}^2], \tag{B.7}$$

a_B being the Bohr radius.

The diagonal elements of the operator $\sigma(v)$ have the physical meaning of cross-sections of so-called optical collisions, i.e., the cross-sections of collisions leading to virtual transitions inside level a between its sublevels and to virtual transitions inside level b between its sublevels, resulting in the broadening of Stark components of the H-line. According to Eq. (B.6), these cross-sections $\langle\alpha|\langle\beta|\sigma(v)|\beta\rangle|\alpha\rangle$ are expressed through diagonal elements of the operator W^2, which in the parabolic coordinates have the form (see, e.g., [3, 4]):

$$\begin{aligned} {}_{\alpha\beta}(W^2)_{\beta\alpha} = [2\hbar^2/(3m_e^2v^2)](9/8)[n^2(n^2+q^2-m^2-1) \\ - 4nqn'q' + n'^2(n'^2+q'^2-m'^2-1)]. \end{aligned} \tag{B.8}$$

Here, n and n' are the principal quantum numbers of the upper and lower levels, respectively:

$q = n_1 - n_2$ and $q' = n_1' - n_2'$ are the electric quantum numbers of the Stark sublevels α and β, respectively, expressed through the corresponding parabolic quantum numbers; m and m' are the quantum numbers of the projection of the angular momentum in the states α and β, respectively. Thus, from Eq. (B.6) it follows:

$${}_{\alpha\beta}(\sigma)_{\beta\alpha} = \int d\rho 2\pi\rho_{\alpha\beta}(W^2)_{\beta\alpha}/\rho^2. \tag{B.9}$$

Obviously, the right side of Eq. (B.9) diverges at both small and large impact parameters. This divergence is one of the primary deficiencies of the CT.

The divergence at large ρ is an intrinsic feature of the long-range Coulomb potential involved. The plasma screens out the electric field of perturbing electrons at the distances larger than the Debye radius

$$\rho_D = [T_e/(4\pi e^2 N_e)]^{1/2}, \tag{B.10}$$

so that the upper cutoff at $\rho_{\max} = \rho_D$ is more or less natural.[1] In contrast, the divergence at small impact parameters is caused exclusively by the utilization of the perturbation expansion. Indeed, according to Eqs. (B.3) and (B.6), the quantity

$$[1 - S_a(\rho, v)S_b{}^*(\rho, v)]_{\text{ang.av}} = (W/\rho)^2 \tag{B.11}$$

goes to infinity when the impact parameter approaches zero. This fact contradicts the unitarity of the S-matrices:

$$|1 - S_a(\rho, v)S_b^*(\rho, v)| = C, \quad C \leq 2. \tag{B.12}$$

To remedy the problem, the CT uses the quantity $(W/\rho)^2$, which has a meaning of a "strength of a collision", to subdivide the collisions into "weak" and "strong" collisions as follows. Collisions having impact parameters $\rho > \rho_{\min}$, where

$$_{\alpha\beta}(W^2)_{\beta\alpha}/\rho_{\min}^2 = C, \tag{B.13}$$

are considered weak. Collisions having impact parameters $\rho < \rho_{\min}$, are considered strong, so that for them the quantity $\langle\alpha|\langle\beta|W^2|\beta\rangle|\alpha\rangle/\rho_{\min}^2$ under the integral in Eq. (B.9) is substituted by the constant C, which is called a "strong collision constant." Then the Eq. (B.9) takes the form:

$$_{\alpha\beta}(\sigma)_{\beta\alpha} = \int_{\rho_{\max}}^{\rho_{\min}} d\rho 2\pi\rho_{\alpha\beta}(W^2)_{\beta\alpha}/\rho^2 + \int_0^{\rho_{\min}} d\rho 2\pi\rho C. \tag{B.14}$$

Obviously, the second integral in Eq. (B.14) is equal to

$$\pi\rho_{\min}^2 C = \pi_{\alpha\beta}(W^2)_{\beta\alpha}, \tag{B.15}$$

where we substituted C by the left side of Eq. (B.13). Then Eq. (B.14) can be represented in the form:

$$_{\alpha\beta}(\sigma)_{\beta\alpha} = 2\pi_{\alpha\beta}(W^2)_{\beta\alpha}\{\ln[\rho_{\max}/\rho_{\min}(C)] + 1/2\}. \tag{B.16}$$

[1]More rigorously, $\rho_{\max} = \min[\rho_D, v/\Delta\omega, v/\delta\omega(F_i)]$, where $\Delta\omega$ is the detuning from the center of the line, $\delta\omega(\mathrm{F}_i)$ is the static Stark splitting in the ion field F_i. Physically, the requirements $\rho < v/\Delta\omega$ and $\rho < v/\delta\omega(F_i)$ are the allowances for incomplete collisions and for the removal of the degeneracy by the ion field, respectively. Typically, $\min[\rho_D, v/\Delta\omega, v/\delta\omega(F_i)] = \rho_D$.

The quantity $\rho_{\min}(C)$, defined by Eq. (B.13), obviously depends on the choice of the strong collision constant $C \leq 2$. Kepple and Griem [1] chose the left side of Eq. (B.15) to be equal to $\pi\rho_{\min}^2$, i.e., to choose $C = 1$. Later, in his book [2] on page 70, Griem changed the choice of the left side of Eq. (B.15) to $3\pi\rho_{\min}^2/2$, what is equivalent to choosing $C = 3/2$. However, under the logarithm in Eq. (B.16), he still kept $\rho_{\min}(1)$ — as in Kepple–Griem paper [3] — what is inconsistent. Indeed, it is clear that the choice of the right side of Eq. (B.15), which is the choice of the value of C, affects the choice of the quantity $\rho_{\min}$ defined by Eq. (B.13).

Further, Eq. (B.13) defines the quantity $\rho_{\min}$ individually for each Stark component of the H-line. The next step in the CT is to re-define this quantity to be the same, universal value for the entire H-line. For this purpose, in the CT one averages the matrix element ${}_{\alpha\beta}(W^2)_{\beta\alpha}$ in Eq. (B.13) over the Stark sublevels in the following way.

First, the quantity

$$W_a^2 = [2\hbar^2/(3m_e^2v^2)]\langle\alpha|\mathbf{R}_a^2|\alpha\rangle/a_B^2 \tag{B.17}$$

is averaged over the Stark sublevels of the upper level a, resulting in

$$(W_a^2)_{av} = 2\hbar^2(\mathbf{R}_a^2/a_B^2)_{av}/(3m_e^2v^2), \tag{B.18}$$

where we denoted for brevity

$$\langle\alpha|\mathbf{R}_a^2|\alpha\rangle_{av} = (\mathbf{R}_a^2)_{av}. \tag{B.19}$$

Then, the corresponding quantity $(W_b^2)_{av}$ for the lower level b is calculated. Next, the square root of the averaged matrix element $(\langle\alpha|\langle\beta|W^2|\beta\rangle|\alpha\rangle)$ is declared to be

$$[{}_{\alpha\beta}(W^2)_{\beta\alpha}]_{av}^{1/2} = [(W_a^2)_{av}^{1/2} - (W_b^2)_{av}^{1/2}] \tag{B.20}$$

with a justification that in this form, it allows for the partial cancellation of terms in ${}_{\alpha\beta}(W^2)_{\beta\alpha}$ when n' approaches n. Finally, the average of the square root of Eq. (B.13) is represented in the form

$$[(W_a^2)_{av}^{1/2} - (W_b^2)_{av}^{1/2}]/C^{1/2} = \rho_W, \tag{B.21}$$

where the average quantity $(\rho_{\min})_{av}$ is denoted ρ_W and called the Weisskopf radius. More explicitly, from Eqs. (B.18) and (B.21) it follows that

$$\rho_W(C) = [2/(3C)]^{1/2}[\hbar/(m_e v)][(\mathbf{R}_a^2)_{av}^{1/2} - (\mathbf{R}_b^2)_{av}^{1/2}]/a_B. \qquad \text{(B.22)}$$

Then Eq. (B.16) is rewritten in the form:

$$_{\alpha\beta}(\sigma)_{\beta\alpha} = 2\pi_{\alpha\beta}(W^2)_{\beta\alpha}\{\ln[\rho_{\max}/\rho_W(C)] + 1/2\}. \qquad \text{(B.23)}$$

Kepple and Griem [1], while choosing $C = 1$, used the following expression for the Weisskopf radius

$$\hbar(n^2 - n'^2)/(m_e v) = \rho_{WG}(1) \qquad \text{(B.24)}$$

(here and below the superscript "G" stands for "Griem"). The same expression for the Weisskopf radius was used also in Griem's book [2]. It corresponds to setting $(\mathbf{R}_a^2)_{av}/a_B^2 = 3n^4/2$ and $(\mathbf{R}_b^2)_{av}/a_B^2 = 3n'^4/2$.

However, a more accurate calculation of $(\mathbf{R}_a^2)_{av}$ and $(\mathbf{R}_b^2)_{av}$ yields a different result. Indeed,

$$\langle\alpha|\mathbf{R}_a^2|\alpha\rangle/a_B^2 = (9/8)n^2(n^2 + q^2 - m^2 - 1), \qquad \text{(B.25)}$$

(see, e.g., [3,4] or Eqs. (B.7) and (B.8)). Since, $(q^2)_{av} = (m^2)_{av}$, then

$$\langle\alpha|\mathbf{R}_a^2|\alpha\rangle_{av})/a_B{}^2 = (9/8)n^2(n^2 - 1). \qquad \text{(B.26)}$$

The same result can be obtained from the well-known expression for the matrix element of $\mathbf{R}_a^2$ in the spherical quantization $(n,\ l,\ m)$ — see, e.g., Landau–Lifshitz's textbook [5]:

$$\langle \mathrm{n}l\mathrm{m}|\mathbf{R}_a^2|\mathrm{n}l\mathrm{m}\rangle/a_B{}^2 = (9/4)\mathrm{n}^2(n^2 - l^2 - l - 1). \qquad \text{(B.27)}$$

Indeed,

$$\begin{aligned}(\langle \mathrm{n}l\mathrm{m}|\mathbf{R}_a^2|\mathrm{n}l\mathrm{m}\rangle)_{av}/a_B^2 &= (9/4)\mathrm{n}^2[(1/\mathrm{n}^2)\sum_{l=0}^{n-1}(n^2 - l^2 - l - 1)(2l+1)\\ &= (9/8)\mathrm{n}^2(\mathrm{n}^2 - 1),\end{aligned} \qquad \text{(B.28)}$$

i.e., the same result as obtained above in the parabolic quantization.

Therefore, the leading term in $(\langle \mathrm{n}l\mathrm{m}|\mathbf{R}_a^2|\mathrm{n}l\mathrm{m}\rangle)_{av}/a_B^2$ is $9n^4/8$ rather than the quantity $3n^4/2$ used by Griem. Substituting this more accurate result in Eq. (B.22) we obtain the following more accurate expression for the Weisskopf radius:

$$\rho_{WA}(C) = (3/C)^{1/2}\hbar(n^2 - n'^2)/(2\, m_e v), \tag{B.29}$$

(here and below the superscript "A" stands for "accurate").

Thus, in Griem's CT, Eq. (B.23) becomes

$$_{\alpha\beta}(\sigma)_{\beta\alpha}, G = 2\pi_{\alpha\beta}(\mathrm{W}^2)_{\beta\alpha}\{\ln[\rho_{\max}/\rho_{WG}(1)] + 1/2\} \tag{B.30}$$

with $\rho_{WG}(1)$ given by Eq. (B.24), while more accurately it should be

$$_{\alpha\beta}(\sigma)_{\beta\alpha}, A = 2\pi_{\alpha\beta}(\mathrm{W}^2)_{\beta\alpha}\{\ln[\rho_{\max}/\rho_{WA}(\mathrm{C})] + 1/2\} \tag{B.31}$$

with $\rho_{WA}(C)$ given by Eq. (B.29).

B.2 Conventional Theory (CT) Allowing for the Scattering of Perturbing Electrons on the Atomic Electric Dipole

In reality, the perturbing electrons do not pass the radiating atom as free particles. Rather, they move in the dipole potential

$$V = e^2\langle \mathbf{R}\rangle \cdot \mathbf{r}/r^3, \tag{B.32}$$

where $\mathbf{r}$ is the radius-vector of the perturbing electrons and $\langle \mathbf{R}\rangle$ is the mean value of the radius-vector of the atomic electron:

$$\langle \mathbf{R}\rangle = -3e^2\mathbf{A}/(4|E_{\mathrm{at}}|), \tag{B.33}$$

where E_{at} is the energy of the atomic electron. Physically, this is a consequence of the fact that hydrogen atoms possess permanent electric dipole moments, what is intimately related to the existence of an additional conserved vector quantity — the Runge–Lenz vector $\mathbf{A}$:

$$\mathbf{A} = -\mathbf{R}/R + (\mathbf{p}\times\mathbf{L} - \mathbf{L}\times\mathbf{p})/(2m_e e^2). \tag{B.34}$$

Here, $\mathbf{R}$, $\mathbf{p}$, $\mathbf{L}$, e and m_e are the radius-vector, momentum, angular momentum, charge, and mass of the atomic electron, respectively.

The consequences of the allowance for this dipole potential were presented in paper [6] and are described below.

In the spherical coordinates in the **r**-space with the polar axis along the mean value $\langle \mathbf{R} \rangle$ of the radius-vector of the atomic electron, the dipole potential takes the form

$$V(r,\theta) = (\mathrm{d}\cos\theta)/r^2, \quad d = e^2|\langle \mathbf{R} \rangle|, \tag{B.35}$$

where θ is the polar angle of the vector **r** and $\langle \mathbf{R} \rangle$ is the mean value of the radius-vector of the atomic electron. In the parabolic coordinates in the **R**-space, the latter quantity is (see, e.g., Landau–Lifshitz's textbook [5]) $\langle \mathbf{R} \rangle = 3nqa_B/2$, so that for the Ly-lines $d = 3n|q|e^2 a_B/2$. More generally, one should use the arithmetic average of the values of d for the upper and lower Stark sublevels — as suggested by Nienhuis [7] and used by Zsudy and Baylis [8]. Therefore, in the present paper, we use the following value of d:

$$d = 3(n|q| + n'|q'|)e^2 a_B/4. \tag{B.36}$$

The motion in a dipole potential, like the one from Eq. (B.35), has been studied in detail by a number of authors [9–11]. The primary feature of this physical system is that, in addition to the conservation of the projection M_z of the angular momentum **M** of the perturbing electron on the axis of symmetry (which is along $\langle \mathbf{R} \rangle$ in our case), there exists an additional conserved quantity B:

$$B = M^2 + 2m_e d\cos\theta. \tag{B.37}$$

(this quantity could be interpreted as the square of a *generalized* angular momentum of the perturbing electron). Since, this quantity is conserved at any instant of time, it can be expressed through its value at $t = -\infty$ as follows,

$$B = (m_e v_0 \rho_0)^2 + 2m_e d\cos\theta_0. \tag{B.38}$$

Here, ρ_0 is the asymptotic impact parameter at $t = -\infty$ and θ_0 is the angle between $\langle \mathbf{R} \rangle$ and $\mathbf{r}(-\infty)$, which is the same angle as between $\langle \mathbf{R} \rangle$ and $-\mathbf{v}_0$, where $\mathbf{v}_0$ is the velocity $\mathbf{v}(-\infty)$ of the perturbing electron at $t = -\infty$.

Another important feature of this physical system is that it allows the separation of the radial motion from the angular motion. The radial motion occurs in an effective potential $U_{\text{eff}} = B/(2m_e r^2)$, resulting in the following time dependence of the absolute value of the radial coordinate

$$r(t) = [B/(2m_e E) + (2E/m_e)t^2]^{1/2}, \tag{B.39}$$

where E is the energy of the perturbing electron. Substituting B from Eq. (B.38) and $E = m_e v_0^2/2$ in Eq. (B.39), we get

$$r(t) = (\rho_{\text{eff}}^2 + v_0^2 t^2)^{1/2}, \tag{B.40}$$

where

$$\rho_{\text{eff}} = [\rho_0^2 + 2d\cos\theta_0/(m_e v_0^2)]^{1/2}. \tag{B.41}$$

Below for brevity of formulas, we use v instead of v_0. After denoting

$$2d/(m_e v^2) = \rho_d^2, \tag{B.42}$$

Equation (B.41) takes the form:

$$\rho_{\text{eff}} = (\rho_0^2 + \rho_d^2 \cos\theta_0)^{1/2}. \tag{B.43}$$

The comparison of Eqs. (B.40) and (B.8) shows that the motion of the perturbing electron occurs with an effective impact parameter ρ_{eff} given by Eq. (B.41). This allows us to apply the standard effective trajectories method used in atomic physics for calculating cross-sections — see, e.g., Lebedev–Beigman's book [12], page 230. According to this method, the actual nonrectilinear trajectory is substituted by an effective rectilinear trajectory characterized by the velocity v_0 and by the effective impact parameter ρ_{eff} from Eq. (B.43). Then instead of Eq. (B.12) we have

$$\begin{aligned} {}_{\alpha\beta}[\sigma(\theta_0)]_{\beta\alpha} &= \int d\rho_0\ 2\pi\rho_0{}_{\alpha\beta}(W^2)_{\beta\alpha}/\rho_{\text{eff}}^2 \\ &= \int d\rho_0\ 2\pi\rho_0[{}_{\alpha\beta}(W^2)_{\beta\alpha}/(\rho_0^2 + \rho_d^2\cos\theta_0)]. \end{aligned} \tag{B.44}$$

The next step is the averaging of $1/(\rho_0^2+\rho_d^2\cos\theta_0)]$ over the angle θ_0:

$$(1/2)\int_{-1}^{1} d(\cos\theta_0)/(\rho_0^2+\rho_d^2\cos\theta_0)$$
$$= [1/(2\rho_d^2)]\ln[(\rho_0^2/\rho_d^2+1)/(\rho_0^2/\rho_d^2-1)]. \tag{B.45}$$

Now, instead of Eq. (B.17), we have

$$_{\alpha\beta}(\sigma)_{\beta\alpha,d}$$
$$= \int^{\rho_{\max}} d\rho_0 2\pi\rho_0[_{\alpha\beta}(W^2)_{\beta\alpha}/(2\rho_d^2)]\ln[(\rho_0^2/\rho_d^2+1)/(\rho_0^2/\rho_d^2-1)]$$
$$+\int^{\rho_{\min}} d\rho_0 2\pi\rho_0 C, \tag{B.46}$$

where $\rho_{\min}$ is defined by the condition:

$$[_{\alpha\beta}(W^2)_{\beta\alpha}/(2\rho_d^2)]\ln[(\rho_{\min}^2/\rho_d^2+1)/(\rho_{\min}^2/\rho_d^2-1)] = C, \tag{B.47}$$

(obviously, $\rho_{\min} > \rho_d$). Here and below, the superscript "*d*" in $_{\alpha\beta}(\sigma)_{\beta\alpha,d}$ signifies that this cross-section was obtained with the allowance for the scattering on the atomic electric dipole.

The integration over ρ_0 in Eq. (B.46) can be also performed analytically, yielding:

$$_{\alpha\beta}(\sigma)_{\beta\alpha,d} = (\pi/2)_{\alpha\beta}(W^2)_{\beta\alpha}\{[(\rho_{\max}^2/\rho_d^2)\ln[(\rho_{\max}^2+\rho_d^2)/(\rho_{\max}^2-\rho_d^2)]$$
$$-(\rho_{\min}^2/\rho_d^2)\ln[(\rho_{\min}^2+\rho_d^2)/(\rho_{\min}^2-\rho_d^2)]$$
$$+\ln[(\rho_{\max}^4-\rho_d^4)/(\rho_{\min}^4-\rho_d^4)]\} + \pi C\rho_{\min}^4. \tag{B.48}$$

After substituting the expression for the strong collision constant C from Eq. (B.47) in formula (B.48), the latter reduces to:

$$_{\alpha\beta}(\sigma)_{\beta\alpha,d} = 2\pi_{\alpha\beta}(W^2)_{\beta\alpha}\{\ln[(\rho_{\max}/\rho_{\min})(\rho_{\max}^4-\rho_d^4)^{1/4}/$$
$$(\rho_{\min}^4-\rho_d^4)^{1/4}] + [\rho_{\max}^2/(4\rho_d^2)]\ln[(\rho_{\max}^2+\rho_d^2)/$$
$$(\rho_{\max}^2-\rho_d^2)]\}. \tag{B.49}$$

In the limit, where $\rho_{\max}/\rho_{\min} \gg 1$, Eq. (B.49) simplifies to:

$$ {}_{\alpha\beta}(\sigma)_{\beta\alpha,d} = 2\pi_{\alpha\beta}(W^2)_{\beta\alpha}[\ln[(\rho_{\max}/\rho_{\min}) \\ +1/2 - (1/4)\ln(1 - \rho_d^4/\rho_{\min}^4)]. \quad \text{(B.50)} $$

The quantity $\rho_{\min}$ in Eqs. (B.49) and (B.50) is the solution of Eq. (B.47) with respect to $\rho_{\min}$:

$$ \rho_{\min} = \rho_d\{[\exp(2C\rho_d^2/{}_{\alpha\beta}(W^2)_{\beta\alpha}) + 1]/ \\ [\exp(2C\rho_d^2/{}_{\alpha\beta}(W^2)_{\beta\alpha}) - 1]\}^{1/2}. \quad \text{(B.51)} $$

In the limit, where $C\rho_d^2/{}_{\alpha\beta}(W^2)_{\beta\alpha} \gg 1$, Eq. (B.51) yields $\rho_{\min} = [{}_{\alpha\beta}(W^2)_{\beta\alpha}/C]^{1/2}$, while in the opposite limit Eq. (B.51) yields $\rho_{\min} = \rho_d$. However, the opposite limit does not correspond to the actual situation for any H-line, as will be shown below.

Equation (B.51) defines the quantity $\rho_{\min}$ individually for each Stark component of the H-line. The next step is the same as in Sec. B.1: to re-define $\rho_{\min}$ to be a universal value for the entire H-line. For the average quantity $[{}_{\alpha\beta}(W^2)_{\beta\alpha})/C]^{1/2}$ we use the same result as in Sec. B.1, i.e., $(3/C)^{1/2}\hbar(n^2 - n'^2)/(2m_e v)$ denoted as $\rho_{WA}(C)$ in Eq. (B.32). As for the quantity d that controls the value of ρ_d, by averaging the right side of Eq. (36) (where d was defined) over Stark sublevels we get $\langle d\rangle_{av} = (n^2 + n'^2)e^2a_B/4$. Substituting this into the definition of ρ_d in Eq. (B.42), we find:

$$ \langle\rho_d\rangle_{av} = [(n^2 + n'^2)/2]^{1/2}\hbar/(\mathrm{m_e v}). \quad \text{(B.52)} $$

Thus, from Eqs. (B.51) and (B.52) we obtain the universal value $\langle\rho_{\min}\rangle_{av}$ for the entire H-line:

$$ \langle\rho_{\min}\rangle_{av} = \langle\rho_d\rangle_{av}\{[\exp(2\langle\rho_d\rangle_{av}^2/\rho_{WA}(C)^2) + 1]/ \\ [\exp(2\langle\rho_d\rangle \mathrm{av}^2/\rho_{WA}(C)^2) - 1]\}^{1/2}. \quad \text{(B.53)} $$

Now, we come back to Eq. (B.49), substitute $\rho_{\min}$ by $\langle\rho_{\min}\rangle_{av}$ and ρ_d by $\langle\rho_d\rangle_{av}$, and by introducing dimensionless parameters

$$ x = \langle\rho_d\rangle_{av}/\rho_{\max}, \quad b = \langle\rho_d\rangle_{av}/\rho_{WA}(C) \\ = (2C/3)^{1/2}(n^2 + n'^2)^{1/2}/(n^2 - n'^2), \quad \text{(B.54)} $$

we finally obtain:

$$ {}_{\alpha\beta}(\sigma)_{\beta\alpha,A,d} = 2\pi {}_{\alpha\beta}(W^2)_{\beta\alpha}\{\ln[(\exp(2b^2) - 1)^{1/2}(1/x^4 - 1)^{1/4}/2^{1/2}] - b^2/2 + [1/(4x^2)]\ln[(1 + x^2)/(1 - x^2)]\}. \quad \text{(B.55)} $$

Since, the parameter b in Eq. (B.55) is defined (by Eq. (B.54)) through the more accurate expression for the Weisskopf radius $\rho_{WA}(C)$, the superscript "A" is added in ${}_{\alpha\beta}(\sigma)_{\beta\alpha,A,d}$.

More importantly, the velocity v of the perturbing electrons cancelled out from the definition of the parameter b, so that the ratio $b/C^{1/2}$ is just a combination of the principal quantum numbers n and n': it takes a fixed value specific for each H-line and it does not depend on the temperature T and the electron density N_e of the plasma. Taking into account that $C \leq 2$, from the definition of b in Eq. (B.54) it is seen that $b < 1$ always, reaching maximum values for $n' - n - 1$, i.e., for the most intense H-line of each spectral series. For example, for H_α line we get $b = 0.59\,C^{1/2}$.

The other dimensionless parameter $x = \langle\rho_d\rangle_{av}/\rho_{\max}$ in Eq. (B.55) significantly depends on plasma parameters. In the typical case, where $\rho_{\max}$ is equal to the Debye radius ρ_D (given in Eq. (B.13)), the parameter x can be re presented in the form:

$$ x = (2e\hbar/T)[(n^2 + n'^2)N_e/m_e]^{1/2} = 2.097 \times 10^{-11}[(n^2 + n'^2)N_e]^{1/2}/T, \quad \text{(B.56)} $$

where in the last, practical part of Eq. (B.56), the temperature T is in eV and the electron density N_e is in cm^{-3}. While deriving Eq. (B.56), the quantity $1/v$ in the expression for $\langle\rho_d\rangle_{av}$ (given by Eq. (52)) was substituted by its average over the Maxwell distribution $\langle 1/v\rangle = [2m_e/(\pi T)]^{1/2}$ — just as in Griem's CT. For warm dense plasmas emitting H-lines, the parameter x can reach values ~0.1 or even slightly greater.

From the ratio of the cross-sections ${}_{\alpha\beta}(\sigma)_{\beta\alpha,A,d}/{}_{\alpha\beta}(\sigma)_{\beta\alpha,A}$, where the denominator is given by Eq. (B.34) and the numerator — by Eq. (B.55), the matrix element ${}_{\alpha\beta}(W^2)_{\beta\alpha}$ cancels out, so that this ratio becomes a universal function of just two dimensionless parameters x and b applicable for any set of the five parameters N_e,

T, n, n' and C:

$$\begin{aligned}\text{ratio} &= {}_{\alpha\beta}(\sigma)_{\beta\alpha,A,d}/{}_{\alpha\beta}(\sigma)_{\beta\alpha,A}\\ &= \{\ln[(\exp(2b^2)-1)^{1/2}(1/x^4-1)^{1/4}/2^{1/2}] - b^2/2\\ &\quad + [1/(4x^2)]\ln[(1+x^2)/(1-x^2)]\}/\{\ln[b/x]+1/2\}. \qquad \text{(B.57)}\end{aligned}$$

Moreover, since the parameter x in Eq. (B.57) is given by Eq. (B.56) obtained by averaging over the Maxwell distribution of the velocities of the perturbing electrons, then the ratio in Eq. (B.57) is essentially the same as the ratio of widths:

$$\text{ratio} = \gamma_{\alpha\beta,A,d}/\gamma_{\alpha\beta,A}. \qquad \text{(B.58)}$$

Here, both the numerator and the denominator are defined by Eqs. (4) and (6), but with substituting ${}_{\alpha\beta}(\sigma)_{\beta\alpha,A,d}$ in Eq. (4) for calculating $\gamma_{\alpha\beta,A,d}$ and with substituting ${}_{\alpha\beta}(\sigma)_{\beta\alpha,A}$ in Eq. (4) for calculating $\gamma_{\alpha\beta,A}$. In other words, this is the ratio of widths calculated with and without the allowance for the scattering of the perturbing electrons on the atomic electric dipole, where both in the numerator and the denominator, we use the more accurate expression for the Weisskopf radius ρ_{WA} given by Eq. (B.32).

For comparison with Griem's CT, a modification has to be made to the corresponding ratio of widths. After taking into account that Griem's choice of the Weisskopf radius ρ_{WG} (given by Eq. (B.27)), the denominator in Eq. (B.57) changes to $\{\ln[b/(xC^{1/2})]+0.356\}$, so that the corresponding ratio becomes

$$\begin{aligned}\text{ratio A to G} &= \gamma_{\alpha\beta,A,d}/\gamma_{\alpha\beta,G}\\ &= \{\ln[(\exp(2b^2)-1)^{1/2}(1/x^4-1)^{1/4}/2^{1/2}]\\ &\quad - b^2/2 + [1/(4x^2)]\ln[(1+x^2)/(1-x^2)]\}/\\ &\quad \{\ln[b/(xC^{1/2})]+0.356\}. \qquad \text{(B.59)}\end{aligned}$$

Figure B.1 presents a three-dimensional plot of this ratio with the quantity $C = 3/2$ suggested on page 70 of Griem's book [2]. It is seen that the above refinements of the CT increase the electron contribution to the width.

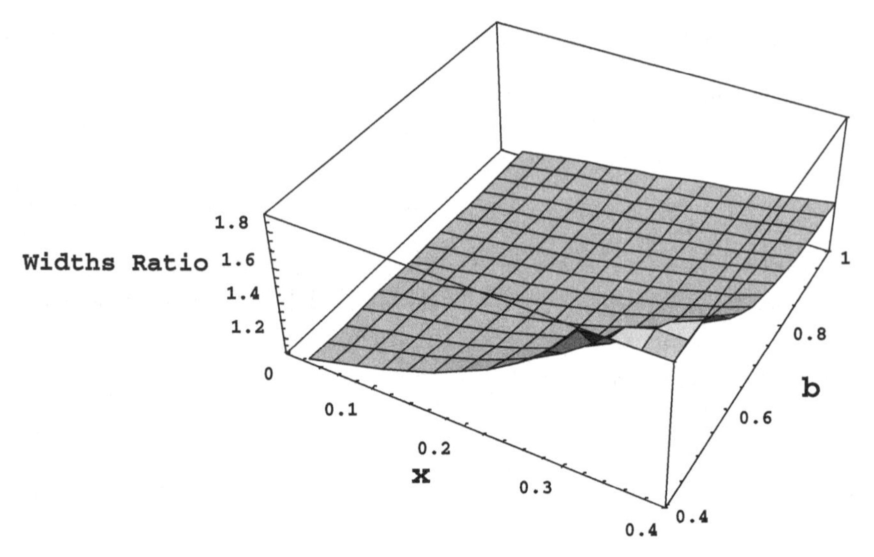

Fig. B.1. A three-dimensional plot of the ratio (given by Eq. (B.59)) of the electron impact width in the CT, refined in the present paper, with $C = 3/2$ to the electron impact width in Griem's CT versus dimensionless parameters x and b given by Eqs. (B.56) and (B.54), respectively.

Figure B.2 presents the same ratio from Eq. (B.59) for the H_α line versus the dimensionless parameter x (given by Eq. (B.56)) for two different choices of the strong collision constant: $C = 3/2$ (solid curve) and $C = 1$ (dashed curve). In the range of $0 < x < 0.56$, where the solid curve is above the dashed curve, the relative difference between the two curves is about 15% at $x = 0.4$ and it diminishes toward both ends of this range.

Figures B.3 and B.4 present the comparison of the experimental widths of the H_α line from two different benchmark experiments with several theories. Namely, in Fig. B.3, the experimental data was obtained by Kunze's group in a gas-liner pinch plasma [13], while in Fig. B.4, the experimental data was obtained by Vitel's group in a flash tube plasma [14]. In both figures, the experimental widths are presented by separated dots. As for the theories, in both figures the solid curve corresponds to the refined CT from the present paper, the dotted curve — to Griem's CT, and the dashed curve — to the Extended Generalized Theory (EGT). (A brief description

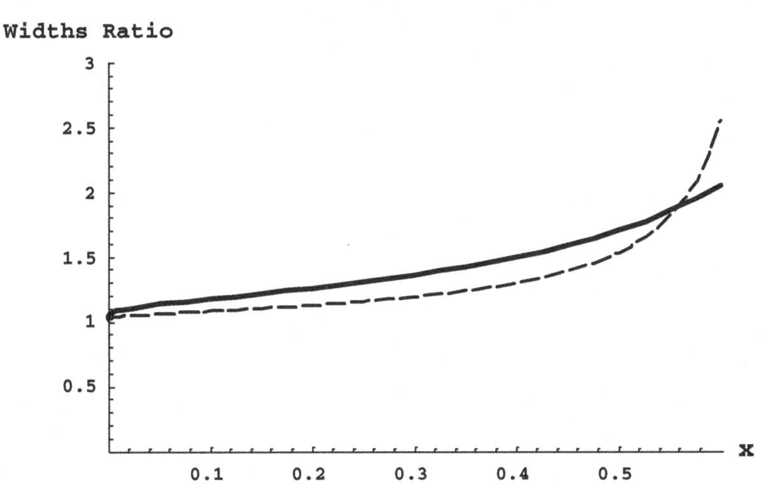

Fig. B.2. The ratio (given by Eq. (B.59)) of the electron impact width in the CT, refined in the present paper, to the electron impact width in Griem's CT, calculated for the H_α line versus the dimensionless parameter x (given by Eq. (B.56)) for two different choices of the strong collision constant: $C = 3/2$ (solid curve) and $C = 1$ (dashed curve).

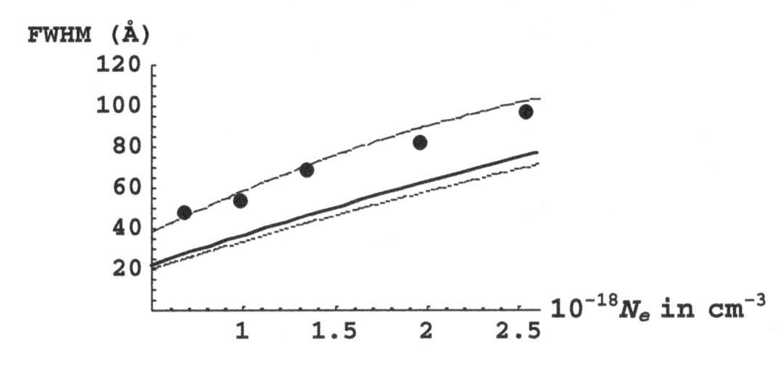

Fig. B.3. Comparison of the experimental widths of the H_α line (separated dots) obtained by Kunze's group in a gas-liner pinch plasma [13] at the temperatures (6–8) eV with the following theories: the refined CT from the present paper (solid line), Griem's CT (dotted line), the EGT (dashed line).

of the GT is presented in Appendix C.) It should be emphasized that the theoretical widths based on the refined CT, developed in the present paper, have been calculated taking into account both diagonal and non-diagonal matrix elements of the electron-impact

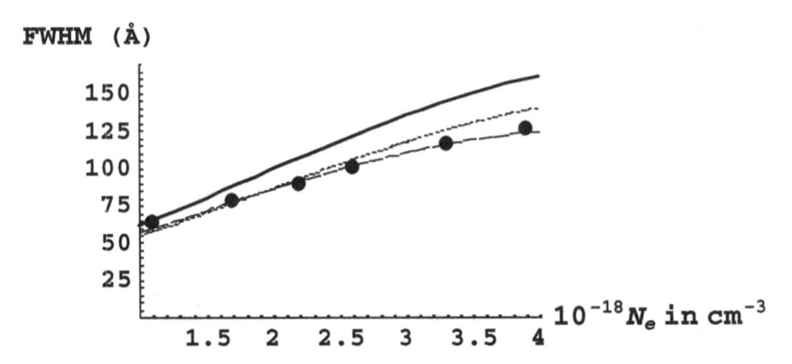

Fig. B.4. Comparison of the experimental widths of the H_α line (separated dots) obtained by Vitel's group in a flash tube plasma [14] at the temperatures (1–1.5) eV and the initial gas pressure 600 Torr with the following theories: the refined CT from the present paper (solid line), Griem's CT (dotted line), the EGT (dashed line).

broadening operator, as well as the quasistatic broadening by ions, i.e., in the same way as in Griem's CT.

The experimental data by Kunze's group [13] presented in Fig. B.3 were obtained at the high end of the T range ((6–8) eV) and the low end of the N_e range ($(0.5\text{–}2.5)\times10^{-18}$ cm^{-3}). It is seen that in this range, where Griem's CT dramatically underestimated the widths compared to the experiment, the employment of the refined CT from the present paper diminished the discrepancy, though only slightly. In this range of plasma parameters, the primary reason for the discrepancy between the experiment and either one of the CT theories is the neglect of the ion dynamics. Thus, since in this range of plasma parameters, the employment of the refined CT brought the theoretical widths closer to the experimental widths, then the role of the ion dynamics might be slightly smaller than previously thought. (For the latest results on the ion dynamical broadening we refer to paper [15] and references therein.) As for the EGT which incorporated the ion dynamics using analytical results obtained within the GT formalism, it shows the best agreement with the experiment.

The experimental data by Vitel's group [14] presented in Fig. B.4 were obtained at the low end of the T range ((1–1.5) eV) and the high end of the N_e range (up to 4×10^{-18} cm^{-3}). It is seen that in this range, where Griem's CT slightly overestimated the widths compared to the experiment, the employment of the refined CT from the present paper brought the theoretical widths further away from the experimental widths. In this range of plasma parameters, the ion dynamics is less important and the primary reason for the discrepancy between the experiment and either one of the CT theories is the neglect of the acceleration of perturbing electrons by the ion field in the vicinity of the radiating atom: this effect diminishes the widths. Thus, since in this range of plasma parameters, the employment of the refined CT brought the theoretical widths further away from the experimental widths, then the role of the acceleration of perturbing electrons by the ion field might be greater than previously thought. As for the EGT which incorporated this effect, it shows again the best agreement with the experiment.[2]

---

[2]For completeness, we note here, other comparisons between a version of the CT and the core GT applied to the dynamical SB by ions in magnetic fusion plasmas. Paper [16] of 1994 presented a study of this situation based on the core GT leading to the following main result. At values of the magnetic field B typical for magnetic fusion plasmas, practically the entire dynamical Stark width due to ions is due only to the so-called *adiabatic* contribution. This is because, as the magnetic field B increases, causing the increase of the separation ω_B between the Zeeman sublevels of hydrogenic atoms, the *non-adiabatic* contribution to the dynamical SB by ions decreases — specifically, it decreases dramatically at magnetic fields typical for magnetic fusion plasmas. The adiabatic contribution was calculated in [16] exactly (no perturbation theory), which is the primary feature of the GT.

In 2009, i.e., 15 years later, a group of authors published a paper [17], where they revisited the subject studied in paper [16]. The authors of paper [17] based their analytical results on the CT.

The comparison of the results of the GT [16] and the CT [17] was made in paper [18], in the review [19], and in paper [20]. There it was shown in detail that *the CT* [17] *overestimated the primary, adiabatic contribution to the broadening by up to an order of magnitude*, and also overestimated the secondary, nonadiabatic contribution.

We also note in passing that works [18–20] clearly demonstrated that for the overwhelming majority of hydrogenic spectral lines, the secondary, non-adiabatic contribution calculated by the GT does not violate the unitarity of the S-matrix and therefore does not require an artificial cutoff at small impact parameters — in distinction to the CT in general and the CT version from [17] in particular.

References

[1] 1.P. Kepple and H.R. *Griem, Phys. Rev.* **173** 317 (1968).
[2] H.R. Griem, *Spectral Line Broadening by Plasmas*, Academic, New York 1974.
[3] V.S. Lisitsa, Sov. Phys. *Uspekhi* **122** 603 (1977).
[4] G.V. Sholin, A.V. Demura and V.S. Lisitsa, *Sov. Phys. JETP* **37** 1057 (1973).
[5] L.D. Landau and E.M. Lifshitz, *Quantum Mechanics*, Pergamon, Oxford (1965).
[6] E. Oks, J.Q.S.R.T. **152** 74 (2015).
[7] G. Nienhuis, *Physica* **66** 245 (1973).
[8] J. Szudy and W.E. Baylis, *Canad. J. Phys.* **54** 2287 (1976).
[9] R.J. Cross Jr., *J. of Chem. Phys.* **46** 609 (1967).
[10] K. Fox, *J. Phys. A (Proc. Phys. Soc.)* **1** 124 (1968).
[11] T. Chandrasekaran and T.W. Wilkerson, *Phys. Rev.* **181** 329 (1969).
[12] V.S. Lebedev and I.L. Beigman. *Physics of Highly Excited Atoms and Ions*, Springer, Berlin (1998).
[13] S. Büscher, T. Wrubel, S. Ferri and H.-J. Kunze, *J. Phys. B: Atom. Mol. Opt. Phys.* **35** 2889 (2002).
[14] S.A. Flih, E. Oks and Y. Vitel, *J. Phys. B: Atom. Mol. Opt. Phys.* **36** 283 (2003).
[15] A.V. Demura and E. Stambulchik, *Atoms* **2** 334 (2014).
[16] A. Derevianko and E. Oks, *Phys. Rev. Lett.* **73** 2059 (1994).
[17] J. Rosato, Y. Marandet, H. Capes, S. Ferri, C. Mosse, L. Godbert-Mouret *et al.*, *Phys. Rev. E* **79** 046408 (2009).
[18] E. Oks, Intern. *Review of Atom. and Mol. Phys.* **1** 169 (2010).
[19] E.Oks, in: V. Shevelko and H. Tawara, Atomic Processes in Basic and Applied Physics, (Eds.) (Springer, Heidelberg) Chap. 15, p. 393 (2012).
[20] E. Oks, *J. Quant. Spectrosc. Rad. Transfer* **171** 15 (2016).

Appendix C

The Generalized Theory of the Stark Broadening of Hydrogen Lines in Non/Weakly-Magnetized Plasmas

C.1 Introduction

Plasma spectroscopy had served as a fertile field for applications of the formalism of Dressed Atomic States (DAS) in plasmas [1]. The theory named in the title of this paper is primarily based on a *generalization* of the formalism of DAS in plasmas.

DAS is the formalism initially designed to describe the interaction of a *monochromatic* (or quasi-monochromatic) field — e.g., laser or maser radiation — with gases. Later, it was applied for the interaction of a laser or maser radiation with plasmas [1]. The employment of DAS led to the enhancement of the accuracy of analytical calculations and to more robust codes.

Our *generalization* of DAS was based on using atomic states dressed by a *broad-band* field of plasma electrons and ions [2]. Therefore, generalized DAS is a more complicated concept than the usual DAS, where the dressing was due to a monochromatic field.

The employment of the generalized DAS allowed to describe analytically a *coupling* of the electron and ion microfields facilitated by the radiating atom — *indirect coupling* [2]. The indirect coupling *increases with the principal quantum number* n and *with the electron density* N_e.Besides, *it increases as the temperature* T *decreases.*

The above employment of the generalized DAS resulted in the Generalized Theory (GT) of the SB of hydrogen (and hydrogenlike) spectral lines in plasmas, also known as the *core GT* [3, 4].

Later extensions of the GT, mentioned in this paragraph and in the next one, together with the core GT, represented the Extended Generalized Theory (EGT). Namely, the employment of the generalized DAS also allowed to obtain *accurate analytical results for the ion-dynamical broadening.* These results were produced in a *non-binary* description and *without involving the impact approximation* [2].

Our theory took into account also the Acceleration of (perturbing) Electrons by the Ion Field (AEIF). This phenomenon, representing a *direct coupling* of the electron and ion microfields, was described in a *non-binary* approach [2].

C.2 Eight Stages in the Design of the Core Generalized Theory (GT)

1. *Separation of Static and Dynamic Ions.* For a given value τ of the argument of the correlation function $C(\tau)$, the ion-dynamical part of $C(\tau)$ originates from the collisions, for which the instants of the closest approach t_0 fall within the interval $(-\tau/2,\tau/2)$. The rest of the perturbing ions are considered static. This is consistent with the fact that at $\tau \to \infty$ all ions are dynamic, while at $\tau \to 0$ all ions are static. Thus, to different values of τ correspond different proportions of dynamic and static ions.

2. *Partition of the Hamiltonian.* The Hamiltonian $H(t)$ is broken down in the following four terms:

$$\begin{aligned} H(t) &= H_0 + V^{\mathrm{IS}} + V_{\mathrm{par}}(t) + V_{\mathrm{perp}}(t), V_{\mathrm{par}}(t) \\ &= V_{||}^{\mathrm{ID}}(t) + V_{||}^{E}(t), V_{\mathrm{perp}}(\mathrm{t}) = V \perp^{\mathrm{ID}} (t) + V \perp^{\mathrm{E}} (t). \end{aligned} \quad \text{(C.1)}$$

Here, H_0 is the Hamiltonian of the unperturbed radiating atom/ion (radiator); superscripts IS, ID and E refer to interactions of the radiator with static ions, dynamic ions, and electrons, respectively; subscripts $||$ and $\perp$ refer, respectively, to parallel and perpendicular components of the electron and dynamic ion microfields with respect to the direction of the quasistatic ion field at the location of the radiator.

3. *Generalized DAS.* The atomic states of the radiator are dressed by the following time-dependent factor:

$$Q(t) = \exp\{(i/\hbar)[(H_0 + V^{\mathrm{IS}})t + \int_{\infty}^{t} dt' V_{\mathrm{par}}(t')]\}. \tag{C.2}$$

In other words, the interactions V^{IS} and $V_{\mathrm{par}}(t')$ are taken into account *exactly*, moreover *analytically*. Then the interaction $V_{\mathrm{perp}}(t)$ is allowed for by Dirac's perturbation theory. The dressing factor Q(t) is controlled by the interaction $V_{\mathrm{par}}(t)$ with a *broadband* field in distinction to the usual DAS dressed by a monochromatic field (examples of the latter can be found in [1]).

4. *Evolution Operator in the MODIFIED Interaction Representation.* It employs the dressing factor $Q(t)$ from Eq. (C.2)

$$U(t_1, t_2) = \mathrm{Texp}[-(i/\hbar) \int_{t1}^{t2} dt Q^*(t) V_{\mathrm{perp}}(t) Q(t)], \tag{C.3}$$

where $T\exp[\cdots]$ is the time-ordered exponential.

5. *Calculation of the M-matrix.* The M-matrix that controls the calculation of the correlation function (defined in stage 6) is calculated as follows:

$$M(t_1, t_2) = \exp[(i/\hbar) \int_{t1}^{t2} dt' V_{\mathrm{par}}(t')] U(t_1, t_2), \tag{C.4}$$

where $\exp[\cdots]$ is calculated *exactly*, analytically, while $U(t_1, t_2)$ — via the perturbation theory. The impact approximation is *not* used.

6. *Correlation Function.* It is defined by the formula

$$C(\tau) = \mathrm{Tr}\{\mathbf{d}\bullet\mathbf{d} \exp[-(i/\hbar)(H_0 + V^{\mathrm{IS}})\tau]\rho \langle M(\tau/2, -\tau/2)\rangle_{\mathrm{av}}\}, \tag{C.5}$$

where $\mathrm{Tr}\{\cdots\}$ stands for the trace, $\langle\cdots\rangle_{\mathrm{av}}$ denotes the ensemble average.

7. *Lineshape at a fixed quasistatic ion field F.* It is calculated as the Fourier-transform of the correlation function $C(\tau)$.

8. *Averaging over the distribution* $W(F)$. The distribution $W(F)$ of the quasistatic ion field can be produced using, e.g., APEX [5].

The central point of the core GT is taking into account *the interaction* $V_{||}(t)$ *on the equal footing with the unperturbed Hamiltonian* H_0 — *without any perturbation expansion.*

Now let us make a general comparison of the GT and the CT. In the most general terms, the GT can be characterized by the following controlling parameter

$$Y \equiv \frac{\langle V^{\mathrm{IS}} \rangle}{\Omega_w(V_{\mathrm{par}})}. \tag{C.6}$$

$\Omega_W(V_{\mathrm{par}})$ is the Weisskopf frequency characterizing the dynamic interaction $V_{\mathrm{par}}(\mathrm{t})$. (The Weisskopf frequency is the ratio of the mean thermal velocity of perturbers to the Weisskopf radius defined in Eqs. (B.21), (B.22).) The important point is that in the limit $Y \to 0$, we recover the CT.

Second, the GT is *convergent* at small impact parameters while the corresponding CT for neutral radiators is divergent. Physically, the difference from the CT can be explained as follows. The GT deals with virtual transitions, caused by the interaction $V_{\mathrm{perp}}(\mathrm{t})$, between atomic sublevels *"dressed"* by the interaction $V_{\mathrm{par}}(\mathrm{t})$. It is the allowance for this "dressing" that eliminates the divergence and enhances the accuracy of the results.

Third, the matrix $M(t_2, t_1)$ in Eq. (B.4), as well as its simplified version — the scattering matrix $S = M(-\infty, \infty)$ — consist of two physically different terms:

$$M = M_a + M_{na}, \tag{C.7}$$

$$S = S_a + S_{na}. \tag{C.8}$$

The first term in Eqs. (C.7, C.8) represents a purely adiabatic contribution, similar to what is usually referred as the "Old Adiabatic Theory" of broadening (see e.g., [6]). In particular, in the scattering matrix, the adiabatic term has the form:

$$S_a \equiv \exp\left[\frac{i}{\hbar}\int_{\infty}^{\infty} dt V_1(t)\right] \quad \text{(C.9)}$$

Here and below, we use shorter notations: $V_1 = V_{\text{par}}$, $V_2 = V_{\text{perp}}$. The second term in Eqs. (C.7, C.8) is non-adiabatic: it would vanish without non-adiabatic virtual transitions caused by the interaction $V_2(t)$ between sublevels dressed by the interaction $V_1(t)$. In particular, in the scattering matrix the adiabatic term has the form:

$$S_{na} \equiv \exp\left[\frac{i}{\hbar}\int_{\infty}^{\infty} dt V_1(t)\right]\left\{T\exp\left[\frac{i}{\hbar}\int_{\infty}^{\infty} dt Q^* V_2(t) Q\right]\right\}, \quad \text{(C.10)}$$

where after the expansion of T exp$[\cdots]$ and the angular averaging, the first non-vanishing contribution is usually of the second order with respect to V_2.

Fourth, it turns out that the GT can be developed *analytically* to the same level as the CT. This is counterintuitive because the starting formulas for the GT are more complicated than for the CT.

C.3 Details of the Core Generalized Theory (GT)

We present the Hamiltonian of a hydrogen or deuterium atom under the action of the quasistatic part $\mathbf{F}$ of the ion microfield and an electron-produced dynamic field $\mathbf{E}$(t) in the form,

$$H = H_0 - \mathbf{dF} - \mathbf{dE}(t), \quad \text{(C.11)}$$

where H_0 is the unperturbed Hamiltonian, $\mathbf{d}$ is the dipole moment operator. We choose the axis Oz of the parabolic quantization along the field $\mathbf{F}$. Then the operator — $\mathbf{dF}$, representing the interaction with this field, is diagonal in any subspace of a fixed principal quantum number n (for brevity, in any n-subspace). Therefore, in the CT this interaction was allowed for "exactly" (neglecting corrections originating from matrix elements of d_z corresponding to $\Delta n \neq 0$); the interaction with the field $\mathbf{E}(t)$ was then treated in the second order of the time-dependent perturbation theory.

The most important point of the GT, in this case is taking into account the entire z-component of the total field $\mathbf{F} + \mathbf{E}$ (t) in the same (or analogous) manner as the field $\mathbf{F}$ was treated in the CT. The idea behind this approach is that the interaction $-d_z[F+E_z(\mathrm{t})]$ (and not only its part $-d_z\mathrm{F}$) is diagonal in any n-subspace. Therefore, the z-component of the electron microfield can be allowed for more accurately than in the CT.

For radiative transitions between upper $(\alpha, \alpha', \alpha'', \ldots)$ and lower $(\beta,\ \beta', \beta'', \ldots)$ Stark sublevels, we write perturbed time-dependent wave functions in the form

$$\Psi_j(t, \mathbf{r}) = \exp\left[-i\omega_j t + i\int_{-\infty}^{t} dt'(d_z)_{jj}E_z(t')\right] \times U(t, -\infty)\varphi_j(\mathbf{r}), \quad \omega_j = \omega_{0j} - (d_z)_{jj}F, \quad \text{(C.12)}$$

where ω_{0j} is the unperturbed energy of the Stark sublevel $j(j = \alpha, \alpha', \ldots$ for upper sublevels and $j = \beta, \beta', \ldots$ for lower sublevels). Here and below, unless specified to the contrary, the units are such chosen that $\hbar = 1$.

The evolution operator $U(t_2, t_1)$ in our modified interaction representation can be written as

$$U(t_2, t_1) = T\exp[-i\int_{t1}^{t2} dtV(t)], \quad \text{(C.13)}$$

$$V(t) = \exp\left[i\int_{-\infty}^{t} dt' H_1(t')\right][-d_\perp E_\perp(t)]\exp\left[-i\int_{-\infty}^{t} dt' H_1(t')\right], \quad \text{(C.14)}$$

where $d_\perp E_\perp = d_x E_x + d_y E_y$ and the operator $H_1(t)$ looks as follows:

$$H_1(t) = H_0 - d_z F - d_z E_z(t). \quad \text{(C.15)}$$

In the lineshape formula:

$$J(\Delta\omega) = (2\pi)^{-1}\int_{-\infty}^{\infty} d\tau C(\tau)\exp(i\Delta\omega\tau), \quad \text{(C.16)}$$

where $\Delta\omega = \omega - \omega_0$, the correlation function can be represented in the form:

$$C(\tau) = \sum_{\alpha,\alpha',\beta,\beta'} \langle\alpha|\mathbf{d}|\beta'\rangle\langle\beta|\mathbf{d}|\alpha'\rangle \exp[-i(\omega_\alpha - \omega_\beta)\tau]$$
$$\langle\alpha'|\langle\beta'|[M(\tau/2, -\tau/2)]_{\rm av}|\beta\rangle|\alpha\rangle. \tag{C.17}$$

Here, the operator M is related to the evolution operator (C.13) as follows:

$$M(\tau/2, -\tau/2) = \exp\left[id_z \int_{-\tau/2}^{\tau/2} dt E_z(t)\right] U(\tau/2, -\tau/2); \tag{C.18}$$

$[\cdots]_{av}$ in Eq. (C.17) stands for the ensemble average.

There are adiabatic and non-adiabatic terms in the operator M and therefore in the correlation function $C(\tau)$ (see Eq. (C.7)). The adiabatic term in the correlation function is proportional to the part $d_z d_z$ of the operator $\mathbf{dd}$ in Eq. (C.17), the operator $d_z d_z$ being diagonal in the line space. The rest of the correlation function corresponds to the non-adiabatic contribution: it is proportional to the operator $d_x d_x + d_y d_y$ that has both diagonal and non-diagonal matrix elements in the line space.

In particular, the "diagonal" part of the correlation function, i.e., the part proportional to the diagonal elements of the operator $\mathbf{dd}$, can be represented as

$$C_{\rm diag}(\tau) = |\mathbf{d}_{\alpha\beta}|^2 \exp[-i(\omega_\alpha - \omega_\beta)\tau] \left\{ \exp\left[i\Delta d_z \int_{\tau/2}^{\tau/2} dt E_z(t)\right] \right.$$
$$\left. \times\, U_{\alpha\alpha}(\tau/2, -\tau/2) U^*_{\beta\beta}(\tau/2, -\tau/2) \right\}_{\rm av}, \tag{C.19}$$

where,

$$\Delta d_z \equiv (d_z)_{\alpha\alpha} - (d_z)_{\beta\beta}. \tag{C.20}$$

After expanding the evolution operator in Eq. (C.19) up to the second order, we obtain the following expression:

$$C_{\text{diag}}(\tau) = |\mathbf{d}_{\alpha\beta}|^2 \exp[-i(\omega_\alpha - \omega_\beta)\tau][\langle f_a(\tau)\rangle_{\text{av}} + \langle f_{na}(\tau)\rangle_{av}], \tag{C.21}$$

where,

$$f_a(\tau) \equiv \exp[i\Delta \mathrm{d_z} \int_{-\tau/2}^{\tau/2} dt E_z(t)], \tag{C.22}$$

$$f_{\text{na}}(\tau) \equiv -f_a(\tau) \left\{ \sum_{\alpha'} \int_{-\tau/2}^{\tau/2} \mathrm{dt}_1 \int_{-\tau/2}^{\tau/2} \mathrm{dt}_2 \exp\left[i(\omega_\alpha - \omega_{\alpha\prime})(t_1 - t_2) \right.\right.$$
$$\left. - i(\delta d_z)_\alpha \int_{t2}^{t1} dt E_z(t) \right] [(d_\perp)_{\alpha\alpha'} E_\perp(t_1)][(d_\perp)_{\alpha'\alpha} E_\perp(t_2)]$$
$$\left. + \sum_{\beta'} [\alpha \to \beta, \alpha\prime \to \beta']^* \right\}, \tag{C.23}$$

$$(\delta d_z)_\alpha \equiv (d_z)_{\alpha\alpha} - (d_z)_{\alpha'\alpha'}. \tag{C.24}$$

Here, the notation $\Sigma_{\beta'}[\alpha \to \beta, \alpha' \to \beta']$ means a term obtained from the $\Sigma_{\alpha\prime}[\cdots]$ in Eq. (C.23) by the substitution $\alpha \to \beta$, $\alpha' \to \beta'$.

In the diagonal part of the correlation function in Eq. (C.21) — just like in the entire correlation function given by Eq. (C.17) — there are two physically different terms. The term proportional to $f_a(\tau)$ represents the adiabatic contribution, similar to what is usually referred to as "The Old Adiabatic Theory of Broadening" (see, e.g., [6]). The term proportional to $f_{na}(\tau)$ is the non-adiabatic one: it is due to non-adiabatic virtual transitions between *dressed* Stark sublevels, the transitions being induced by x- and y-components of the electron microfield $\mathbf{E}(t)$.

It should be emphasized that the Stark sublevels are *dynamically dressed* by the entire z-component of the total microfield $\mathbf{F} + \mathbf{E}(t)$. This is the primary feature of the GT that leads to its numerous advantages over the CT.

Equations (C.13)–(C.24) represent the GT in its most general form. They allow calculating the correlation function and then the lineshape (see Eq. (C.16)) without further approximations. It is this general form of the GT that was used for most numerical calculations presented in book [2], including the Tables of Stark widths of hydrogen and deuterium lines in plasmas.

In order to have a better physical insight into the GT and for tracing more easily the relation between the GT and the CT, we present below a simplified version of the GT as well. The simplified version of the GT is obtained by introducing the impact approximation.

Under the impact approximation, the operator $[\mathrm{M}(\tau/2, -\tau/2)]_{\mathrm{av}}$ in Eq. (C.17) is approximated as

$$[M(\infty, -\infty)]_{\mathrm{av}} \equiv \Phi. \tag{C.25}$$

In distinction to the operator $[M(\tau/2, -\tau/2)]_{\mathrm{av}}$, the operator Φ does not depend on the argument τ of the correlation function, but it still depends on the quasistatic part $\mathbf{F}$ of the ion microfield. The operator $\Phi(F)$ is called the Electron Broadening Operator (EBO) — a.k.a. the electron impact operator.

Under the impact approximation, the correlation function $C(\tau)$ and the line shape $J(\Delta\omega)$ simplify to the form:

$$C(\tau) = \sum_{\alpha,\alpha',\beta,\beta'} \langle\alpha|\mathbf{d}|\beta'\rangle\langle\beta|\mathbf{d}|\alpha'\rangle \exp[-i(\omega_\alpha - \omega_\beta)\tau]\langle\alpha'|\langle\beta'|\Phi|\beta\rangle|\alpha\rangle, \tag{C.26}$$

$$J(\Delta\omega) = -\pi^{-1}\mathrm{Re}\left\{ \sum_{\alpha,\alpha'\beta,\beta'} \langle\alpha|\mathbf{d}|\beta'\rangle\langle\beta|\mathbf{d}|\alpha'\rangle\langle\alpha'|\langle\beta'| \right.$$
$$\left. \times\ [i(\Delta\omega + \omega_\beta - \omega_\alpha) + \Phi)]^{-1}|\beta\rangle|\alpha\rangle \right\}. \tag{C.27}$$

If the correlation function would not have non-diagonal elements of the operator $\mathbf{dd}$, then the line shape given by Eq. (C.27) would consist of the sum of Lorentzians characterized by the electron impact Halfwidth at the Half Maximum (HWHM) $\Gamma_{\alpha\beta}$ and by the electron impact shift $D_{\alpha\beta}$ as follows:

$$\Gamma_{\alpha\beta} = -\mathrm{Re}\langle\alpha|\langle\beta|\Phi|\beta\rangle|\alpha\rangle, \quad D_{\alpha\beta} = -\mathrm{Im}\langle\alpha|\langle\beta|\Phi|\beta\rangle|\alpha\rangle. \tag{C.28}$$

In reality, the would-be-Lorentzians are coupled by non-diagonal elements of the operator **dd**. Therefore, the actual calculation of the line profile based on (C.27) is much more complicated: it involves the inversion of the matrix of the rank $(n_\alpha n_\beta)^2$, e.g., of the rank 36 for the H_α line and of the rank 64 for the H_β line. This can be done only numerically (except for the L_α and L_β lines).

The *adiabatic* part of the EBO in the GT has only diagonal elements as follows:

$$\langle\langle\alpha\beta|\Phi_{ad}|\alpha\beta\rangle\rangle = \frac{4\pi\hbar^2 N_e}{3\,m_e^2\,a_0^2}\int_0^\infty dv W_M(v)\frac{1}{v} \times [z_{\alpha\alpha}z_{\alpha\alpha} - 2_{z_{\alpha\alpha}}z_{\beta\beta} + z_{\beta\beta}z_{\beta\beta}]\,I(u_{\alpha\beta}). \quad \text{(C.29)}$$

Here,

$$I(u) = u^2/2 - [u^2\cos(1/u) + (2u^3 - u)\sin(1/u) + Ci(1/u)]/6; \quad \text{(C.30)}$$
$$u_{\alpha\beta} = m_e v \rho_D/[3\hbar(n_\alpha q_\alpha - n_\beta q_\beta)],$$

where $Ci(x)$ is the integral cosine function, ρ_D is the Debye radius, $q = n_1 - n_2$ is the electric quantum number (expressed via the parabolic quantum numbers).

We emphasize that the adiabatic part of the EBO given by Eqs. (C.29)–(C.30) corresponds to the summation of all orders of the Dyson expansion of the time-ordered exponential in Eq. (C.13). This is one of the most important distinctions from the CT, where all terms in the EBO are calculated only in the first non-vanishing (namely, the second) order of the Dyson expansion.

The *non-adiabatic* part of the EBO in the GT has the form:

$$\langle\langle\alpha\beta|\Phi_{na}|\alpha'\beta'\rangle\rangle = \frac{4\pi\hbar^2 N_e}{3m_e^2 a_0^2}\int_0^\infty dv W_M(v)\frac{1}{v}\left[\sum_{\alpha''}(x_{\alpha\alpha''}x''_{\alpha}{}_{\alpha'} + y_{\alpha\alpha''}y_{\alpha''\alpha'}) \times \int_0^\infty \frac{dZ_\alpha}{Z_\alpha}c\pm(X_{\alpha'}, Y_\alpha, Z_\alpha) + \sum_{\beta''}[(\alpha\alpha'\alpha'') \to (\beta\beta'\beta'')]\right] \quad \text{(C.31)}$$

where the expression under the summation over β'' should be obtained from the expression under the summation over α'' via the substitution $(\alpha\alpha'\alpha'') \rightarrow (\beta\beta'\beta'')$ and the complex conjugation. In (C.31) the upper signs correspond to the non-diagonal elements ($\alpha' \neq \alpha$ or $\beta' \neq \beta$) of the xx- and yy-operators, the lower signs correspond to diagonal elements ($\alpha' = \alpha$ in the first term or $\beta' = \beta$ in the second term).

The dimensionless variables of integrations Z_α and Z_β are proportional to the impact parameter ρ as follows:

$$Z_{\alpha,\beta} = 3n_{\alpha,\beta}\hbar F\rho/(2m_e ev), \tag{C.32}$$

where F is the quasistatic part of the ion field.

The generalized broadening functions $C_\pm$ in (C.31) are defined as follows:

$$\begin{aligned} C_\pm(\chi, Y, Z) = &-\frac{3}{4}\int_\infty^\infty dx_1 \int_\infty^{x_1} dx_2 [w(x_1)w(x_2)]^3 exp[iZ(x_1 \pm x_2)] \\ &- \left\{ j_0(\varepsilon) + (2x_1x_21)\frac{j_1(\varepsilon)}{\varepsilon} \right. \\ &\left. + [(1x_1x_2)\sigma_1^2(x_1+x_2)\sigma_1\sigma_2]\varepsilon^2 j_2(\varepsilon) \right\}, \\ \varepsilon \equiv \sqrt{\sigma_1^2+\sigma_2^2}, \quad &\sigma_1 \equiv \frac{Y}{Z}[x_1w(x_1) \pm x_2w(x_2) + 1 \pm 1 - 2\chi], \\ \sigma_2 \equiv \frac{Y}{Z}[w(x_1) \pm w(x_2)], \quad &w(x) \equiv 1/\sqrt{1+x^2}. \end{aligned} \tag{C.33}$$

Spherical Bessel functions $j_0(\varepsilon), j_1(\varepsilon), j_2(\varepsilon)$ in (C.33) can be represented in terms of the elementary functions as follows:

$$\begin{aligned} j_0(\varepsilon) &= \frac{1}{\varepsilon} sin\varepsilon, \quad j_1(\varepsilon) = \frac{1}{\varepsilon^2}(sin\varepsilon - \varepsilon cos\varepsilon), \\ j_2(\varepsilon) &= \frac{1}{\varepsilon^3}(3sin\varepsilon - 3\varepsilon cos\varepsilon - \varepsilon^2 sin\varepsilon). \end{aligned} \tag{C.34}$$

In distinction to the CT, there are two new parameters that enter the generalized broadening functions. The first one χ stands for:

$$\chi_a \equiv (n_\alpha q_\alpha \delta_{\alpha\alpha'} - n_\beta q_\beta \delta_{\beta\beta'})/(n_\alpha(q_\alpha - q_{\alpha''})),$$
$$\chi_b \equiv (n_\alpha q_\alpha \delta_{\alpha\alpha'} - n_\beta q_\beta \delta_{\beta\beta'})/(n_\beta(q_\beta - q_{\beta''})). \tag{C.35}$$

The second new parameter Y is expressed as:

$$Y_k = \left(\frac{3n_k \hbar}{2Z_k m_e v}\right)^2 \frac{F}{e}, \quad k = \alpha, \beta. \tag{C.36}$$

The parameter Y provides a connection between the non-adiabatic parts of the EBO in the GT and in the CT. In the limiting case $Y \to 0$, corresponding to plasmas of either to relatively low density, or relatively high temperature, the generalized broadening function (C.33), we recover the broadening function of CT:

$$C_\pm(\chi, 0, Z)$$
$$= -\frac{1}{2}\int_{-\infty}^{+\infty} dx_1 \int_{-\infty}^{x1} \frac{dx_2(1 + x_1 x_2)}{[(1 + x_1^2)(1 + x_2^2)]^{\frac{3}{2}}} \exp[iZ(x_1 \pm x_2)]. \tag{C.37}$$

On the other hand, *the higher the electron density and/or the principal quantum number (or the lower the temperature), the greater become the parameter Y and, consequently, the inaccuracy of the CT.*

One of the most characteristic features of the GT is that the integrals over Z in (C.31) converge for any finite Y (while in the CT they diverge at low Z). Physically, this is due to the allowance for an average Stark splitting produced by a z-component of electron microfield (in addition to the splitting produced by the ion microfield $\mathbf{F}$). It is just this splitting that prevents the integrals over Z in (C.31) from diverging at small Z (i.e. at small ρ).

We note that both in the CT and in the GT, the EBO contains also a so-called "interference term." The interference term originates from the averaging of the product of two first order terms in the expansion of $S_a S_b^*$ (and therefore is also called "cross term"). The interference/cross term $\Phi_\times$ differs from zero only for non-diagonal elements of the EBO:

$$\langle\langle\alpha\beta|\Phi_\times|\alpha'\beta'\rangle\rangle = 2\frac{4\pi\hbar^2 N_e}{3m_e^2 a_0^2}(x_{\alpha\alpha'}x^*_{\beta\beta'} + y_{\alpha\alpha'}Y^*_{\beta\beta'}$$

$$\times \int_0^\infty \frac{dZ_\alpha}{Z_\alpha} C_\times(\chi_\alpha, Y_\alpha, Y_\beta, Z_\alpha, Z_\beta). \quad \text{(C.38)}$$

The broadening function $C_\times$ in the interference term (C.38) has the form:

$$\begin{aligned} &C_\times(\chi_\alpha, Y_\alpha, Y_\beta, Z_\alpha, Z_\beta) \\ &\quad = \frac{3}{4}\int_\infty^\infty dx_1 \int_\infty^\infty dx_2 [w(x_1)w(x_2)]^3 \exp[i(Z_\beta x_2 - Z_\alpha x_1)] \\ &\quad - \left\{ j_0(\varepsilon) + (2_{x_1x_2} - 1)\frac{j_1(\varepsilon)}{\varepsilon} + [(1-_{x_1+x_2})\sigma_1^2 \right. \\ &\quad \left. - (x_1 + x_2)\sigma_1\sigma_2]\varepsilon^{-2} j_2(\varepsilon) \right\}, \end{aligned} \quad \text{(C.39)}$$

$$\varepsilon \equiv \sqrt{\sigma_1^2 + \sigma_2^2}, \quad \sigma_1 \equiv 2\frac{\chi_\alpha}{Z_\alpha} + (1 + x_2 w(x_2))\frac{Y_\beta}{Z_\beta} - (1 + x_1 w(x_1))\frac{Y_\alpha}{Z_\alpha},$$

$$\sigma_2 \equiv w(x_1)\frac{Y_\alpha}{Z_\alpha} - w(x_2)\frac{Y_\beta}{Z_\beta}, \quad w(x) \equiv 1/\sqrt{1 + x^2}.$$

As in the CT, we represent $C_\pm(\chi, Y, Z)$ as a sum of two terms:

$$C_\pm(\chi, Y, Z) \equiv A_\pm(\chi, Y, Z) + iB_\pm(\chi, Y, Z), \quad \text{(C.40)}$$

where $A_\pm(\chi, Y, Z)$ and $B_\pm(\chi, Y, Z)$ are real and are called width and shift functions, respectively.

In Fig. C.1, we present the comparison of the width functions $A_-(Z)$ of the GT and of the CT. It is seen that for large impact parameters they asymptotically merge together. However, for small impact parameters, there is a very significant difference. Namely, the CT width function is finite at $Z = 0$, so that after the multiplication by the factor of $1/Z$ (see (C.31)) the integral over impact parameters diverges at $Z = 0$. In distinction, the width function in the GT oscillates at small Z, becomes zero at $Z = 0$, and, being multiplied by $1/Z$, converges under the integration over impact parameters.

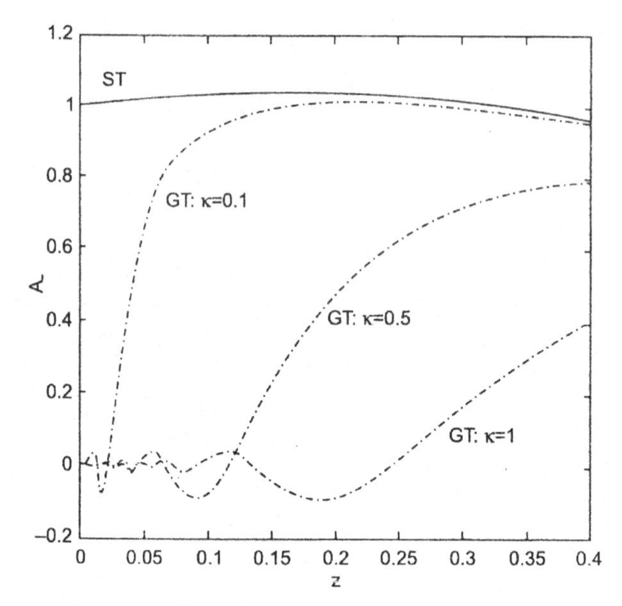

Fig. C.1. The width function $A_-(Z)$ of the GT and of the CT (labeled inside the figure as the ST — the Standard Theory) versus the scaled impact parameter $Z = 3n\hbar F\rho/(2m_e \mathrm{ev})$ for several values of the coupling parameter denoted as $\chi = [9n(n_\alpha q_\alpha - n_\beta q_\beta)F/2][\hbar/(\mathrm{m_e}v)]^2$.

References

[1] E. Oks, *Plasma Spectroscopy: the Influence of Microwave and Laser Fields*, Springer, Berlin, (1995).

[2] E. Oks, *Stark Broadening of Hydrogen and Hydrogenlike Spectral Lines in Plasmas: The Physical Insight*, (Alpha Science International, Oxford, UK) (2006).

[3] Ya. Ispolatov and E. Oks, *J. Quant. Specrosc. Rad. Transfer* **51** 129 (1994).

[4] E. Oks, A. Derevianko and Ya. Ispolatov, *J. Quant. Specrosc. Rad. Transfer* **54** 307 (1995).

[5] C.A. Iglesias, J. Lebowitz and D. McGowan, *Phys. Rev. A* **32** 1667 (1983).

[6] V.S. Lisitsa, *Sov. Phys. Uspekhi* **122** 603 (1977).

Appendix D

The Generalized Theory of the Stark Broadening of Hydrogenlike Lines in Non/Weakly-Magnetized Plasmas

We start from the same Hamiltonian as in Eq. (C.1) of Appendix C. In distinction to Sec. 4.2, we now use hyperbolic trajectories of perturbing electrons with the usual parametrization [1, 2]:

$$\begin{aligned}
\mathbf{r}(\tau) &= \frac{\rho}{K}\{f_v(\tau)\mathbf{e}_v + f_\rho(\tau)\mathbf{e}_\rho\},\\
r(\tau) &= \rho(K\cosh(\tau) - \alpha),\\
f_v(\tau) &= \alpha(K - \alpha\cosh(\tau)) + \sinh(\tau),\\
f_\rho(\tau) &= K - \alpha(\cosh(\tau) + \sinh(\tau)),\\
\mathbf{E}(\tau) &= \frac{e\mathbf{r}(\tau)}{r^3(\tau)},\\
\alpha &= \frac{(Z_r - 1)e^2}{\rho}/(m_v^2) = \frac{\rho_c}{\rho},\\
K &= \sqrt{1+\alpha^2},\\
t &= \frac{\rho}{v}K\sinh(\tau) - \alpha\tau).
\end{aligned} \tag{D.1}$$

In Eq. (D.1), the first and fifth lines contain vector $\mathbf{r}(t)$ defined via the unit vector $\mathbf{e}_\mathrm{v}$ in the $\mathbf{v}$-space ($\mathbf{v} = \mathrm{v}\mathbf{e}_\mathrm{v}$) and the unit vector $\mathbf{e}_\rho$ in the $\boldsymbol{\rho}$-space ($\boldsymbol{\rho} = \rho\mathbf{e}_\rho$).

Concerning the *adiabatic* part of the EBO, we obtain basically Eq. (C.29) with the only difference that $I(u_{\alpha\beta})$ should be now substituted by $I(u_{\alpha\beta}) - I(w_{\alpha\beta})$, where,

$$w_{\alpha\beta} = m_e \mathrm{v}\rho_c/[3\hbar(n_\alpha q_\alpha - n_\beta q_\beta)], \quad \rho_c \equiv (Z_r - 1)e^2/[mv^2]. \quad \text{(D.2)}$$

The expressions for $I(u)$ and $u_{\alpha\beta}$ are given by Eq. (C.30).

As for the *non-adiabatic* part of the EBO, the result can be expressed by Eq. (C.31), but with more complicated generalized broadening functions $C_\pm(\chi, Y, \xi, Z)$. Of course, hyperbolic trajectories complicate the calculations compared to the case of rectilinear trajectories. Despite this complication, it is possible, just as in Appendix C, to perform all three angular integrations analytically and to obtain the generalized broadening functions $C_\pm(\chi, Y, \xi, Z)$ expressed in terms of elementary functions:

$$\begin{aligned} C_\pm(\chi, Y, \xi, Z) = \frac{3}{4K^2} \int_\infty^\infty dx_1 \int_\infty^{x_1} dx_2 \\ \times \frac{\exp\{iZ(K(\sinh x_1 \pm \sinh x_2) - \alpha(x_1 \pm x_2)\}}{((K\cosh x_1 - \alpha)(K\cosh x_2 - \alpha))^2} \\ \times \left\{ f_{\rho 1} f_{\rho 2} j_0(\varepsilon) + (2f_{v1}f_{v2} - f_{\rho 1} f_{\rho 2}) \frac{j_1(\varepsilon)}{\varepsilon} \right. \\ - ((f_{v1}f_{v2} - f_{\rho 1}f_{\rho 2})\eta_1(x_1, x_2) \\ \left. + (f_{v1}f_{\rho 2} + f_{\rho 1}f_{v2})\eta_2(x_1, x_2))\eta_2(x_1, x_2)\frac{j_2(\varepsilon)}{\varepsilon^2} \right\}, \end{aligned}$$

$$\begin{aligned} \varepsilon \equiv \sqrt{\eta_1(x_1, x_2)^2 + \eta_2(x_1, x_2)^2}, \\ \eta_1(x_1, x_2) = \frac{Y}{Z}\frac{1}{K}\left\{ \Theta_1(\tanh(x_2/2)) - \Theta_1(\tanh(x_1/2)) \right. \\ \left. + 2\frac{\chi}{K} + \frac{1}{K} \pm \frac{1}{K} \right\}, \\ \eta_2(x_1, x_2) = \frac{Y}{Z}\frac{1}{K}\left\{ \Theta_2(\tanh(x_2/2)) - \Theta_2(\tanh(x_1/2)) + 2\frac{\alpha\chi}{K} \right. \\ \left. + \frac{\alpha}{K} \pm \frac{\alpha}{K} \right\}, \end{aligned}$$

$$\Theta_1(u) = \frac{2u + \frac{\alpha}{K}(1u^2)}{(K+\alpha)u^2 + (K-\alpha)}, \quad \Theta_2(u) = \frac{2\alpha u - \frac{1}{K}(1-u^2)}{(K+\alpha)u^2 + (K-\alpha)},$$
$$\alpha \equiv \frac{\xi}{Z}, \quad K = \sqrt{1+\alpha^2}. \tag{D.3}$$

We remind that spherical Bessel functions $j_0(\varepsilon)$, $j_1(\varepsilon)$, $j_2(\varepsilon)$ in (4.3.4) are actually elementary functions given by Eq. (C.34). The parameter ξ in the last line of Eq. (D.3), which is the only one new parameter compared to the case of rectilinear trajectories, is defined as:

$$\xi_k \equiv \frac{\rho_c}{(\rho_s)_k}, \quad \rho_c \equiv \frac{(Z_r - 1)e^2}{mv^2}, \quad k = \alpha, \beta, \tag{D.4}$$

where,

$$(\rho_s)_k \equiv \frac{\hbar v}{[(\delta d)_k F]}$$
$$(\delta d)_\alpha \equiv (d_z)_{\alpha\alpha} - (d_z)_{\alpha'\alpha'}, \quad (\delta d)_\beta \equiv (d_z)_{\beta\beta} - (d_z)_{\beta'\beta'}. \tag{D.5}$$

In the limit, where the coupling parameter Y (defined by Eq. (C.36)) is relatively small ($Y \ll 1$), we recover the broadening functions of the CT, for example

$$\begin{aligned} &C_-(\chi, 0, \xi, Z) \\ &= \frac{1}{2}\int_\infty^\infty dx_1 \int_\infty^{x_1} dx_2 \frac{\exp\{iZ(K(\sinh x_1 - \sinh x_2) - \alpha(x_1 - x_2)\}}{((K\cosh x_1 - \alpha)(K\cosh x_2 - \alpha))^2} \\ &\quad \times \{K^2 + \sinh x_1 \sinh x_2 + \alpha^2 \cosh x_1 \cosh x_2 \\ &\quad - \alpha K(\cosh x_1 + \cosh x_2)\}. \end{aligned} \tag{D.6}$$

It should be emphasized, just as in Appendix C, that *the higher the electron density and/or the principal quantum number (or the lower the temperature), the greater become both the parameter Y and, consequently, the inaccuracy of the CT.*

Figure D.1 shows the real part of the generalized broadening function Re C_- from Eq. (D.3), calculated for $n = 2$, $Z_r = 2$, $\chi = 0$, $\xi = 0.37$, and $Y = 0.5$. The abscissa scale is logarithmic with respect

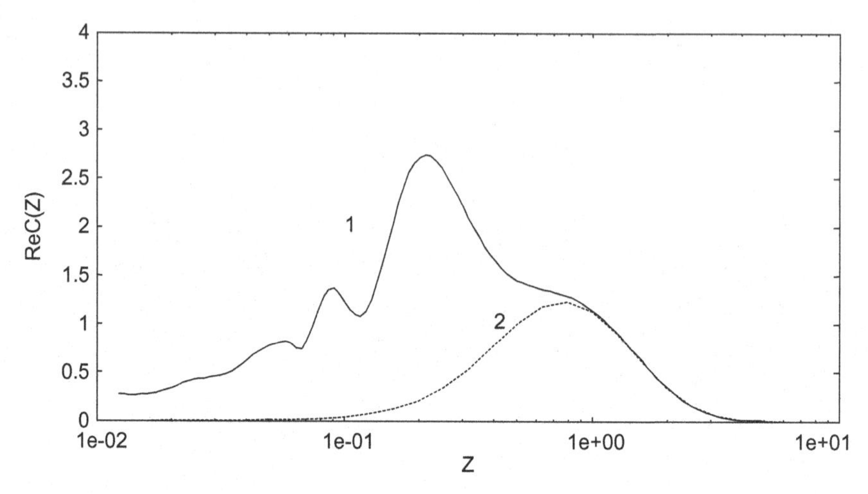

Fig. D.1. The real part of the generalized broadening function Re C_- from Eq. (D.3), calculated for $n = 2$, $Z_r = 2$, $\chi = 0$, $\xi = 0.37$, and $Y = 0.5$. The abscissa scale is logarithmic with respect to the reduced impact parameter Z, so that the non-adiabatic width is controlled by the area under curve 1 in Fig. D.1. A comparison with the corresponding result of the CT (curve 2) shows that the CT significantly underestimates the non-adiabatic contribution to the electron-impact broadening.

to the reduced impact parameter Z, so that the non-adiabatic width is controlled by the area under curve 1 in Fig. D.1. The comparison with the corresponding result of the CT (curve 2) shows that the CT significantly underestimates the non-adiabatic contribution to the electron-impact broadening.

In this appendix, we presented a simplified version of the GT for HL-lines — the version employing the impact approximation — for the purpose of having a better physical insight into the GT and for tracing more easily the relation between the GT and the CT. However, it should be noted that most of the numerical calculations of the SB of HL-lines presented in book [3] were based on equations analogous to Eqs. (C.1)–(C.7) representing the GT in its most general form. They allowed calculating the correlation function and then the lineshape without the impact approximation.

References

[1] H.R. Griem, *Spectral Line Broadening by Plasmas*, Academic, New York (1974).
[2] S. Sahal-Brechot, *Astron. & Astrophys.* **1** 91 (1969).
[3] E. Oks, *Stark Broadening of Hydrogen and Hydrogenlike Spectral Lines in Plasmas: The Physical Insight*, Alpha Science International, Oxford, UK (2006).

Appendix E

Stark Broadening of Hydrogen Lines under Strong Magnetic Fields in Laboratory and Astrophysical Plasmas

E.1 Approach Based on Rectilinear Trajectories of the Perturbers: Intense (low-n) Hydrogen Lines

Mechanisms of spectral line broadening are also divided into two types: *homogeneous* and *inhomogeneous*. A particular broadening mechanism is *homogeneous* when it is the same for all radiators. A typical example is the Stark broadening (SB) by the electron microfield (within the Conventional Theory (CT)). In distinction, the SB by the quasistatic part F_{qs} of the ion microfield is *inhomogeneous* because different radiators are subjected to generally different values of F_{qs}.

In the conditions typical for the edge plasmas of magnetic fusion devices (e.g., in the divertor region of tokamaks), the homogeneous Stark width of Hydrogen/Deuterium Spectral Lines (HDSL) is controlled by the dynamic part of the ion microfield. A strong magnetic field B (up to $\sim$10 T) in such plasmas significantly affects shapes of HDSL and thus provides opportunities for spectroscopic diagnostics. Physically, this is because a relatively large Zeeman splitting of hydrogen energy levels diminishes the range of ion impact parameters ρ, for which the characteristic frequency of the variation v_i/ρ of the electric field of perturbing charged particles exceeds the Zeeman splitting.

Paper [1] presented a study of this situation based on the most advanced semiclassical theory of the SB: the Generalized Theory (GT). In the semiclassical theories of the dynamical SB of *non-hydrogenic* spectral lines by electron or ion microfields, the broadening is controlled by *virtual transitions* between different sublevels characterized by the same principal quantum number n (see, e.g., book [2]). This is a *non-adiabatic* contribution — since, it involves different sublevels. For hydrogenic spectral lines (including HDSL), under the usual assumption that the fine structure can be neglected, in addition to the non-adiabatic contribution, there is also a significant *adiabatic* contribution — see, e.g., review [3] and book [4]. Physically, the adiabatic contribution is due to the fact that for hydrogenic spectral lines, the overwhelming majority of sublevels have permanent electric dipole moments (which is especially clear in the parabolic quantization). Even an isolated sublevel having a permanent electric dipole moment would produce some adiabatic contribution — in other words, the latter does not require virtual transitions between different sublevels. Classically the adiabatic contribution originates from the phase modulation of the atomic oscillator by the dynamic microfield of electrons or ions.

In the GT, contributions to the dynamical Stark width due to ions are explicitly separated into adiabatic and non-adiabatic. The main result of the GT for magnetic fusion plasmas is the following. At values of the magnetic field B typical for magnetic fusion plasmas, practically the entire dynamical Stark width due to ions is due only to the *adiabatic* contribution. This is because, as the magnetic field B increases, it causes the increase of the separation

$$\omega_B = eB/(2m_e c) \tag{E.1}$$

between the Zeeman sublevels of hydrogenic atoms, the non-adiabatic contribution to the dynamical SB by ions decreases — specifically, it decreases dramatically at magnetic fields typical for magnetic fusion plasmas.

In accordance to [4, 5], the adiabatic contribution to the dynamical Stark width due to ions γ_{ad} has the following form, which is the *exact, non-perturbative* analytical result:

$$\gamma_{\text{ad}} = 18(\hbar/m_e)^2(X_{\alpha\beta}/Z_r)^2 Z_i^2 N_i (2\pi M/T_i)^{1/2} I(R_i),$$
$$X_{\alpha\beta} = |n_\alpha q_\alpha - n_\beta q_\beta|. \qquad \text{(E.2)}$$

Here, Z_i, N_i and T_i are the charge, the density, and the temperature of the plasma ions, respectively; Z_r is the nuclear charge of the radiator; M is the reduced mass of the pair "radiator — perturbing ion"; n_γ and $q_\gamma = (n_1 - n_2)_\gamma$ are, respectively, the principal quantum number and the electric quantum number, the latter being expressed via the parabolic quantum numbers n_1 and n_2; the subscript $\gamma = \alpha$ (the upper Stark sublevel) or β (the lower Stark sublevel). The function $I(R_i)$ in Eq. (E.2) is defined as follows:

$$I(R_i) = \{R_i^2[3 - \cos(1/R_i)] + (R_i - 2R_i^3)\sin(1/R_i) - ci(1/R_i)\}/6, \qquad \text{(E.3)}$$

In Eq. (E.3), $ci(1/R_i)$ is the cosine integral function, the quantity R_i being:

$$R_i = r_{\text{D}}/r_{\text{WA}}, \qquad \text{(E.4)}$$

where,

$$r_{\text{D}} = [T_e/(4\pi e^2 N_e)]^{1/2} = 743.40[T_e(\text{eV})/N_e(\text{cm}^{-3})]^{1/2}, \text{cm} \qquad \text{(E.5)}$$

is the Debye radius[1] and

$$r_{\text{WA}} = 3X_{\alpha\beta}\hbar/(Z_r m_e v_i)$$
$$= 3.5486 \times 10^{-6}(X_{\alpha\beta}/Z_r)(M/M_p)^{1/2}/[T_i(\text{eV})]^{1/2}, \text{cm} \qquad \text{(E.6)}$$

is the adiabatic Weisskopf radius (M_p is the proton mass). The quantity r_{WA} naturally arises in the GT with the *exact* coefficient given in Eq. (E.6) — in distinction to the Weisskopf radius of the ST defined only by the order of magnitude. We note that the

[1] If the screening by ions would be taken into account (in addition to the screening by electrons), the Debye radius from Eq. (E.5) should be divided by $2^{1/2}$.

quantity $r_{\rm WA}$ in Eq. (E.6) coincides with the corresponding quantity arising from the *exact* solution for the SB of hydrogen/deuterium spectral lines by just one sort of perturbing charges — in the binary approximation [6] and in the multi-particle description [7]. The practical part of Eq. (E.6) was obtained using the fact that the averaging over ion velocities is performed with the effective statistical weight factor $W_M(v_i)/v_i$, where $W_M(v_i)$ is the Maxwell distribution, and that $W_M(v_i)/v_i$ has the maximum at $v_i = (T_i/M)^{1/2}$.

In 2009, Rosato *et al.* published a paper [8], where they revisited the subject studied in paper [1] in 1994: the dynamical SB of hydrogen/deuterium lines by ions in magnetized plasmas. Rosato *et al.* [19] based their analytical results on the CT — despite they knew that a more advanced study (based on the GT) has been already published 15 years earlier. Essentially, Rosato *et al.* [8] repeated Sholin–Demura–Lisitsa's version of the CT [9] with the substitution of the magnetic splitting ω_B of hydrogen/deuterium energy levels instead of the electric splitting ω_F.

Then in paper [10], in review [11], and in paper [12], we showed that the analytical results from [8] yield a very dramatic inaccuracy — up to an order of magnitude. Below, we present the corresponding arguments and illustrations.

For comparing the adiabatic Stark widths of the GT and of the CT, it is convenient to introduce the adiabatic broadening cross-section $\sigma_a(v_i)$ related to the adiabatic width γ_a as follows:

$$\gamma_a = N_i \int_0^\infty dv_i W(v_i) v_i \sigma_a(v_i), \tag{E.7}$$

where $W(v_i)$ is the velocity distribution. In the GT, the adiabatic broadening cross-section $\sigma_{aGT}(v_i)$ is

$$\sigma_{aGT}(v_i) = 2\pi[r_{\rm WA}(v_i)]^2 I[R_i(v_i)], \tag{E.8}$$

where $I[R_i(v_i)]$ is given by Eq. (E.3) and $r_{\rm WA}(v_i)$ is given by the first equality in Eq. (E.6). This is the *exact* analytical result equivalent to the summation of *all orders* of the Dyson perturbation expansion.

In paper [8], based on the *second order of the Dyson perturbation expansion* of the ST, the adiabatic broadening cross-section $\sigma_{\text{aRos},\alpha}(v_i)$ for the upper level involved in the radiative transition, was (the subscript Ros is the abbreviated name of the first author, Rosato, of paper [8])

$$\sigma_{\text{aRos},\alpha}(v_i) = \pi[r_{\text{str},\alpha}(v_i)]^2(1 + 2\ln[r_{\text{D}}/r_{\text{str},\alpha}(v_i)], \tag{E.9}$$

where $r_{\text{str},\alpha}(v_i)$ is a so-called "strong collision radius" (i.e., the boundary between weak and strong collisions). For the adiabatic contribution from [8], $r_{\text{str},\alpha}(v_i)$ is the solution of the following equation:

$$r_{\text{str},\alpha} = (2/3)^{1/2}[\hbar n_\alpha^2/(m_e v_i)]K_{\|\alpha}^{1/2}, \quad K_{\|\alpha} = (1/n_\alpha^4)\sum |z_{\alpha\alpha}|^2, \tag{E.10}$$

Here, $z_{\alpha\alpha}$ are matrix elements (in atomic units) of the projection of radius-vector of the atomic electron on the magnetic field **B**. (We note that in the left side of the similar Eq. (12) from [8], there was erroneously $\langle \text{S}_{n\alpha}^{(2)} \rangle$ instead of $\langle \text{S}_{n\alpha}^{(2)} - 1 \rangle$.) Using the parabolic quantization, one obtains:

$$K_{\|\alpha} = q_\alpha^2/n_\alpha^2. \tag{E.11}$$

In the CT, the impact parameter, at which perturbers contribute most effectively to the dynamical SB, is called Weisskopf radius. It is defined only by the order of magnitude (which is one of the major sources of the inaccuracy of the CT): it is $\sim n_\alpha^2\hbar/(m_e v_i)$ for $Z_r = Z_i = 1$. The authors of paper [8] arbitrarily chose the following numerical coefficient in the Weisskopf radius of the CT, which enters their Eq. (E.9) for the cross-section:

$$r_{W,\text{Ros},\alpha}(v_i) = (2/3)^{1/2} n_\alpha^2 \hbar/(m_e v_i). \tag{E.12}$$

The ratio of the adiabatic broadening cross-sections, denoted as $\kappa = \sigma_{a\text{Ros}}/\sigma_{a\text{GT}}$, is the following:

$$\kappa = \sigma_{a\text{Ros}}/\sigma_{a\text{GT}} = q_\alpha^2 n_\alpha^2\{1/2 + \ln[6^{1/2}2|X_{\alpha\beta}|r_{\text{D}}/(|q_\alpha|n_\alpha r_{Wa})]\}. \tag{E.13}$$

Below, we provide examples of the values of the ratio κ for some hydrogen lines in a hydrogen plasma (so that $M = M_p/2$)

at the conditions typical for tokamak divertors. The components of a particular line are identified by the parabolic quantum numbers: $(n_1 n_2 m)_\alpha - (n_1 n_2 m)_\beta$; we also indicate the polarization of the component (π or σ). The ratio κ is calculated at $T = 4\,\mathrm{eV}$ and $N_e = 10^{13}\,\mathrm{cm}^{-3}$.

First, we consider the Paschen-alpha line, specifically its component (102)–(101), which is one of the two most intense σ-components. The calculation of the ratio, using Eq. (E.13) yields: $\kappa = 14$.

Second, we consider the Paschen-beta line, specifically its component (121)–(020), which is the σ-component closest to the line center. The calculation of the ratio, using Eq. (E.13) yields: $\kappa = 22$.

Third, we consider the Balmer-alpha line, specifically the component (101)–(100), which is one of the two most intense σ-components. The calculation of the ratio, using Eq. (E.13) yields: $\kappa = 9$.

Thus, "the authors of paper [8] overestimated the primary, adiabatic contribution to the dynamical SB by ions in magnetic fusion plasmas by up to an order of magnitude."

For completeness, let us now discuss also the secondary, minor, non-adiabatic contribution. The non-adiabatic contribution to the dynamical Stark width due to ions γ_{na}, calculated for magnetic fusion plasmas using the GT is controlled by the integral of a so-called width function $A_-(\chi, Y, Z)$ over scaled (dimensionless) impact parameters Z:

$$a_-(\chi, Y, Z_\mathrm{D}) = \int_0^{Z_\mathrm{D}} dZ A_-(\chi, Y, Z)/Z. \tag{E.14}$$

The scaled impact parameter Z is defined as:

$$Z(\rho) = \rho/\rho_B = 2m_e cv\rho/(eB). \tag{E.15}$$

The upper limit of the integration in (E.14) is $Z_\mathrm{D} = Z(r_\mathrm{D})$. Typically, $r_\mathrm{D} > r_{\mathrm{Wa}}$, which is assumed in Eq. (E.14).

Compared to the CT, there are two new parameters that enter the width function. The first one χ stands for

$$\chi = (n_\alpha q_\alpha - n_\beta q_\beta)/n_\gamma = X_{\alpha\beta}/n_\gamma, \quad \gamma = \alpha \text{ or } \beta. \tag{E.16}$$

The second new parameter Y is physically the most important: it is a coupling parameter defined as:

$$Y = 3n_\gamma Z_i \hbar e B/(2m_e^2 c v_i^2)$$
$$= 0.31885\; n_\gamma Z_i (M/M_p) B(\text{Tesla})/T_i(\text{eV}), \quad \gamma = \alpha \text{ or } \beta, \quad \text{(E.17)}$$

where Z_i is the charge of the plasma ions.

For example, for the D_α or L_β line ($n_\nu = 3$) from a deuterium plasma ($M/M_p = 1$, $Z_i = 1$), Eq. (E.17) yields: $Y = 0.95655\; B(T)/T_i$ (eV). For the typical parameters of tokamak divertors, the ratio $B(\text{Tesla})/T_i$ (eV) is greater or of the order of unity, so that Y is also greater or of the order of unity. At these values of the coupling parameter, first, the ST becomes quite inaccurate and second (but most importantly), there occurs a dramatic decrease of the non-adiabatic contribution to the dynamical Stark width due to ions. Thus, the total contribution to the dynamical Stark width due to ions can be well represented by the adiabatic contribution given by Eq. (E.2).

The finding, that the non-adiabatic contribution significantly decreases with the increase of the magnetic field, was quite clear already in 1994: from the results of the paper [1] (where Eqs. (E.14–E.17) were first presented) complemented by the results of the paper [5] (where it was shown that the function $a_-(\chi, Y, Z_\text{D})$, controlling the non-adiabatic contribution, significantly decreases with the increase of the coupling parameter Y). Therefore, the claim by Rosato *et al.* [8] that they were the first to "discover" this effect in their paper [8] published in 2009 is without merit.

Now let us discuss the relation between the unitarity of the S-matrix and the non-adiabatic contribution calculated by the GT or by the CT. In both theories, the non-adiabatic contribution is calculated via $\{1 - S_\text{na}\}_\text{ang}$, which is the angular average of the non-adiabatic part of the S-matrix. It is calculated up to the second order of the Dyson perturbation expansion, but using the different basis: the basis of the DAS in the GT as opposed to the usual atomic basis in the CT. At small impact parameters, the CT would violate the unitarity of the S-matrix. To avoid the violation, the CT has to separate collisions into "weak" and "strong", the boundary between them being defined by the condition,

$$|\{1 - S_{\text{na}}\}_{\text{ang}}| = C, \quad 0 \le C \le 2. \tag{E.18}$$

The uncertainty in the choice of the constant C in (E.18) is yet another major source of inaccuracy of the CT.

In paper [13], it was rigorously shown analytically that for the overwhelming majority of hydrogen/deuterium spectral lines, the non-adiabatic contribution calculated by the GT does not violate the unitarity of the S-matrix — in distinction to the CT. Therefore, for the overwhelming majority of hydrogen/deuterium spectral lines the lower limit of the integration in Eq. (E.14) can remain to be zero.

As an illustration of this important distinction between the GT and CT, we present Figs. E.1–E.3. For the most intense π-component (400)–(100) of the Balmer-gamma line, Fig. E.1 shows the dependence of the integrand A_-/Z in Eq. (E.14) versus the

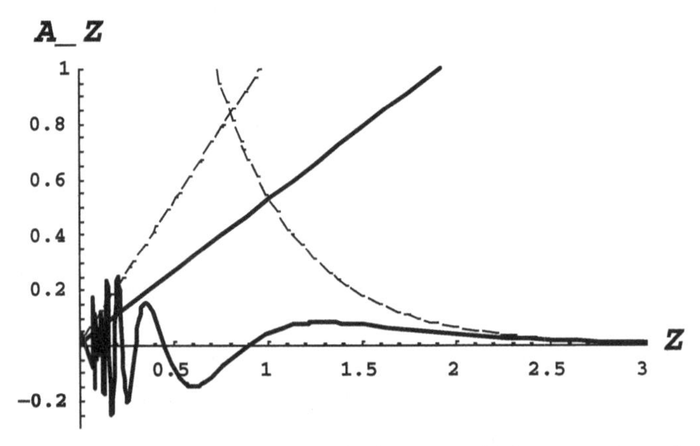

Fig. E.1. Dependence of the integrand A_-/Z in Eq. (14) versus the scaled impact parameter Z: by the GT, shown by the solid curve, and by the CT, shown by the dashed curve. The solid curve is calculated by the GT for the coupling parameter $Y = 0.85$, which corresponds, e.g., to $B = 4T$ and $T = 1.5\,\text{eV}$, or $B = 6T$ and $T = 2.25\,\text{Ev}$, or $B = 8T$ and $T = 3\,\text{eV}$. Two possible unitarity restrictions are presented by straight lines. The solid straight line and the dashed straight line correspond to the choice $C = 1$ or $C = 2$ in Eq. (E.18), respectively. The entire illustration is for the most intense π-component (400)–(100) of the Balmer-gamma line.

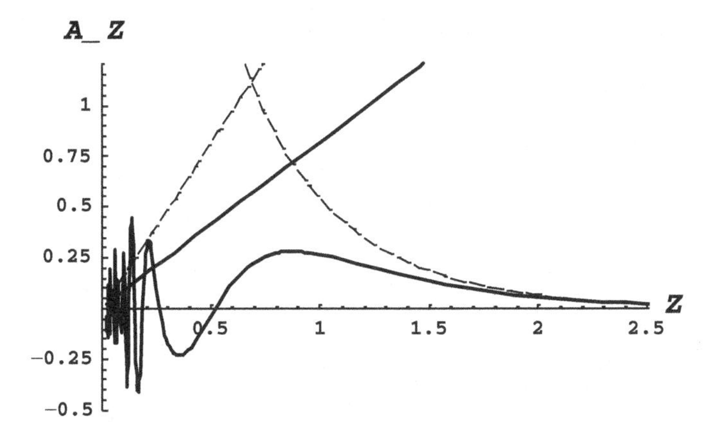

Fig. E.2. The same as Fig. E.1, but for the most intense π-component (300)–(000) of the Lyman-gamma line.

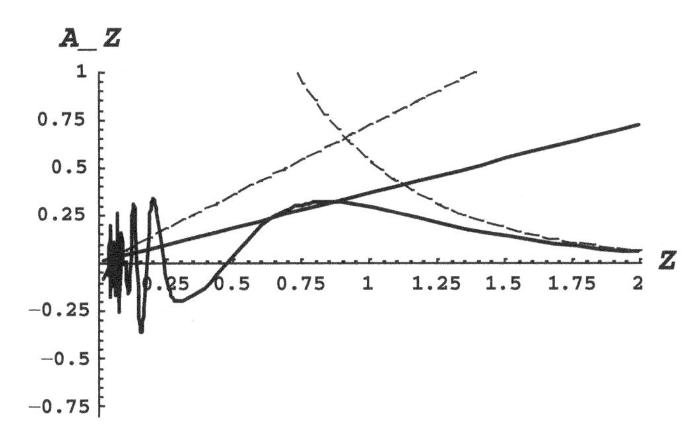

Fig. E.3. The same as Fig. E.1, but for the only one (and thus, the most intense) π-component (200)–(000) of the Lyman-beta line.

scaled impact parameter Z: by the GT (solid curve) and by the CT (dashed curve). The solid curve is calculated by the GT for the coupling parameter $Y = 0.85$, which corresponds, e.g., to $B = 4T$ and $T = 1.5\,\text{eV}$, or $B = 6T$ and $T = 2.25\,\text{eV}$, or $B = 8T$ and $T = 3\,\text{eV}$. Two possible unitarity restrictions are presented by straight lines. The solid straight line and the dashed straight line correspond to the choice $C = 1$ or $C = 2$ in Eq. (E.18), respectively.

Figure E.2 shows the same as Fig. E.1, but for the most intense π-component (300)–(000) of the Lyman-gamma line.

Figure E.3 shows the same as Fig. E.1, but for the only one (and thus, the most intense) π-component (200)–(000) of the Lyman-beta line.

Figures E.1–E.3 clearly demonstrates the following:

1. The CT violates the unitarity of the S-matrix and has to separate collisions into weak and strong at the value of Z somewhere between 0.7 and 1.1.
2. The GT does not need to engage the unitarity cutoff: the integrand A_{-}/Z strongly oscillates at small Z and thus practically "kills" the contribution from the small impact parameters to the integral.
3. Even after engaging the unitarity cutoff, the CT considerably overestimates the non-adiabatic contribution — by several times (in addition to significantly overestimating the adiabatic contribution by up to an order of magnitude).

The fact that at the presence of an additional static field, the width function of the GT exhibits oscillations at small impact parameters just like in the exact solution at the absence of the additional static field can be explained as follows. At small impact parameters, at the closest approach of the perturber to the radiator, the interaction with the electric field of the perturber is much stronger than the interaction with an additional static field, which makes the situation similar to the case where the additional static field is absent (the exact solutions for the latter case are well-known in the binary approach [6] and in the multi-particle description [7]). Physically, the oscillatory character of the exact solution in the region of small impact parameters in both situations is due to the fact that the closer/stronger the collision, the more transitions between atomic sublevels it causes.

We note that for the Lyman-alpha line, *as an exception*, the GT might need engaging the unitarity restriction and therefore separate collisions into weak and strong. Figure E.4 presents the plot for the same conditions as in Fig. E.1, but for the σ-components of the Lyman-alpha line: (001)–(000), (00-1)–(000).

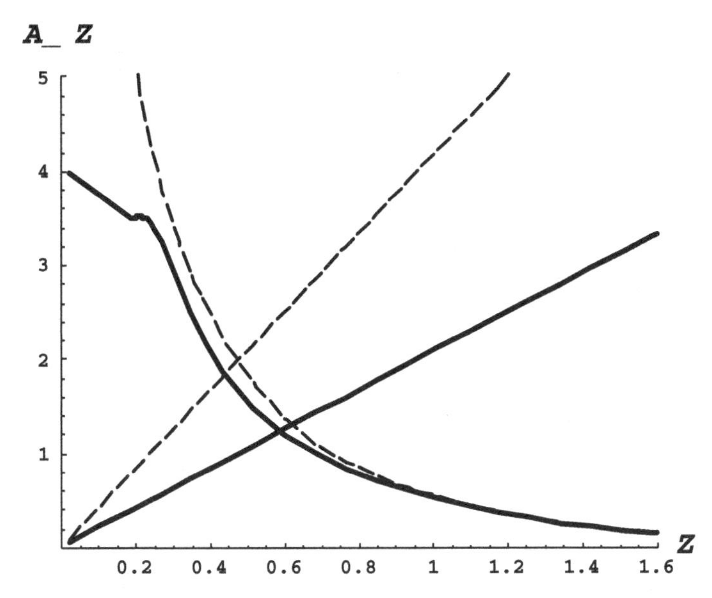

Fig. E.4. The same as in Fig. E.1, but for the σ-components of the Lyman-alpha line: (001)–(000), (00-1)–(000).

Figure E.4 shows that for the chosen plasma conditions, both the CT and the GT need engaging the unitarity cutoff. It also shows that, after engaging the unitarity cutoff for both theories, the CT still overestimates the non-adiabatic contribution to the broadening, though only slightly.

We note that in the later paper [14], the authors of the paper [8] made the second attempt to show that for magnetized plasmas, the CT is allegedly better than the GT. The results from the paper [14] cannot be trusted (just like those from paper [8]) for the following reasons detailed below.

Speaking of the adiabatic contribution, Rosato *et al.* [14] complained that the cross-section $\sigma_{a\mathrm{GT}}$ automatically took into account the contributions from both the upper and lower levels, as well as the interference term, while $\sigma_{a\mathrm{Ros}}$ did not. However, we compared with the GT the CT version as it was presented by Rosato *et al.* in Ref. [8]. The fact that Rosato *et al.* were unwilling or unable to include the contributions from both the upper and lower levels, as well as the interference term, is their problem — not ours: it would not make sense for us to try to improve their results in any case (because this should have been the task for Rosato *et al.*), but especially given that their results are based on the obsolete theory, such as the CT, which is less advance than the GT by the design.

We also note that even in the later paper [14], Rosato *et al.* were still unwilling or unable to include the contributions from both the upper and lower levels, as well as the interference term. Instead, they undertook a desperate (but clumsy) attempt to save their inaccurate, obsolete theory by *intentionally messing up the results of the GT*, while comparing the two theories. After messing up the results of the GT, Rosato *et al.* [14] essentially pronounced the ratio of the results of the two theories to be a factor of two. Then, by insulting the common sense, they assigned this discrepancy to "the typical uncertainty range expected by the use of the GT." On the part of Rosato *et al.*, this was a classical example of dumping their own blame on others. We have to remind the readers, including Rosato *et al.*, that in the GT the adiabatic contribution was calculated *exactly, i.e., in all orders of the Dyson perturbation expansion*, while in their obsolete theory, it was calculated only up to the second order of the Dyson perturbation expansion. The assault on the common sense by Rosato *et al.* [14] is equivalent to comparing $\exp(x)$ at $x = 2.5$ to its expansion up to the second order $1 + x + x^2/2$, where the ratio $(1 + x + x^2/2)/\exp(x)$ is about 0.5, and to state that the factor of two difference is due to the uncertainty in $\exp(x)$.

Trying to distract the attention from the primary, undisputable deficiency of their CT theory concerning the adiabatic contribution (as shown above), the Rosato *et al.* [14] focused on the much smaller non-adiabatic contribution. After attempting to numerically calculate the non-adiabatic contribution by the GT, they claimed that it violates unitarity at small impact parameters. However, the violation of the unitarity at small impact parameters for both the adiabatic and non-adiabatic contributions is actually the characteristic feature of the ST for *all* hydrogen/deuterium lines (resulting, after some rather arbitrary cutoff, in the inability of the CT from [8, 14] to calculate the adiabatic contribution better than by the order of magnitude). As for the GT, in paper [13] it was rigorously shown analytically that for the overwhelming majority of hydrogen/deuterium spectral lines, the non-adiabatic contribution does not violate unitarity at small impact parameters — as we already noted above. Also, this fact is illustrated above in Figs. E.1–E.3.

As for the Ly_α line, where the secondary, non-adiabatic contribution of GT does require the unitarity cutoff at small impact parameters ρ (which was mentioned already in 1995 in paper [15] by the authors of the GT), the width function for the σ-components $A_-(\rho)$ is proportional to ρ at small ρ, which is a rigorous analytical result — the proof presented in paper [12] is reproduced below at the end of this section. Also, this fact is illustrated in the above Fig. E.4.

In distinction, the corresponding attempted "GT computations" by Rosato *et al.* in paper [14] yielded that $A_-(0)$ is infinite. This is an absurd, which just by itself already shows that the results by Rosato *et al.* from paper [14] cannot be trusted — just like their results the from the paper [8].

There is an even more striking absurd in simulations by Rosato *et al.* [14]. It is well-known since the publication of paper [9] in 1973, that the impact shift of the spectral components of hydrogen/deuterium lines of the Ly series is proportional to the following combination of the parabolic quantum numbers $n(n_1 - n_2)$, so that for the σ-components of the Ly_α line $n_1 - n_2 = 0$ and this shift vanishes, which is a rigorous analytical result for rectilinear trajectories of the perturbers. Despite this, computations of the authors of the paper [14] yielded a non-zero impact shift of the σ-components of the Ly_α line (for rectilinear trajectories of the perturbers), which is preposterous.

One of the possible reasons for the failure of simulations by Rosato *et al.* [14] was actually noted by the authors of the paper [14]: they wrote that "the broadening function used in the GT involves a multiple integrals of strongly oscillating functions, not suitable for fast numerical evaluation." So, it seems that their computations failed to properly handle the multiple integrals of strongly oscillating functions.

Finally, here is the analytical proof that the width function for the σ-components $A_-(\rho)$ is proportional to ρ at small ρ. For the σ-components, the GT parameter $\chi = 0$ according to Eq. (E.16). Therefore, we consider the width function $A_-(\chi, Y, Z)$ of the GT for $\chi = 0$:

$$A_-(0,Y,Z) = -\frac{3}{4}\int_{\infty}^{\infty} dx_1 \int_{\infty}^{x_1} dx_2 [w(x_1)w(x_2)]^3 \cos[Z(x_1 - x_2)]$$

$$\left\{j_0(\varepsilon) + (2x_1x_2 - 1)\frac{j_1(\varepsilon)}{\varepsilon} + [(1 - x_1x_2)\sigma_1^2 - (x_1 + x_2)\sigma_1\sigma_2]\varepsilon^2 j_2(\varepsilon)\right\},$$

$$\varepsilon \equiv \sqrt{\sigma_1^2 + \sigma_2^2}, \quad \sigma_1 \equiv \tfrac{Y}{Z}[x_1 w(x_1) - x_2 w(x_2)],$$

$$\sigma_2 \equiv \tfrac{Y}{Z}[w(x_1) - w(x_2)], \quad w(x) \equiv 1/\sqrt{1 + x^2}. \tag{E.19}$$

Spherical Bessel functions $j_0(\varepsilon)$, $j_1(\varepsilon)$, $j_2(\varepsilon)$ in Eq. (E.19) can be expressed in terms of the elementary functions as follows:

$$\begin{aligned} j_0(\varepsilon) &= \frac{1}{\varepsilon}\sin\varepsilon, \quad j_1(\varepsilon) = \frac{1}{\varepsilon^2}(\sin\varepsilon - \varepsilon\cos\varepsilon), \\ j_2(\varepsilon) &= \frac{1}{\varepsilon^3}(3\sin\varepsilon - 3\varepsilon\cos\varepsilon - \varepsilon^2\sin\varepsilon). \end{aligned} \tag{E.20}$$

Since, the integrand in Eq. (E.19) is symmetric with respect to the interchange of x_1 and x_2, the upper limit of the integration over x_2 can be extended to infinity, while multiplying the result by 1/2. After the substitution,

$$x_1 = u_1 + u_2, \quad x_2 = u_1 - u_2, \tag{E.21}$$

the integral from Eq. (E.19), denoted below for brevity as I, becomes of the following type:

$$I = \int_{-\infty}^{\infty}\int_{-\infty}^{\infty} du_1 du_2 f(u_1, u_2) S[\lambda g(u_1, u_2)], \tag{E.22}$$

where S is an oscillatory function and λ is a large parameter:

$$\lambda = 2^{1/2} Y/Z \gg 1 \tag{E.23}$$

(so that S is a strongly oscillatory function). The function $g(u_1, u_2)$ has the following properties at $u_1 = u_2 = 0$:

$$g = 0, \quad \partial g/\partial u_2 = |u_2|, \quad \partial^{(n)} g/\partial u_2^n = 0 \text{ for any } n. \tag{E.24}$$

Therefore, according to the well-known stationary phase method for evaluating integrals containing strongly oscillatory functions (see, e.g., book [16]), the integral in Eq. (E.22) can be approximated as follows:

$$\begin{aligned} I &\approx 2\left[\int_{-\infty}^{\infty} du_1 f(u_1, 0)\right] \int_0^{\infty} du_2 S(\lambda u_2) \\ &= (2/\lambda)\left[\int_{-\infty}^{\infty} du_1 f(u_1, 0)\right] \int_0^{\infty} dw S(w). \end{aligned} \tag{E.25}$$

Here,

$$f(u_1, 0) = 3/[4(1 + u_1^2)^3], \quad S(w) = j_0(w) + j_1(w) + 2j_0(w). \tag{E.26}$$

Performing analytically the integrations in Eq. (E.25), we obtain:

$$I \approx 3Z/Y. \tag{E.27}$$

Thus, the integral controlling the width

$$a_- = \int_0^{\infty} dZ A_-(0, Y, Z)/Z \tag{E.28}$$

does not diverge at small (scaled) impact parameters Z — in distinction to the attempt by Rosato *et al.* [14] to simulate the corresponding GT result. Since their simulations yielded an absurd result, (the divergence of the integral in Eq. (B.28) at small impact parameters) contradicting to the rigorous analytical calculations, it seems that their simulations of anything cannot be trusted.

E.2 Approach Based on Rectilinear Trajectories of the Perturbers: Lorentz–Doppler Profiles of Highly-excited Hydrogen Lines

In magnetized plasmas, radiating hydrogen atoms moving with the velocity **v** across the magnetic field **B** experience a Lorentz electric field $\mathbf{E}_\mathrm{L} = \mathbf{v}\mathrm{x}\mathbf{B}/\mathrm{c}$ in addition to other electric fields. In plasmas, the atomic velocity $\mathbf{v}$ has a distribution — therefore, so does the Lorentz field, thus making an additional contribution to the *broadening* of spectral lines. In 1970, Galushkin [17] considered a joint effect of Lorentz and Doppler broadenings (including also Zeeman splitting) and produced many interesting analytical results. He noted, in particular, that Lorentz and Doppler broadenings cannot be accounted for via a convolution, but rather they intertwine in a more complicated manner.

In the intervening decades, other authors also contributed to the subject [3–7]. However, the scope of the papers [17–22] was limited to some spectral lines originating from levels of the principal quantum number $n = 2$, 3, or 4, namely Ly_α, H_α and H_β lines of hydrogen in Refs. [17–19, 21, 22] and Ly_α of C VI in Ref. [20].

Here, we present analytical results for the Lorentz–Doppler broadening of *highly-excited hydrogen/deuterium lines* obtained in papers [23, 24]. We consider weakly coupled (a.k.a. ideal) plasmas, i.e., plasmas where there is a large number of ions ν in the sphere of the Debye radius:

$$\nu = [T_e^3/(36\pi N_e \mathrm{e}^6)]^{1/2} = 1377[T_e^3/N_e \mathrm{e}^6]^{1/2} \gg 1. \tag{E.29}$$

Here, T_e is the electron temperature in Kelvin and N_e is the electron density in cm^{-3}. In these plasmas, the ion microfield can be described by the Holtsmark distribution [25] characterized by the so-called Holtsmark field

$$E_0 = 2.603 e N_e^{2/3} = 1.25 \times 10^{-9} N_e^{2/3}. \tag{E.30}$$

In this case, the most probable ion field is,

$$E_{i\max} = 1.608 E_0. \tag{E.31}$$

The average Lorentz field,

$$E_\mathrm{LT} = B v_T/c = 4.28 \times 10^{-3} B[T(K)]^{1/2} \tag{E.32}$$

($v_T = (2T/M)^{1/2}$ is the atomic thermal velocity) can exceed the most probable ion microfield E_i when the magnetic field B exceeds the following critical value:

$$B_c = 4.69 \times 10^{-7} N_e^{2/3}/[T(K)]^{1/2}, \tag{E.33}$$

(in Eqs. (E.32) and (E.33), B is in Tesla). For example, in the solar chromosphere, the typical plasma parameters are $N_e \sim 10^{11}\,\text{cm}^{-3}$ and $T \sim 10^4\,\text{K}$ (except solar flares where N_e can be higher by two orders of magnitude) — see, e.g., [26,27]. In this case, from Eq. (E.33) we get $B_c = 0.2$ T. A more accurate estimate for this example can be obtained by taking into account that non-thermal velocities v_{nonth} in the solar chromosphere can be approximately several tens of km/s, so that the total velocity $v_{\text{tot}} = (v_T^2 + v_{\text{nonth}}^2)^{1/2} \sim (15\text{–}30)$ km/s. Then, $E_L = E_{i\max}$ already at $B \sim 0.05$ Tesla, while B can reach 0.4 T in sunspots.

Another example is edge plasmas of tokamaks. For a low-density discharge in Alcator C-Mod [28], where $N_e \sim 3 \times 10^{13}\,\text{cm}^{-3}$ and $T \sim 5 \times 10^4\,\text{K}$, we get $B_c = 4$ T, while the actual magnetic field was 8 T.

For completeness, we analyze the ratio of the Zeeman width of hydrogen lines $\Delta\omega_Z$ to the corresponding "halfwidth" of the n-multiplet due to the Lorentz broadening $\Delta\omega_L$:

$$\Delta\omega_Z = eB/(2m_e c), \quad \Delta\omega_L = 3n(n-1)\hbar B v_T/(2m_e ec),$$
$$\Delta\omega_Z/\Delta\omega_L = 5680/[n(n-1)T^{1/2}], \tag{E.34}$$

where n is the principal quantum number of the upper level involved in the radiation transition and the atomic temperature T is in Kelvin. For example, for the typical temperature at the edge plasmas of tokamaks $T \sim 5 \times 10^4\,\text{K}$, Eq. (E.34) yields $\Delta\omega_Z/\Delta\omega_L = 25.4/[n(n-1)]$: so that the Lorentz width exceeds the Zeeman width for hydrogen lines of $n > 5$, while Balmer lines up to $n = 16$ were observed, e.g., at Alcator C-Mod [28]. Another example: for the typical temperature in the solar chromosphere $T \sim 10^4\,\text{K}$, Eq. (6) yields $\Delta\omega_Z/\Delta\omega_L = 56.8/[n(n-1)]$. So, the Lorentz width exceeds the Zeeman width for hydrogen lines of $n > 8$, while Balmer lines up to $n \sim 30$ were observed [26, 27].

Thus, we encounter the situations where the primary broadening mechanism of hydrogen lines is the Lorentz broadening. The Lorentz broadening intertwines with the Doppler broadening, as mentioned above. Indeed, the most prominent feature of the combined Lorentz–Doppler broadening is that it does not reduce to the convolution of Lorentz and Doppler broadenings, as first noted by Galushkin [17]. These two broadening mechanisms entangle in a more complicated way because in the laboratory reference frame, the Lorentz–Doppler profile of a Stark component of a hydrogen line is proportional to the following δ-function (in the frequency scale) $\delta[\Delta\omega - (\omega_0 v/c)\cos\alpha - (kX_{\alpha\beta}\mathrm{B}v/c)\sin\vartheta]$, where α is the angle between the direction of observation and the atomic velocity $\mathbf{v}$, and ϑ is the angle between vectors $\mathbf{v}$ and $\mathbf{B}$.

For obtaining the results in the most universal way, we use the following dimensionless notations,

$$w = c\Delta\omega/(v_T\omega_0) = c\Delta\lambda/(v_T\lambda_0), \quad \mathbf{b} = kX_{\alpha\beta}\mathbf{B}/\omega_0, \quad \mathbf{u} = \mathbf{v}/v_T, \tag{E.35}$$

where w is the scaled detuning from the unperturbed frequency ω_0 or from the unperturbed wavelength λ_0 of a hydrogen spectral line, $\mathbf{b}$ is the scaled magnetic field, and $\mathbf{u}$ is the scaled atomic velocity. The quantities k and $X_{\alpha\beta}$ in Eq. (E.35) are as follows

$$k = 3\hbar/(2m_e e), \quad X_{\alpha\beta} = n_\alpha(n_1 - n_2)_\alpha - n_\beta(n_1 - n_2)_\beta, \tag{E.36}$$

where n_1, n_2 are the parabolic quantum numbers, and n is the principal quantum numbers of the upper (subscript α) and lower (subscript β) Stark sublevels involved in the radiative transition.

We start from the situation where the direction of the observation is along $\mathbf{B}$. In this case, the Lorentz–Doppler profile of a Stark component of a hydrogen line can be expressed as follows:

$$I_{\mathrm{par}}(w, b) = \int_0^\infty du f(u) \int_0^1 \mathrm{d}(\cos\vartheta)\delta(w - u\cos\vartheta - b\sin\vartheta), \tag{E.37}$$

where the velocity distribution $f(u)$ is usually taken as the Maxwell distribution

$$f(u) = (4u^2/\pi^{1/2})\exp(-u^2). \tag{E.38}$$

For brevity, we denote $\cos\vartheta = x$. The root x_0 of the argument of the δ-function in Eq. (9) is

$$x_0(u,b) = [u - b(1 + b^2 - u^2)^{1/2}]/(1 + b^2). \tag{E.39}$$

Using the properties of the δ-function, Eq. (E.37) can be reduced to

$$I_{\text{par}}(w,b) = (2/\pi^{1/2})\int_{ymin}^{\infty} \times dyy\exp(-y^2)/\{1 - bx_0(w/y,b)/[1 - x_0^2(w/y,b)]^{1/2}\}, \tag{E.40}$$

where y_{min} is given as follows. For $b > 0$:

$$y_{\text{min}} = w/(1 + b^2)^{1/2} \quad \text{for } w > 0, \quad y_{\text{min}} = -w \quad \text{for } w < 0. \tag{E.41}$$

For $b < 0$:

$$y_{\text{min}} = w/(1 + b^2)^{1/2} \quad \text{for } w < 0, \quad y_{\text{min}} = -w \quad \text{for } w > 0. \tag{E.42}$$

Figures E.5 and E.6 show the Lorentz–Doppler profile from Eq. (E.40) for the scaled magnetic field $b = 0.2$ and $b = 1$, respectively. For comparison, the corresponding pure Doppler profile is also shown in these figures. Figure E.7 shows the Lorentz–Doppler profile from Eq. (E.40) for a larger scaled magnetic field $b = 10$.

It is seen that as the magnetic field increases, the maximum of the profile shifts further away from the unperturbed frequency of the spectral line and the profile becomes more and more asymmetric. However, to avoid any confusion, we emphasized that the profiles in Figs. E.4–E.7 are for a Stark component shifted to the blue side, i.e., for a Stark component, for which $X_{\alpha\beta} > 0$ in Eq. (E.36) so that $b > 0$. For a pair of Stark components, corresponding

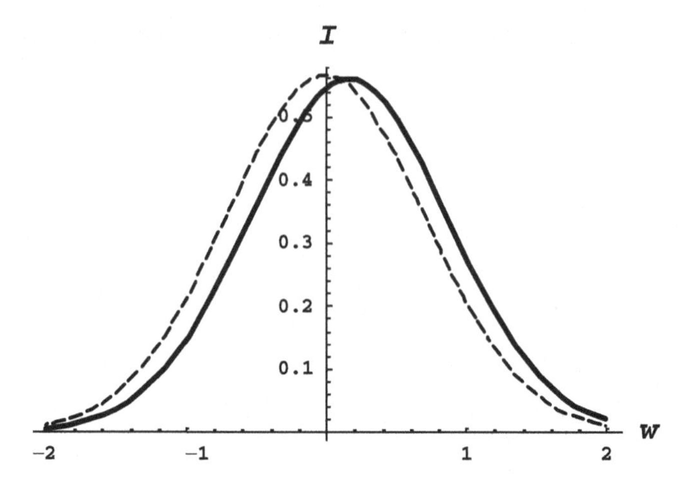

Fig. E.5. The Lorentz–Doppler profile from Eq. (E.40) for the scaled magnetic field $b = 0.2$ (solid line). For comparison, the corresponding pure Doppler profile is also shown (dashed line). The direction of observation is parallel to the magnetic field $\mathbf{B}$.

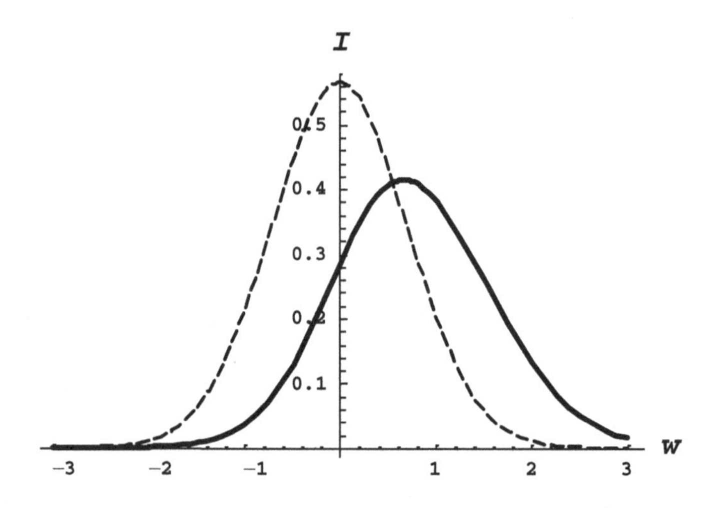

Fig. E.6. Same as in Fig. E.5, but for the scaled magnetic field $b = 1$.

to two equal by magnitude and opposite by sign values of $X_{\alpha\beta}$ (and consequently of b), the resulting profile $I(w, b) = I_{\text{par}}(w, b) + I_{\text{par}}(w, -b)]$ is symmetric — see Fig. E.8 combining the cases of $b = 10$ and $b = -10$.

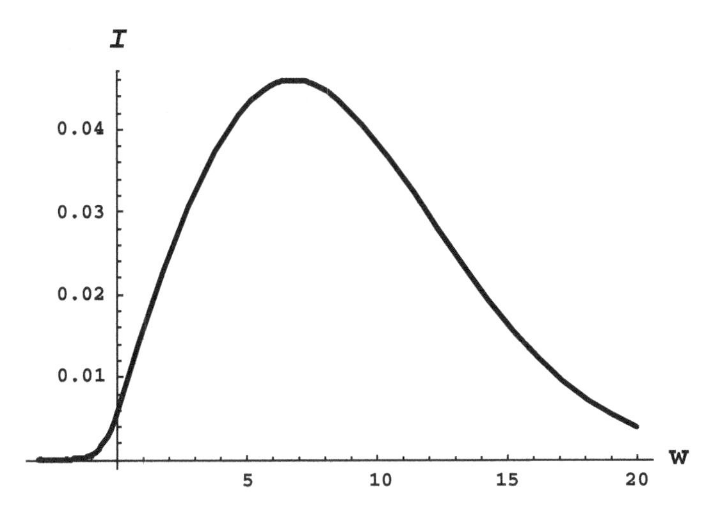

Fig. E.7. The Lorentz–Doppler profile from Eq. (40) for the scaled magnetic field $b = 10$. The direction of observation is parallel to the magnetic field $\mathbf{B}$.

Based on the universal function $I_{\text{par}}(w, b)$ from Eq. (E.40), the profiles $S(\Delta\omega)$ of hydrogen lines in the frequency scale can be calculated as follows (for the observation along $\mathbf{B}$):

$$S(\Delta\omega) = S_\pi(\Delta\omega) + S_\sigma(\Delta\omega), \tag{E.43}$$

$$S_\pi(\Delta\omega) = [c/(v_T\omega_0)] \sum_{(\alpha,\beta)\pi} J_{\alpha\beta} f_\pi I_{\text{par}}[c\Delta\omega/(v_T\omega_0)], \tag{E.44}$$

$$S_\sigma(\Delta\omega) = [c/(v_T\omega_0)] \sum_{(\alpha,\beta)\sigma} J_{\alpha\beta} f_\sigma I_{\text{par}}[c\Delta\omega/(v_T\omega_0)], \tag{E.45}$$

where $f_\pi = 1$, $f_\sigma = 1/2$; $J_{\alpha\beta}$ is the relative intensity of the corresponding Stark component. In Eq. (E.44), the summation is performed over π-components, in Eq. (E.45) — over σ-components.[2]

---

[2]For hydrogen lines having the component of $X_{\alpha\beta} = 0$ (which is usually the σ-component), the sum in (E.45) should be complemented by $J_0\delta(\Delta\omega)$, where J_0 is the relative intensity of the component. Physically, this means that the contribution of this component to the overall profile is controlled by other broadening mechanisms (typically, by the Doppler broadening). However, first, for 50% of hydrogen lines, such as those with even values of $(n_\alpha - n_\beta)$, there is no component of $X_{\alpha\beta} = 0$. Second, for high-n hydrogen lines ($n_\alpha \gg n_\beta$) with odd

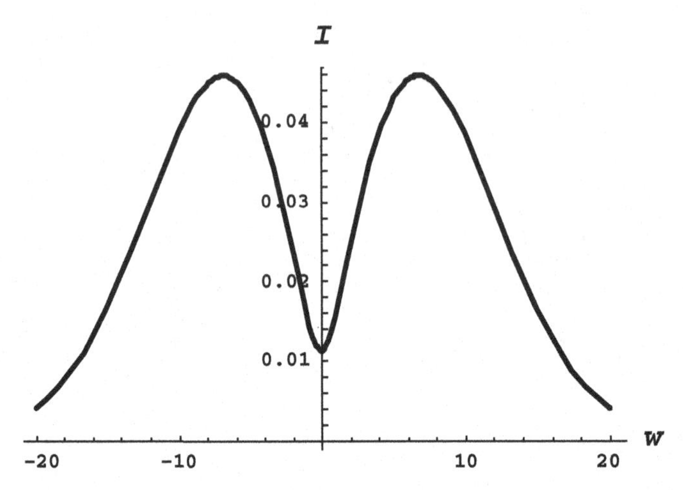

Fig. E.8. The resulting Lorentz–Doppler profile for a pair of Stark components, corresponding to two equal by magnitude and opposite by sign values of $X_{\alpha\beta}$ (in Eq. (E.36)). Specifically, it is the combined profile corresponding to the cases of $b = 10$ and $b = -10$.

For the case of a non-zero angle ψ between the direction of observation and the magnetic field, the relative configuration of the vectors $\mathbf{B}$, $\mathbf{E}_L$ and $\mathbf{v}$, as well as the choice of the reference frame is shown in Fig. E.9.

If we would disregard for a moment the angular dependence of the intensities of π- and σ-components of hydrogen lines, then the Lorentz–Doppler profile of a Stark component can be expressed as follows

$$I(w, b, \psi) = \int_0^\infty du_z f_z(u_z) \int_0^\infty du_R f_R(u_R) \times \int_0^\pi (d\varphi/\pi)\delta[w - u_z \cos\psi - u_R(b + \sin\psi\cos\varphi)]. \tag{E.46}$$

Here,

$$f_z(u_z) = \pi^{-1/2}\exp(-u_z^2) \tag{E.47}$$

values of $(n_\alpha - n_\beta)$, the contribution of this component to the overall profile is relatively small and can be neglected.

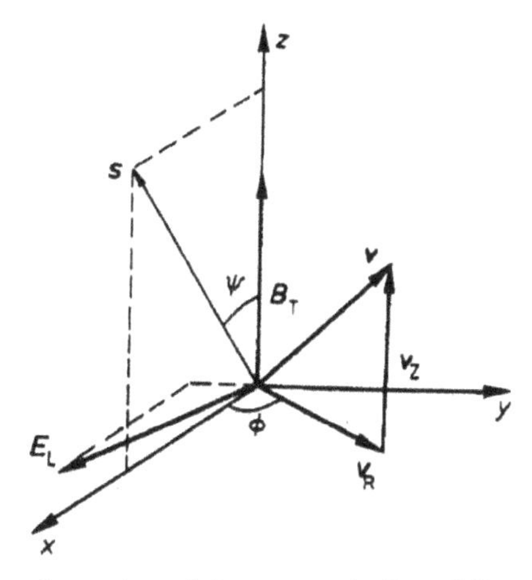

Fig. E.9. Relative configuration of the magnetic **B** and Lorentz $\mathbf{E}_L$ fields and of the direction of the observation **s** ("s" stands for "spectrometer"). The z axis is along **B**. The direction of the observation **s** constitutes a non-zero angle ψ with **B**. The xz plane is spanned on vectors **B** and **s**. The atomic velocity **v** has a component v_z along **B** and a component $\mathbf{v}_R$ perpendicular to **B**. The component $\mathbf{v}_R$ constitutes an angle φ with the x axis.

is the 1-D Maxwell (i.e., Bolzmann) distribution of the scaled component $u_z = v_z/v_T$ of the atomic velocity parallel to **B**, and

$$f_R(u_R) = 2u_R \exp(-u_R^2) \tag{E.48}$$

is the 2D Maxwell distribution of the scaled component $u_R = v_R/v_T$ of the atomic velocity perpendicular to **B**. In Eq. (E.46) we corrected some typographic errors compared to the corresponding Eq. (18) from [24].

For allowing for the angular dependence of the intensities of π- and σ-components, the corresponding angular factors (as functions of ψ) should be entered in Eq. (E.46). For example, for the simplest case, where only the component of the radiation polarized along **B** is observed, this factor is $\sin^2\psi$. Of course, such component of the radiation is only one part of the corresponding σ-Stark-component of the hydrogen line.

For obtaining more specific results, we consider below the case of the observation perpendicular to **B**, i.e., the case of $\psi = \pi/2$. In this case, Eq. (E.46) reduces to:

$$I_{\text{per}}(w,b) = \int_0^\infty du_R f_R(u_R) \int_0^\pi (d\varphi/\pi)\delta[w - u_R(b + \cos\varphi)]. \quad \text{(E.49)}$$

For the subcase of $|b| < 1$, using the properties of the δ-function, Eq. (E.49) can be simplified to:

$$I_{\text{per}}(w,b) = [\pi(1-b)^{1/2}]^{-1} \int_{u_{\text{Rmin}}}^\infty du_R 2u_R \times \exp(-u_R^2)/\{[u_R - w/(1+b)][u_R + w/(1-b)]\}^{1/2}. \quad \text{(E.50)}$$

For $0 < b < 1$:

$$u_{\text{Rmin}} = w/(1+b) \quad \text{for } w > 0, \quad u_{\text{Rmin}} = -w/(1-b) \quad \text{for } w < 0. \quad \text{(E.51)}$$

For $-1 < b < 0$:

$$u_{\text{Rmin}} = -w/(1-b) \quad \text{for } w > 0, \quad u_{\text{Rmin}} = w/(1+b) \quad \text{for } w > 0. \quad \text{(E.52)}$$

We note that an expression equivalent to our Eq. (E.50) was previously obtained by Breton *et al.* [19]. However, in Ref. [19] the lower limit of the integration was set (in our notations) to $u_{\text{Rmin}} = w/(1+b)$ regardless of the sign of b and of the sign of the scaled detuning w from the unperturbed position of the spectral line. Therefore, their calculations of the profiles of the H_α and H_β lines seem to contain errors.

Equation (E.50) yields the spectral line profile for the simplest case, where only the component of the radiation polarized along **B** is observed. Such component of the radiation is only one part of the corresponding σ-Stark-component of the hydrogen line, as mentioned above. For allowing for the angular dependence of the intensities of π- and σ-components, the following factors should be entered in Eq. (E.50) into the integral over du_R:

$$(b - w/u_R)^2 \text{ for } \pi - \text{components},$$
$$[(1 - b^2/2)u_R^2 + bwu_R - w^2/2]/u_R^2 \text{ for } \sigma - \text{components}. \quad \text{(E.53)}$$

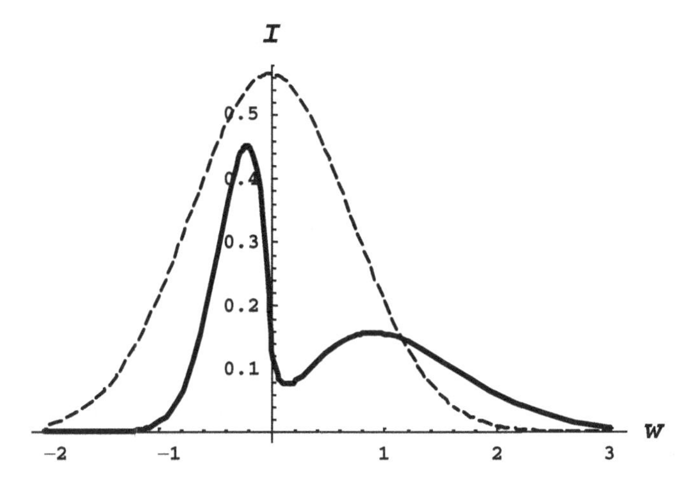

Fig. E.10. The Lorentz–Doppler profile of a π-component of a hydrogen line according to Eqs. (E.50), (E.53) for the scaled magnetic field $b = 0.5$ (solid line). For comparison, the corresponding pure Doppler profile is also shown (dashed line). The direction of observation is perpendicular to the magnetic field **B**.

Figure E.10 shows the Lorentz–Doppler profile of a π-component of a hydrogen line calculated according to Eqs. (50), (53) for the scaled magnetic field $b = 0.5$ (solid line). For comparison, the corresponding pure Doppler profile is also shown (dashed line). Figure E.11 shows the same, but for a σ-component.

Once again, to avoid any confusion, we emphasized that the profiles in Figs. E.10 and E.11 are for a Stark component shifted to the blue side, i.e., for a Stark component, for which $X_{\alpha\beta} > 0$ in Eq. (E.36) so that $b > 0$. For a pair of Stark components, corresponding to two equal by magnitude and opposite by sign values of $X_{\alpha\beta}$ (and consequently of b), the resulting profile $I(w, b) = [I_{\mathrm{per}}(w, b) + I_{\mathrm{per}}(w, -b)]$ is symmetric — see Fig. E.12 combining the cases of $b = 0.5$ and $b = -0.5$ for both π- and σ-components.

Now we proceed to the subcase of $|b| > 1$. Using the properties of the δ-function, Eq. (E.49) can be simplified as follows.

For $b > 1$ and $w > 0$:

$$I_{\mathrm{per}}(w, b) = [\pi(1 - b)^{1/2}]^{-1} \int_{u_{\mathrm{Rmin}}}^{u_{\mathrm{Rmax}}} du_R 2u_R \times \exp(-u_R^2)/\{[w/(b+1) - u_R][u_R - w/(b-1)]\}^{1/2}, \tag{E.54}$$

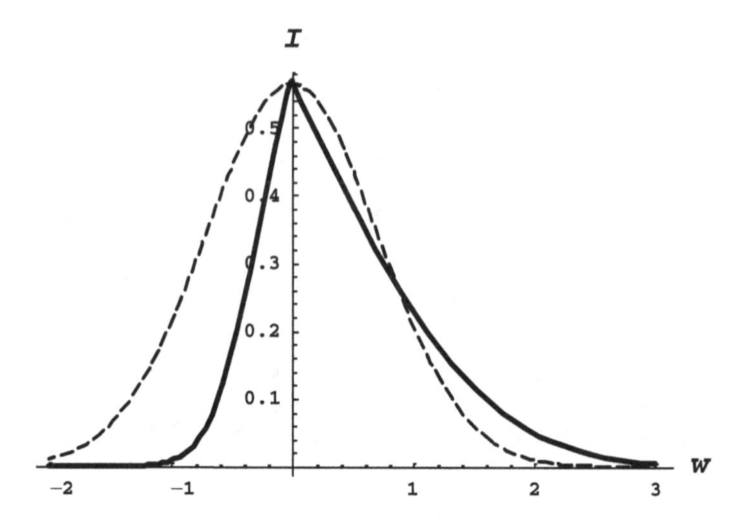

Fig. E.11. Same as in Fig. E.10, but for a σ-component.

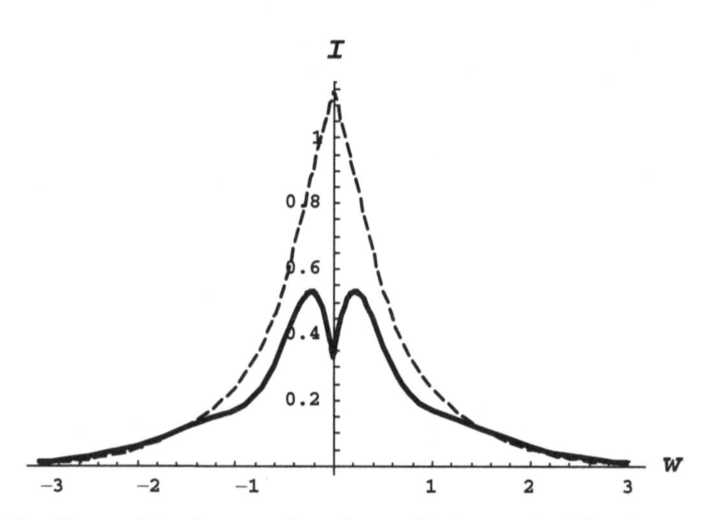

Fig. E.12. The resulting Lorentz–Doppler profile for a pair of Stark components, corresponding to two equal by magnitude and opposite by sign values of $X_{\alpha\beta}$ (in Eq. (E.36)). Specifically, it is the combined profile corresponding to the cases of $b = 0.5$ and $b = -0.5$ for the observation perpendicular to the magnetic field **B**. Solid line: the pair of π-components. Dashed line: the pair of σ-components.

where

$$u_{\text{Rmin}} = w/(1+b), \quad u_{\text{Rmax}} = w/(b-1). \tag{E.55}$$

For $b > 1$ and $w < 0$, or for $b < -1$ and $w > 0$:

$$I_{\text{per}}(w,b) = 0. \tag{E.56}$$

For $b < -1$ and $w < 0$:

$$I_{\text{per}}(w,b) = [\pi(1-b)^{1/2}]^{-1} \int_{u_{\text{Rmin}}}^{u_{\text{Rmax}}} du_R 2u_R \times \exp(-u_R^2)/\{[w/(b+1) - u_R][u_R - w/(b-1)]\}^{1/2}, \tag{E.57}$$

where

$$u_{\text{Rmin}} = w/(b-1), \quad u_{\text{Rmax}} = w/(1+b). \tag{E.58}$$

Similarly to the subcase of $|b| < 1$, in the subcase of $|b| > 1$, Eqs. (E.54), (E.57) yield the spectral line profile for the simplest situation, where only the component of the radiation polarized along **B** is observed. For allowing for the angular dependence of the intensities of π- and σ-components, the corresponding factors given by Eq. (E.53) should be entered in Eqs. (E.54), (E.57) into the integral over du_R.

Figure E.13 shows the Lorentz–Doppler profile of a π-component (solid line) and of a σ-component (dashed line) of a hydrogen line calculated according to Eqs. (E.54), (E.57) for the scaled magnetic field $b = 3$. We emphasize that, while Fig. E.13 shows the Lorentz–Doppler profile of a π-component and of a σ-component, corresponding to $X_{\alpha\beta} > 0$ (in Eq. (E.36)), only for positive values of the scaled detuning w from the unperturbed position of a hydrogen line, at negative values of w the Lorentz–Doppler profile is identically zero — in distinction to the case of $|b| < 1$, illustrated in Figs. E.10 and E.11.

For a pair of Stark components, corresponding to two equal by magnitude and opposite by sign values of $X_{\alpha\beta}$ (and consequently

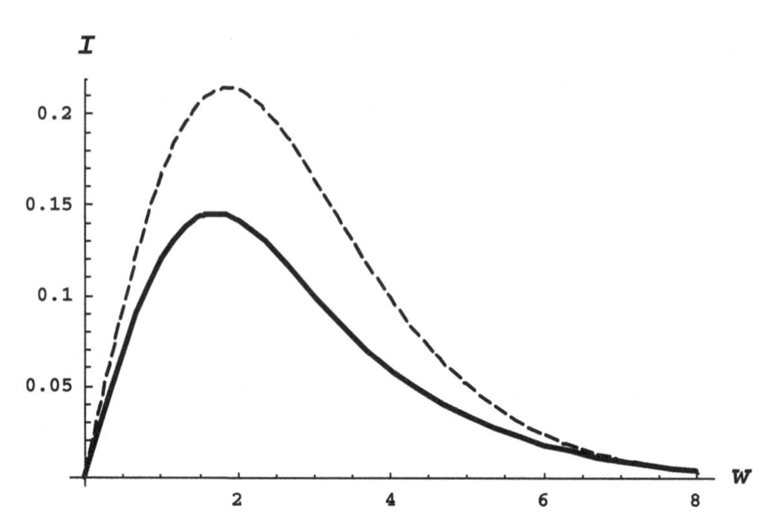

Fig. E.13. The Lorentz–Doppler profile of a π-component (solid line) and of a σ-component (dashed line) of a hydrogen line calculated according to Eqs. (26), (29) for the scaled magnetic field $b = 3$. The direction of observation is perpendicular to the magnetic field $\mathbf{B}$.

of b), the resulting profile $I(w, b) = [I_{\text{per}}(w, b) + I_{\text{per}}(w, -b)]$ is symmetric — see Fig. E.14 combining the cases of $b = 3$ and $b = -3$ for both π- and σ-components.

Finally, we consider the remaining subcase of $|b| = 1$. Using the properties of the δ-function, Eq. (E.49) can be simplified as follows.

For $b = 1$ and $w > 0$, or for $b = -1$ and $w < 0$:

$$I_{\text{per}}(w) = \pi^{-1}|2w|^{-1/2}\int_{|w|/2}^{\infty} du_R 2u_R \exp(-u_R^2)/(u_R - |w|/2)^{1/2}. \tag{E.59}$$

The integral in Eq. (E.59) can be calculated analytically, so that Eq. (E.59) takes the form

$$\begin{aligned} I_{\text{per}}(w) = {} & [\Gamma(1/4)\Gamma(-1/4)]^{-1}|w|^{-1/2}[\Gamma(-1/4)F(3/4, 1/2; -w^2/4) \\ & - \Gamma(1/4)|w|F(5/4, 3/2; -w^2/4)], \end{aligned} \tag{E.60}$$

where $\Gamma(z)$ is the gamma-function, $F(\alpha, \gamma; z)$ is the confluent hypergeometric function.

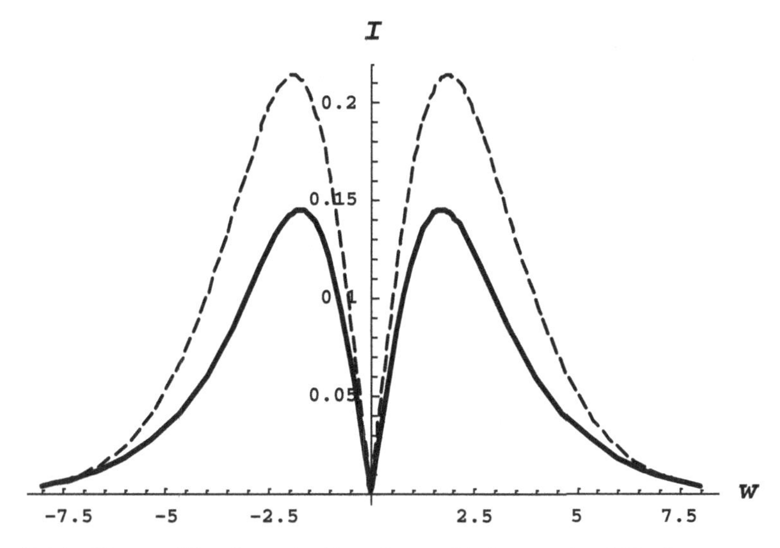

Fig. E.14. The resulting Lorentz–Doppler profile for a pair of Stark components, corresponding to two equal by magnitude and opposite by sign values of $X_{\alpha\beta}$ (in Eq. (E.36)). Specifically, it is the combined profile corresponding to the cases of $b = 3$ and $b = -3$ for the observation perpendicular to the magnetic field **B**. Solid line: the pair of π-components. Dashed line: the pair of σ-components.

For $b = 1$ and $w < 0$, or $b = -1$ and $w > 0$:

$$I_{\text{per}}(w) = 0. \tag{E.61}$$

Similarly to the subcases of $|b| < 1$ and $|b| > 1$, in the subcase of $|b| = 1$, Eqs. (E.59) yields the spectral line profile for the simplest situation, where only the component of the radiation polarized along **B** is observed. For allowing the angular dependence of the intensities of π- and σ-components, the corresponding factors are given by Eq. (E.53), with the substitution of $b = 1$ or $b = -1$, should be entered in Eqs. (E.59) into the integral over du_R.

Figure E.15 shows the Lorentz–Doppler profile of a π-component (solid line) and of a σ-component (dashed line) of a hydrogen line calculated according to Eq. (E.60) for the scaled magnetic field $b = 1$. Just like in the case of $|b| > 1$, while Fig. 11 shows the Lorentz–Doppler profile of a π-component and of a σ-component, corresponding to $X_{\alpha\beta} > 0$ (in Eq. (E.36)), only for positive values of

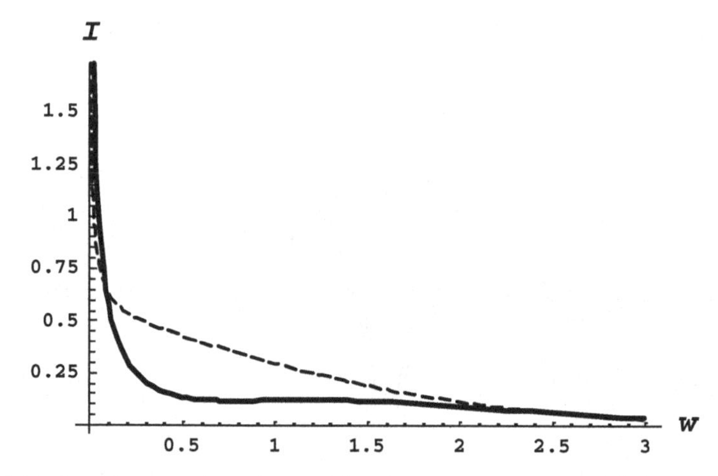

Fig. E.15. The Lorentz–Doppler profile of a π-component (solid line) and of a σ-component (dashed line) of a hydrogen line calculated according to Eq. (E.60) for the scaled magnetic field $b = 1$ for the observation perpendicular to the magnetic field **B**.

the scaled detuning w from the unperturbed position of a hydrogen line, at negative values of w the Lorentz–Doppler profile is identically zero — in distinction to the case of $|b| < 1$, illustrated in Figs. E.10 and E.11.

For a pair of Stark components corresponding to two equal by magnitude and opposite by sign values of $X_{\alpha\beta}$ (and consequently of b), the resulting profile $I(w, b) = [I_{\text{per}}(w, b) + I_{\text{per}}(w, -b)]$ is symmetric — see Fig. E.16 combining the cases of $b = 1$ and $b = -1$ for both π- and σ-components.

Based on the universal functions $I_{\text{per}}(w, b)$ from Eqs. (E.50), (E.54), (E.57), (E.60) and using Eq. (E.53) for the additional factors for π- and σ-components, the profiles $S(\Delta\omega)$ of hydrogen lines in the frequency scale can be calculated by formulas analogous to Eqs. (E.43)–(E.45).

As an example, Fig. E.17 presents calculated Lorentz–Doppler profile of the Balmer line H_{18} for the observations parallel to **B** for $B = 4$ Tesla (solid line). For comparison, the corresponding pure Doppler profile is also shown (dashed line). We remind that

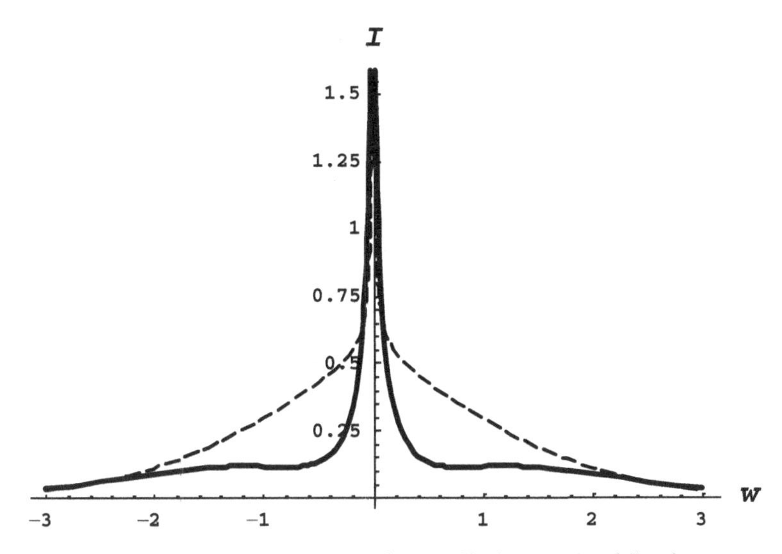

Fig. E.16. The resulting Lorentz–Doppler profile for a pair of Stark components corresponding to two equal by magnitude and opposite by sign values of $X_{\alpha\beta}$ (in Eq. (E.36)). Specifically, it is the combined profile corresponding to the cases of $b = 1$ and $b = -1$ for the observation perpendicular to the magnetic field **B**. Solid line: the pair of π-components. Dashed line: the pair of σ-components.

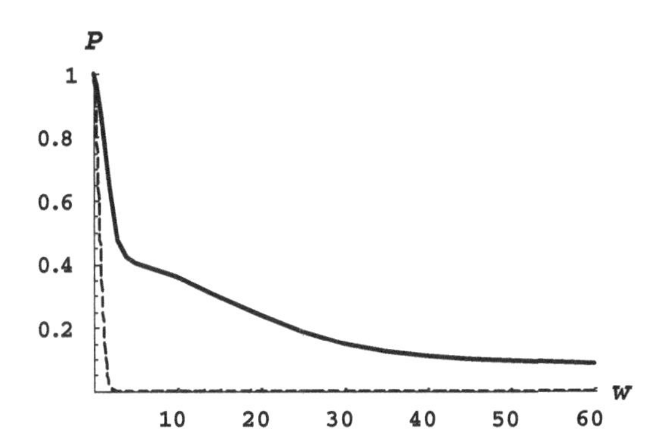

Fig. E.17. Calculated Lorentz–Doppler profile $P(w)$ of the Balmer line H_{18} for the observations parallel to **B** for $B = 4$ Tesla (solid line). For comparison, the corresponding pure Doppler profile is also shown (dashed line). The argument $w = c\Delta\omega/(v_T\omega_0) = c\Delta\lambda/(v_T\lambda_0)$ is the scaled detuning from the unperturbed frequency ω_0 or from the unperturbed wavelength λ_0. The intensity P is relative to the intensity of the maximum.

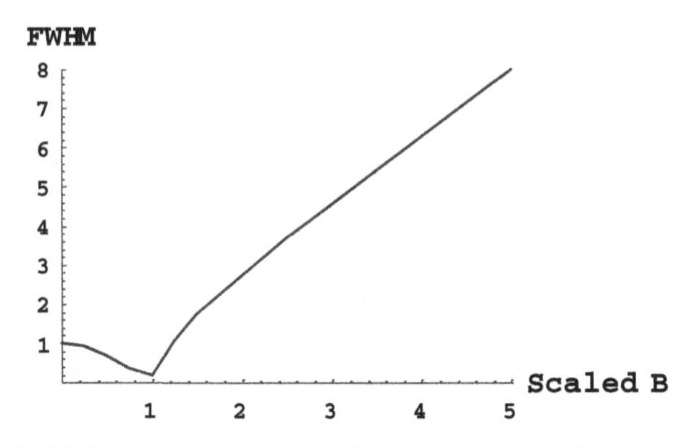

Fig. E.18. The FWHM of the hydrogen/deuterium Ly-beta line observed perpendicular to the magnetic field **B** with a polarizer along **B**. The scaled magnetic field is the ratio of the Lorentz-field shift to the Doppler shift. The FWHM is in units of the Doppler HWHM. The narrowing effect is the most pronounced when the Lorentz-field shift is equal to the Doppler shift.

the argument w defined in Eq. (E.35) is the scaled detuning from the unperturbed frequency ω_0 or from the unperturbed wavelength λ_0.

The most interesting results are the following. The halfwidth of the Lorentz–Doppler profiles is a non-monotonic function of the scaled magnetic field b for observations perpendicular to **B**. As $|b|$ increases from zero, the halfwidth first decreases, then reaches a minimum at $|b| = 1$ (i.e., when the shift in the Lorentz field is equal to the Doppler shift), and then increases — as presented in Fig. E.18 using the Ly-beta line as an example. This is a counterintuitive result.[3]

For completeness, we note that in the far wings, the essentially-exponential decline of the Lorentz–Doppler profiles would switch to a power-law decline caused by the ion and electron microfields. This would happen at the critical detuning $\Delta\omega_{cr} \sim \Delta\omega_{LT}\,[\ln(E_{LT}/E_0)]^{1/2}$, where $E_{\rm LT} \gg E_0$ are defined by Eqs. (E.32) and (E.30), respectively.

[3]In paper [24], where the above results were presented, there were some printing errors in Eqs. (31), (32) and Figs. 9, 10. These printing errors have been corrected here in the corresponding Eqs. (E.27), (E.28) and Figs. E.13, E.14.

Further, if $\Delta\omega_{\rm cr} > T_e/(n^2\hbar)$, then in these far wings, the profile would be proportional to $1/\Delta\omega^{5/2}$ because not only the ion microfield, but also the electron microfield would cause the quasistatic broadening described by the same power law$1/\Delta\omega^{5/2}$.

Finally, let us consider the situation where the Lorentz broadening can predominate over other broadening mechanisms for highly-excited hydrogen lines. Previously in this section, we showed that the Lorentz broadening can significantly exceed both the SB by the plasma microfield and the Zeeman splitting for high-n hydrogen lines. Now let us estimate the ratio of the Lorentz and Doppler broadenings.

The Doppler Full Width and Half Maximum (FWHM) is

$$(\Delta\omega_{\rm D})_{1/2} = 2(\ln 2)^{1/2}\omega_0 v_T/c = 1.665\omega_0 v_T/c, \tag{E.62}$$

where ω_0 is the unperturbed frequency of the spectral line. For highly-excited hydrogen lines, where $n_\alpha \gg n_\beta$, one can use the expression $\omega_0 = m_e e^4/(2n_\beta^2\hbar^3)$.

The Lorentz field $\mathbf{E}_L$ is confined in the plane perpendicular to $\mathbf{B}$, where it has the following distribution:

$$W_L(E_L)dE_L = (2E_L/E_{\rm LT}^2)\exp(-E_L^2/E_{\rm LT}^2)dE_L, E_{\rm LT} = v_T B/c. \tag{E.63}$$

Here, $E_{\rm LT}$ is the average Lorentz field expressed via the thermal velocity v_T of the radiating atoms of mass M. The distribution W_L actually reproduces the shape of the two-dimensional Maxwell distribution of atomic velocities in the plane perpendicular to $\mathbf{B}$. This is because the absolute value of the Lorentz field $\mathbf{E}_L = \mathbf{v}x\mathbf{B}/$c is $E_L = v_R B/c$, where v_R is the component of the atomic velocity perpendicular to $\mathbf{B}$.

The Lorentz-broadened profile of a Stark component of a hydrogen line reproduces the shape of the Lorentz field distribution from Eq. (E.41)

$$S_{\alpha\beta}(\Delta\omega) = (2\Delta\omega/\Delta\omega_{L\alpha\beta}^2)\exp(-\Delta\omega^2/\Delta\omega_{L\alpha\beta}^2),$$
$$\Delta\omega_{L\alpha\beta} = kX_{\alpha\beta}Bv_T/c. \tag{E.64}$$

Its FWHM is

$$(\Delta\omega_{L\alpha\beta})_{1/2} = 2.715\Delta\omega_{L\alpha\beta}. \tag{E.65}$$

The corresponding FWHM $(\Delta\omega_L)_{1/2}$ of the entire hydrogen line can be estimated by using in Eqs. (E.64), (E.65) the average value $\langle k|X_{\alpha\beta}|\rangle = (n_\alpha^2 - n_\beta^2)\hbar/(m_e e)$, which for $n_\alpha \gg n_\beta$ becomes $\langle k|X_{\alpha\beta}|\rangle = n_\alpha^2\hbar/(m_e e)$. Therefore, for the ratio of the FWHM by these two broadening mechanisms, we get:

$$(\Delta\omega_L)_{1/2}/(\Delta\omega_{\mathrm{D}})_{1/2} = n_\alpha^2 n_\beta^2 B(\mathrm{Tesla})/526. \tag{E.66}$$

We note that this ratio does not depend on the temperature.

For Balmer lines ($n_\beta = 2$), Eq. (E.66) becomes

$$(\Delta\omega_L)_{1/2}/(\Delta\omega_{\mathrm{D}})_{1/2} = n_\alpha^2 B(\mathrm{Tesla})/131. \tag{E.67}$$

So, e.g., for the edge plasmas of tokamaks, where Balmer lines of $n_\alpha \sim (10-16)$ have been observed, the Lorentz broadening dominates over the Doppler broadening when the magnetic field exceeds the critical value $B_c \sim 1$ Tesla. This condition is fulfilled in the modern tokamaks and will be fulfilled also in the future tokamaks.

Another example: in solar chromosphere, where Balmer lines of $n_\alpha \sim (25-30)$ have been observed, the Lorentz broadening dominates over the Doppler broadening when the magnetic field exceeds the critical value $B_c \sim (0.15 - 0.2)$ Tesla. This condition can be fulfilled in sunspots where B can be as high as 0.4 Tesla.

Therefore, it is practically useful to calculate pure Lorentz-broadened profiles of highly-excited Balmer lines. To ensure that for each particular hydrogen line, the profile will be *universal*, i.e., applicable for any B and T, we chose the argument γ of the profiles as

$$\gamma = c\Delta\omega/(kBv_T). \tag{E.68}$$

We calculated the corresponding profiles as follows:

$$P(\gamma) = P_\pi(\gamma) + P_\sigma(\gamma), \tag{E.69}$$

$$P_\pi(\gamma) = \sum_{(\alpha,\beta)\pi} (J_{\alpha\beta}/|X_{\alpha\beta}|) f_\pi W(\gamma/|X_{\alpha\beta}|), \tag{E.70}$$

$$P_\sigma(\gamma) = \sum_{(\alpha,\beta)\sigma} (J_{\alpha\beta}/|X_{\alpha\beta}|) f_\sigma W(\gamma/|X_{\alpha\beta}|), \tag{E.71}$$

where

$$W(u) = 2u \exp(-u^2), \tag{E.72}$$

(Eq. (E.72) is the renormalized Eq. (E.63)). The quantities f_π, f_σ in Eqs. (E.70), (E.71) are as follows:

for the observation parallel to **B**,

$$f_\pi = 1, \quad f_\sigma = 1/2, \tag{E.73}$$

while for the observation perpendicular to **B**,

$$f_\pi = 1/2, \quad f_\sigma = 3/4. \tag{E.74}$$

As an example, Fig. E.19 presents calculated universal Lorentz-broadened profiles of the Balmer line H_{18} for the observations perpendicular to **B** (solid line) and parallel to **B** (dashed line).

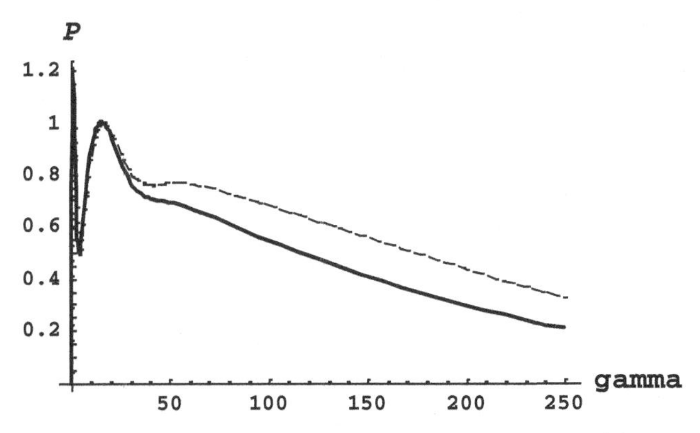

Fig. E.19. Calculated universal Lorentz-broadened profiles $P(\gamma)$ of the Balmer line H_{18} for the observations perpendicular to **B** (solid line) and parallel to **B** (dashed line). The argument $\gamma = c\Delta\omega/(kBv_T)$, where $k = 3\hbar/(2m_e e)$. The intensities of both profiles are relative to the intensity of the second maximum (whose intensity is set as unity). The first maximum would be smeared out by the Doppler broadening.

E.3 Allowance for Spiraling Trajectories of the Perturbers

In Secs. E.2 and E.3 above, we considered rectilinear trajectories of plasma electrons, which is the commonly accepted for the case of neutral radiating atoms (hereafter, radiators). However, in many types of plasmas — such as tokamak plasmas (see, e.g., [28]), laser-produced plasmas (see, e.g., [29]), capacitor-produced plasmas [30], astrophysical plasmas [31, 32] — there are strong magnetic fields. In a strong magnetic field $\mathbf{B}$, perturbing electrons basically spiral along magnetic field lines. Therefore, their trajectories are not rectilinear in the case of neutral radiators or not hyperbolic in the case of charged radiators. Here, we take this into account the following paper [33]. We start from the general framework for calculating shapes of hydrogen (or deuterium) spectral lines in strongly-magnetized plasmas allowing for spiraling trajectories of perturbing electrons.

In a plasma containing a strong magnetic field $\mathbf{B}$, we choose the z-axis along $\mathbf{B}$. For the case of neutral radiators, the radius-vector of a perturbing electron can be represented in the form

$$\mathbf{R}(t) = \rho \mathbf{e}_x + v_z t \mathbf{e}_z + r_{Bp}[\mathbf{e}_x \cos(\omega_B t + \varphi) + \mathbf{e}_y \sin(\omega_B t + \varphi)], \quad \text{(E.75)}$$

where the x-axis is chosen along the impact parameter vector $\boldsymbol{\rho}$. Here, v_z is the electron velocity along the magnetic field and

$$r_{Bp} = v_p/\omega_B, \quad \omega_B = eB/(m_e c), \quad \text{(E.76)}$$

where v_p is the electron velocity in the plane perpendicular to $\mathbf{B}$; ω_B is the Larmor frequency.

For the atomic electron to experience the spiraling nature of the trajectories of perturbing electrons, it requires $\rho_{De}/r_{Bp} > 1$, where ρ_{De} is the electron Debye radius. Taking into account that the average over the 2D-Maxwell distribution $\langle 1/v_p \rangle = \pi^{1/2}/v_{Te}$, where $v_{Te} = (2T_e/m_e)^{1/2}$ is the mean thermal velocity of perturbing electrons, the condition $\rho_{De}/r_{Bp} > 1$ can be rewritten in the form $\pi^{1/2}\omega_B/\omega_{pe} > 1$ (where ω_{pe} is the plasma electron frequency) or

$$B > B_{\text{cr}} = c(4m_e N_e)^{1/2}, \quad B_{\text{cr}}(\text{Tesla}) = 1.81 \times 10^{-7}[N_e(\text{cm}^{-3})]^{1/2}, \quad \text{(E.77)}$$

where N_e is the electron density. For example, for the edge plasmas in tokamaks, at $N_e = 10^{14}\,\mathrm{cm}^{-3}$, the condition (E.77) becomes $B > 1.8$ Tesla, which is fulfilled in modern tokamaks. Another example is DA white dwarfs: at $N_e = 10^{17}\,\mathrm{cm}^{-3}$, condition (E.77) becomes $B > 50$ Tesla, which is fulfilled in many DA white dwarfs. Also, the condition (E.77) is easily fulfilled in capacitor-produced plasmas [30] and in plasmas produced by high-intensity lasers [31] (in the latter case, radiators would be charged rather than neutral).

We consider the situation where the temperature T of the radiators satisfies the condition

$$T \ll (11.12\,\mathrm{keV}/n)(M/M_H)^2, \tag{E.78}$$

where M is the radiator mass, M_H is the mass of hydrogen atoms, n is the principal quantum number of the energy levels, from which the spectral line originates. Under this condition, the Lorentz field effects can be disregarded compared to the "pure" magnetic field effects.

To get the message across in a relatively simple form, we limit ourselves by the Lyman lines. Then matrix elements of the EBO in the impact approximation have the form

$$\begin{aligned}\Phi_{\alpha\alpha'} = N_e \int_0^{\infty} dv_p W_p(v_p) \int_{-\infty}^{\infty} dv_z W_z(v_z) v_z \\ \times \int_0^{\rho\max} d\rho\, 2\pi\rho \langle S_{\alpha\alpha'} - 1\rangle,\end{aligned} \tag{E.79}$$

where $W_p(v_p)$ is the 2D-Maxwell distribution of the perpendicular velocities, $W_z(v_z)$ is the 1D-Maxwell (Boltzmann) distribution of the longitudinal velocities, $\rho_{\max} = \rho_{De}$ is the maximum impact parameter, S is the scattering matrix, $\langle \ldots \rangle$ stands for averaging; α and α' label sublevels of the upper energy level involved in the radiative transition.

In the first order of the Dyson expansion, we have,

$$S_{\alpha\alpha'} - 1 = -i(e^2/\hbar) \int_{-\infty}^{\infty} dt \{\mathbf{r}_{\alpha\alpha'} \mathbf{R}(t)/[R(t)]^3\} \exp(i\omega_{\alpha\alpha'} t), \tag{E.80}$$

where $\mathbf{r}$ is the radius-vector of the atomic electron, $\mathbf{r}_{\alpha\alpha'}$ $\mathbf{R}$ is the scalar product (also known as the dot-product). Here, $\omega_{\alpha\alpha'}$ is the energy difference between the energy sublevels α and α' divided by $\hbar$; for the adjacent energy sublevels we have:

$$|\omega_{\alpha\alpha'}| = \omega_B/2. \tag{E.81}$$

We introduce the following notations:

$$s = \rho/r_{Bp}, \quad g = v_z/v_p, \quad w = v_z t/r_{Bp}. \tag{E.82}$$

In these notations, Eq. (E.80) can be rewritten as:

$$\begin{aligned} S_{\alpha\alpha'} - 1 = -i[e^2/(\hbar r_{Bp} v_z)] \int_{-\infty}^{\infty} dw \\ \{z_{\alpha\alpha'} w \delta_{\alpha\alpha'} + \exp[\pm iw/(2g)][y_{\alpha\alpha'} \sin(w/g + \varphi) \\ + x_{\alpha\alpha'}(\cos(w/g + \varphi) + s)]\}/[1 + w^2 + 2s\cos(w/g + \varphi) + s^2]^{3/2}, \end{aligned} \tag{E.83}$$

where $\delta_{\alpha\alpha'}$ is the Kronecker-delta (we use the parabolic quantization).

In the integrand in Eq. (E.83), the terms containing $z_{\alpha\alpha'}$ is the odd function of w, so that the corresponding integral vanishes. The term containing $y_{\alpha\alpha'}$ vanishes after averaging over the phase φ. As for the term containing $x_{\alpha\alpha'}$, after averaging over the phase φ, it becomes as follows (it should be noted that in this setup the angular averaging of vector ρ is irrelevant — in distinction to the rectilinear trajectories)

$$\begin{aligned} \langle S_{\alpha\alpha'} - 1 \rangle \\ = -i[e^2/(\hbar r_{Bp} v_z)] \int_{-\infty}^{\infty} dw \cos[w/(2g)]\{\mathbf{K}[4s/j(w,s)] \\ - \mathbf{E}[4s/j(w,s)](w^2 + 1 - s^2)/[w^2 + (s-1)^2]\}/\{\pi s[j(w,s)]^{1/2}\}, \end{aligned} \tag{E.84}$$

where $\mathbf{K}(u)$ and $\mathbf{E}(u)$ are the elliptic integrals, and

$$j(w,s) = w^2 + (1+s)^2. \tag{E.85}$$

We denote,

$$u = \rho_{De}/r_{Bp}. \tag{E.86}$$

Then combining Eqs. (E.79) and (E.84), the first order of the *non-diagonal* elements of the EBO $\Phi^{(1)}_{\alpha\alpha'}$ can be represented in the form (the diagonal elements of $\Phi^{(1)}$ vanish because the diagonal elements of the x-coordinate are zeros):

$$\Phi^{(1)}_{\alpha\alpha'} = -ix_{\alpha\alpha'}(e^2/\hbar)\rho_{De}N_e \int_0^\infty dv_p W_p(v_p) \int_{-\infty}^\infty dv_z W_z(v_z) f_0(g,u), \tag{E.87}$$

where

$$f_0(g,u) = (1/u)\int_0^u ds \int_{-\infty}^\infty dw$$
$$\cos[w/(2g)]\{\mathbf{K}[4s/j(w,s)] - \mathbf{E}[4s/j(w,s)]$$
$$\times (w^2+1-s^2)/[w^2+(s-1)^2]\}/\{\pi s[j(w,s)]^{1/2}\}. \tag{E.88}$$

We note that $\Phi^{(1)}_{\alpha\alpha'}$ scales with the electron density as $N_e^{1/2}$.

For obtaining results in the simplest form, we calculate the average $\langle g\rangle = \langle v_z/v_p\rangle$

$$\langle g\rangle = \int_0^\infty dv_p W_p(v_p) \int_{-\infty}^\infty dv_z W_z(v_z) v_z/v_p = 1 \tag{E.89}$$

and denote $f_0(1,u) = f(u)$. The argument can be represented as $u = B/B_{\text{cr}}$, where B_{cr} is the critical value of the magnetic field defined in Eq. (E.77). Figure E.20 shows this universal function $f(u)$ that controls the phenomenon under consideration.

It is seen that $f(u)$ starts growing at $B/B_{\text{cr}} > 1$, reaches the maximum at $B/B_{\text{cr}} = 3$, and then gradually diminishes as B/B_{cr} further increases.

The presence of the maximum of $f(u)$ can be physically understood as follows. At $\rho_{De} < r_{Bp}$, the effect is absent because the radius of the spirals of the trajectory of perturbing electrons exceeds the Debye radius. In the opposite limit, i.e., at $\rho_{De} \gg r_{Bp}$, the radius of the spirals of the trajectory is so small (compared to the Debye radius), that the atomic electron perceives the trajectory almost as a straight line (the straight line of a "width" $\sim r_{Bp}$), so that $\Phi^{(1)}_{\alpha\alpha'}$

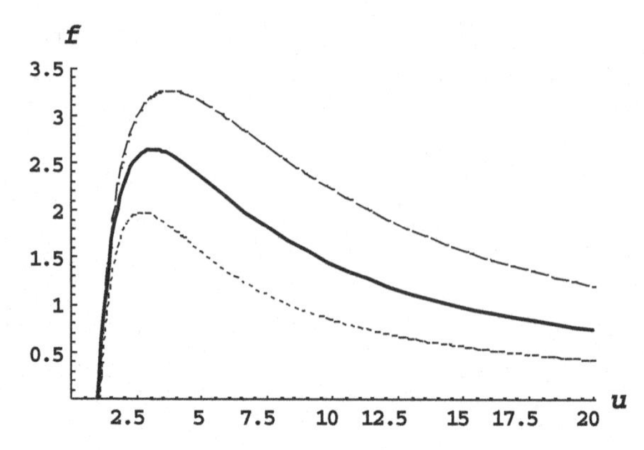

Fig. E.20. The universal function $f(u)$, controlling the first order of the EBO, shown at three different values of the ratio $g = v_z/v_p$: $g = 1$ (solid curve), $g = 3/2$ (dashed curve), $g = 2/3$ (dotted curve). The argument $u = B/B_{\rm cr}$, where $B_{\rm cr}$ is the critical value of the magnetic field defined in Eq. (E.77).

gradually goes to the limit of zero, i.e., to the limit corresponding to rectilinear trajectories of perturbing electrons.

The real and imaginary parts of $\Phi_{\alpha\alpha'}$ relate to the width and shift of spectral components of hydrogen/deuterium lines, respectively, as it is well-known from the fundamentals of the line broadening theory (see, e.g., books [2, 4]). Since, the above first order contribution $\Phi^{(1)}_{\alpha\alpha'}$ is purely imaginary, it is clear in advance that it should translate primarily into an *additional shift* of Zeeman components of hydrogen/deuterium lines. While the above analytical results were obtained for Lyman lines, it is obvious that for *any* hydrogen/deuterium line of *any* spectral series (Balmer, Paschen, etc.), there should be an additional shift of the Zeeman components of the line caused by the spiraling trajectories of perturbing electrons.

For presenting details on the additional shift of the Zeeman components, we use below the Ly_α line, just as an example, because it allows obtaining those detailed results in the analytical form. We show that this additional shift is experienced by the σ-components of the line.

Further, we provide below two different types of applications: to the edge plasmas of tokamaks and to white dwarfs. In the case of the edge plasmas of tokamaks, we consider situations where the width of Zeeman components is controlled by additional factors/mechanisms — other than the broadening by electrons (such as the broadening by a high-frequency electrostatic plasma turbulence or the ion dynamical broadening). However, while the additional broadening factors increase the width of the Zeeman components, they do not affect the shift of the σ-components, as we show below. Therefore, the additional shift of the σ-components, caused by the spiraling trajectories of perturbing electrons, can be detected experimentally at the edge plasmas of tokamaks despite the presence of additional broadening mechanisms.

For calculating the shapes of the Ly_α line, we use the parabolic quantization, i.e., the quantum numbers $(n_1 n_2 m)$. We denote the four sublevels of the upper level $(n = 2)$ as follows: (001) as 1, (100) as 2, (00-1) as 3, and (010) as 4. According to the linear Zeeman effect, states 1 and 3 are shifted (in the frequency scale) by $\omega_B/2$ and $-\omega_B/2$, respectively, while the states 2 and 4 remain unshifted. The shifts are counted from the unperturbed position of the $n = 2$ energy level.

The non-zero matrix elements of the x-coordinates within the $n = 2$ space are the following:

$$x_{12} = x_{21} = x_{23} = x_{32} = x_{34} = x_{43} = x_{41} = x_{14} = -3a_B/2, \quad \text{(E.90)}$$

where a_B is the Bohr radius. We denote

$$a = [3\hbar/(2m_e)]\rho_{De} N_e f[B/B_{\text{cr}}], \quad d = \omega_B/2. \quad \text{(E.91)}$$

The lineshapes of the Ly-line are controlled by matrix elements of operator G^{-1}, where the operator G is defined as follows:

$$G_{\alpha\alpha'} = \langle\alpha|i[\omega - \Delta_{\alpha\alpha}(B)] + \Phi]|\alpha'\rangle, \quad \text{(E.92)}$$

where $\Delta_{11} = d$, $\Delta_{33} = -d$, $\Delta_{22} = \Delta_{44} = 0$. Here and below, ω is counted from the unperturbed frequency of the spectral line. In order to present in the purest form how the non-vanishing first order of the EBO $\Phi^{(1)}$ affects the lineshapes, specifically the shift of the Zeeman components, we consider two situations where there is no need to include the effect of the spiraling trajectories of perturbing electron on $\Phi^{(2)}$.

The first situation is where there is a relatively strong high-frequency electrostatic turbulence in a plasma: the Langmuir turbulence or the electron cyclotron turbulence. For example, in the experiment presented in paper [34], a strong electron cyclotron turbulence was found at the edge plasma of tokamak T-10. In this situation, by extending the results from [11, 35] to the relatively strong Langmuir or electron cyclotron turbulence, we find that the real part of all diagonal elements of the operator G is practically equal to γ_p, which is the largest of the characteristic frequencies of various nonlinear processes involving the Langmuir or electron cyclotron turbulence in the plasma — the processes such as, the generation of the Langmuir or the electron cyclotron turbulence, the induced scattering on the charged particles, the nonlinear decay into ionic sound (in case of the Langmuir turbulence), and so on: the frequency γ_p is assumed to control the width of the power spectrum of the Langmuir or electron cyclotron turbulence.[4] Then matrix G takes the form (we omitted the suffix "p" of γ_p)

$$G = \begin{bmatrix} i(\omega - d) + \gamma & ia & 0 & ia \\ ia & i\omega + \gamma & ia & 0 \\ 0 & ia & i\omega + \gamma & ia \\ ia & 0 & ia & i(\omega + d) + \gamma \end{bmatrix}. \tag{E.93}$$

The general expression for the lineshape $I(\omega)$ of Ly-lines is

$$I(\omega) = \text{const.Re}\left[\sum_{\alpha\alpha'} \mathbf{r}^*_{\alpha 0}\mathbf{r}_{\alpha' 0} < \alpha|G^{-1}|\alpha' >\right] \tag{E.94}$$

where suffix "0" denotes the ground state. For the Ly_α line, the non-zero matrix elements of the operator $\mathbf{r}$ in Eq. (E.94) are as follows

[4] We note in passing that in the case of the Langmuir turbulence, the electron velocity distribution at $v_e \gg v_{Te}$ might deviate from the Maxwell distribution because of the possible appearance of so-called "suprathermal electrons" — see, e.g., review [16]. However, the electron broadening operator is inversely proportional to v_e, so that the non-Maxwellian contribution from the region $v_e \gg v_{Te}$ should be just a very small correction. Even more important is that in the case of the electron cyclotron turbulence, there are no suprathermal electrons, to the best of our knowledge.

(see, e.g., [36]):

$$\langle 4|x|0\rangle = \langle 2|x|0\rangle = \langle 3|z|0\rangle = -\langle 1|z|0\rangle = 128a_B/243,$$
$$\langle 4|y|0\rangle = -\langle 2|y|0\rangle = 128ia_B/243. \quad \text{(E.95)}$$

After inverting the matrix G we obtain the following normalized profiles for each component of the Ly_α line. For the π-component:

$$I(\omega) = (\gamma/\pi)/(\gamma^2 + \omega^2). \quad \text{(E.96)}$$

For the σ-component originating from state (001):

$$I(\omega) = (\gamma/\pi)[\omega^4 + 2d\omega^3 + (d^2 - 2a^2 + 2\gamma^2)\omega^2 + 2d\gamma^2\omega + 8a^4 + 2a^2d^2 + \gamma^2d^2 + 6\gamma^2a^2 + \gamma^4]/\{(\gamma^2 + \omega^2)[\omega^4 - 2(d^2 + 4a^2 - \gamma^2)\omega^2 + (d^2 + 4a^2 + \gamma^2)^2]\}. \quad \text{(E.97)}$$

For the σ-component originating from the state (00–1), the normalized profile can be obtained from Eq. (E.97) by changing d to $-d$.

It is seen that, while the profile of the π-component remains unaffected by the allowance for the spiraling trajectories of perturbing electrons, the profiles of the σ-components are significantly affected by the allowance for the spiraling trajectories of perturbing electrons — both qualitatively and quantitatively: *each* of the two σ-components could generally become a triplet. Indeed, the first factor in the denominator in Eq. (E.97), i.e., the factor $(\gamma^2 + \omega^2)$ has the minimum at $\omega = 0$. This translates into the appearance of the unshifted σ-subcomponent (i.e., the component peaked at $\omega = 0$).

The second factor in the denominator in Eq. (E.97) has the minima at $\omega = \pm(d^2 + 4a^2 - \gamma^2)^{1/2}$. This translates into two additional new features. For example, for the σ-component originating from the state (001), first, the position of the subcomponent shifted to the blue side changes from d to $(d^2 + 4a^2 - \gamma^2)^{1/2}$. At relatively small γ, the new position of this subcomponent is approximately $(d^2 + 4a^2)^{1/2}$, where d linearly depends on the magnetic field B (according to Eqs. (E.76), (E.91)), while a depends on B in a more complicated way: $a(B)$ is proportional to the function $f[B/B_{cr}]$, whose plot was presented in Fig. E.20. Second, in addition to the blue-shifted component, there also appears an unshifted component, as well as a component shifted to the opposite (red) wing of the line.

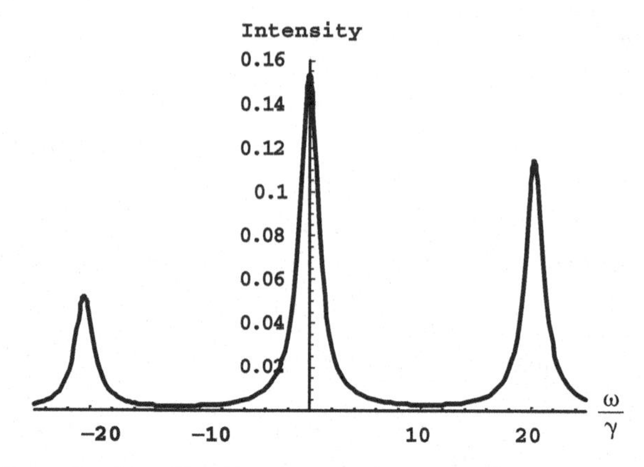

Fig. E.21. Normalized profile of just one σ-component (originating from the state of the parabolic quantum numbers $n_1 = 0$, $n_2 = 0$, $m = 1$) of the Ly_α line for the case of $d/\gamma = 4$, $a/\gamma = 10$, where d and a are defined in Eqs. (E.91) and (E.76). The abscissa is in units ω/γ, where ω is the distance (in the frequency scale) from the unperturbed position of the spectral line and $\gamma = \gamma_r$ is the "width" due to a relatively strong Langmuir or electron cyclotron turbulence (more details are in the paragraph after Eq. (E.92)). The profile of the σ-component originating from the state of $n_1 = 0$, $n_2 = 0$, $m = -1$ can be obtained by reflecting the above profile with respect to the ordinate axis.

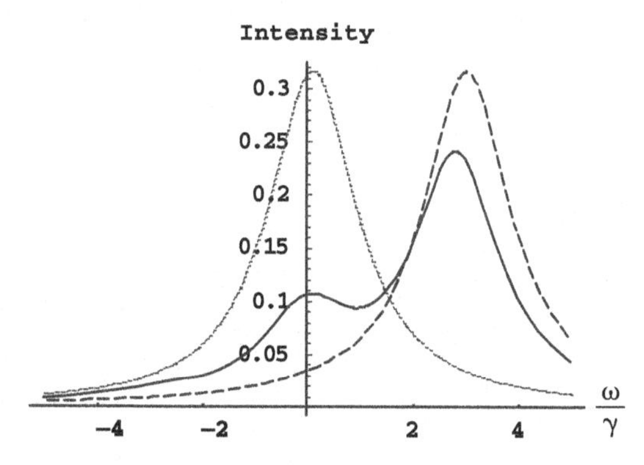

Fig. E.22. Same as in Fig. E.20, but for $d/\gamma = 2$ and $a/\gamma = 1$ (solid curve), $d/\gamma = 3$ and $a/\gamma = 0.2$ (dashed curve), $d/\gamma = 0.1$ and $a/\gamma = 0.05$ (dotted curve).

This illustrated in Fig. E.21 for the case where γ is relatively small compared to d and a.

Figure E.22 demonstrates the same as Fig. E.21, but for three other pairs of $d/\gamma = 2$ and a/γ. The solid curve shows that for a value

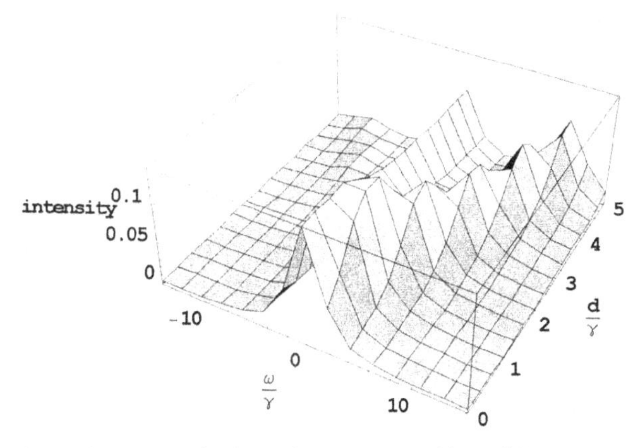

Fig. E.23. Three-dimensional plot of various profiles of just one σ-component (originating from the state of the parabolic quantum numbers $n_1 = 0$, $n_2 = 0$, $m = 1$) of the Ly_α line for the case of $a = d$ and different values of d/γ.

of γ larger than in Fig. E.20 (relative to the values of d and a), the red peak practically disappeared. The dashed curve demonstrates that when $d/\gamma > 1$, but $a/\gamma < 1$, the unshifted subcomponent disappears: *each* of the two σ-components looks like a shifted singlet. The dotted curve shows that if $\gamma > (d^2 + 4a^2)^{1/2}$, both the blue- and red-subcomponents of a particular σ-component collapse to the position practically coinciding with the unperturbed position of the spectral line (so that the entire spectral line, including the π- and σ-components, would look like a singlet structure).

Figure E.23 shows a three-dimensional plot of various profiles of just one σ-component (originating from the state of the parabolic quantum numbers $n_1 = 0$, $n_2 = 0$, $m = 1$) of the Ly_α line for the case of $a = d$ and different values of d/γ. It is seen that as the ratio d/γ increases, there also appears an unshifted component. As the ratio d/γ further increases, there also appears a component shifted to the opposite (red) wing of the line.

Thus, the role of the allowance for the spiraling trajectories of perturbing electrons in the shift of thc σ-components is controlled by the ratio $2a/d$. In the spectrum of the entire Ly_α line, this ratio controls both the new features caused by this allowance: the increase of the ratio of the intensity of the central peak to the intensity of any of the two lateral peaks (which otherwise would be equal to one) and the additional shift of the lateral components. Both new features in the spectrum of the entire Ly_α line are illustrated below.

As an example, we consider parameters corresponding to the edge plasmas of tokamak T-10 in the experiment described in [34]: $B = 1.65$ Tesla, $N_e = 2 \times 10^{13}\,\text{cm}^{-3}$, $T = (10 - 15)\,\text{eV}$ (the latter was the temperature range found experimentally for the region of the maximum intensity of the deuterium lines). In this situation, the ratio is $2a/d = (0.6\text{–}0.7)$, so that the allowance for the spiraling trajectories of perturbing electrons can be important for magnetic fusion plasmas.

It is worth emphasizing the following. For the above parameters of the experiment [34], the spectral line broadening by electrons is smaller than the ion dynamical broadening, which in its turn is smaller than the broadening by the high-frequency electrostatic turbulence.[5] However, the allowance for the spiraling trajectories of the electrons affects primarily the shift (rather than the width) of the Zeeman components of hydrogen/deuterium spectral lines. The contribution to the shift of spectral components of hydrogen/deuterium spectral lines by both the ion impacts and the high-frequency turbulence is controlled by the following combination of the parabolic quantum numbers (see, e.g, papers [9,35], or Eq. (6.1) from book [4], or Eqs. (15.65–66) from review [11]): $n(n_1 - n_2) - n^{(0)}(n_1^{(0)} - n_2^{(0)})$, where the superscript (0) labels parabolic quantum numbers of the lower state involved in the radiative transition. Consequently, for the σ-components of the Ly_α line, the contribution to the shift by both the ion impacts and the high-frequency turbulence is zero (because $n_1 = n_2 = n_1^{(0)} = n_2^{(0)} = 0$). Therefore, there are only two contributions to the shift of the σ-components of the Ly_α line: from

[5] A relatively strong magnetic field $B = 3$ Tesla in experiment [34] partially inhibits the ion dynamical broadening compared to the $B = 0$ case. Physically, this is because a relatively large Zeeman splitting $\omega_B/2$ diminishes the range of ion impact parameters ρ, for which the characteristic frequency of the variation of the ion field v_i/ρ exceeds $\omega_B/2$. This effect was described analytically in the year 1994 in paper [18] using the GT of the SB and presented in Sec. E.1 above.

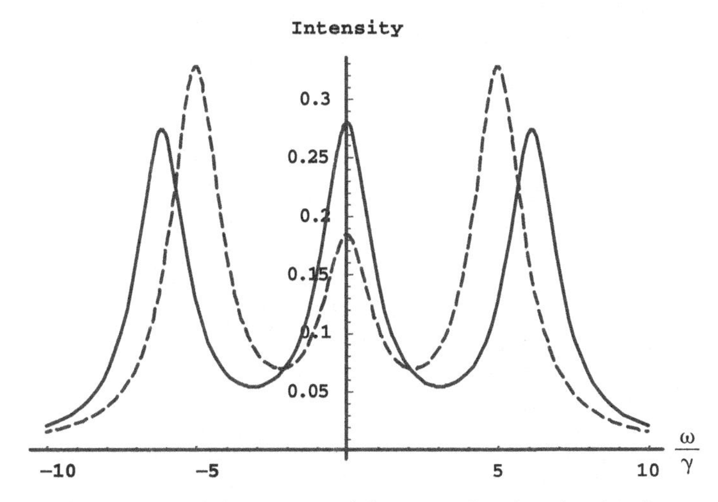

Fig. E.24. Comparison of the spectra of the entire Ly_α line for the above example of tokamak T-10 with the allowance for the spiraling trajectories of the perturbing electrons (solid line) and without this allowance (dotted line). The spectra were calculated for the observation at the angle of 55° with respect to the magnetic field.

the Zeeman effect and from the effect of spiraling trajectories of electrons, the latter being a significant addition to the former. (For clarity, the above two contributions to the shift of the spectral components do not affect the shift of the spectral line as a whole; the latter might be caused by other factors beyond the scope of the present paper).

Figure E.24 shows the comparison of the spectra of the entire Ly_α line for the above example of tokamak T-10 with the allowance for the spiraling trajectories of the perturbing electrons (solid line) and without this allowance (dotted line). The spectra were calculated for the observation at the angle of 55 degrees with respect to the magnetic field. The primary effect of the allowance for the spiraling trajectories of the perturbing electrons is that the ratio of the intensity of the central peak to the intensity of any of the two lateral peaks increased by 82% (from 0.56 to 1.02). The secondary effect is that the shift of the lateral (σ-) components increased by 23%.

As another example, we consider parameters corresponding to the edge plasmas of tokamak EAST: $B = 2$ Tesla, $N_e = 5 \times 10^{13}\,\mathrm{cm}^{-3}$,

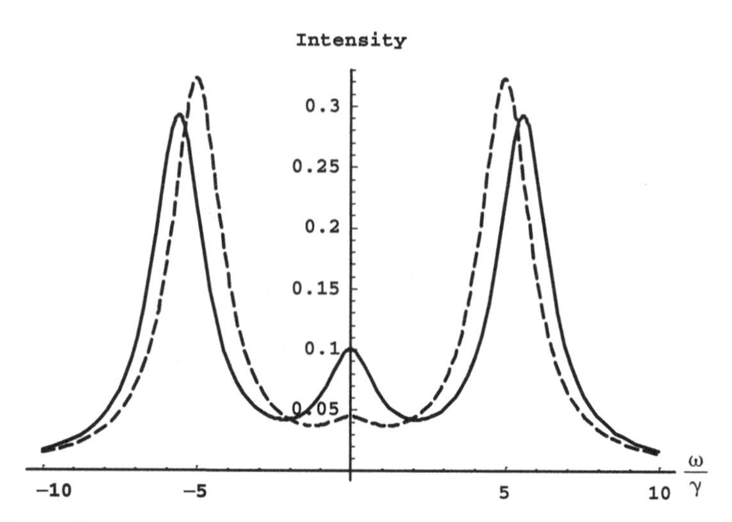

Fig. E.25. Comparison of the spectra of the entire Ly_α line for the above example of tokamak EAST with the allowance for the spiraling trajectories of the perturbing electrons (solid line) and without this allowance (dotted line). The spectra were calculated for the observation at the angle of 20° with respect to the magnetic field, which is the actual angle of the diagnostic window at EAST.

$T = 8\,\text{eV}$. In this situation, the ratio $2a/d = 0.5$, thus demonstrating again that the allowance for the spiraling trajectories of perturbing electrons can be important for magnetic fusion plasmas.

Figure E.25 shows the comparison of the spectra of the entire Ly_α line for the above example of tokamak EAST with the allowance for the spiraling trajectories of the perturbing electrons (solid line) and without this allowance (dotted line). The spectra were calculated for the observation at the angle of 20° with respect to the magnetic field, which is the actual angle of the diagnostic window at EAST. The primary effect of the allowance for the spiraling trajectories of the perturbing electrons is that the ratio of the intensity of the central peak to the intensity of any of the two lateral peaks increased by 150% (from 0.137 to 0.341). The secondary effect is that the shift of the lateral (σ-) components increased by 11%.

As the next example, we consider parameters relevant to the DA white dwarfs (i.e., the white dwarfs emitting hydrogen lines): $B = 100$ Tesla, $N_e = 10^{17}\,\text{cm}^{-3}$, $T_e = 5\,\text{eV}$. In this situation, the ratio is $2a/d = 0.4$. The primary effect of the allowance for the spiraling trajectories of the perturbing electrons for the above example of white dwarfs is that the ratio of the intensity of the central peak to the intensity of any of the two lateral peaks increased by 23% (from 1 to 1.23). The secondary effect is that the shift of the lateral (σ-) components increased by 8%.

So, the allowance for the spiraling trajectories of perturbing electrons can play a noticeable role for these astrophysical objects. We note that at these parameters, the strong magnetic field practically completely inhibits the ion dynamical broadening. We also note that up to now, there was no evidence of the development of a high-frequency electrostatic turbulence in white dwarfs. Therefore, while for obtaining more accurate quantitative results for white dwarfs, it would be necessary in the future to include the effect of the spiraling trajectories of perturbing electron on the second order of the EBO $\Phi^{(2)}$, the results of the present paper still provide a good estimate of the effect of spiraling electron trajectories.

Finally, we describe another situation allowing to present the effect of the non-zero $\Phi^{(1)}$ in the purest form. This is the scenario where the real part of the diagonal elements of the operator G is controlled by the ion dynamical broadening. In practice, for low-n hydrogen/deuterium spectral lines, this relates, e.g., to the electron density range at the edge plasmas of tokamaks in discharges where there is no high-frequency electrostatic turbulence. For the corresponding matrix elements, we can basically use the results from paper [37] obtained for the Ly_α broadening by electrons, "translate" them into the Ly_α broadening by ions, and superimpose these matrix elements with the matrix G from Eq. (E.93) setting $\gamma_p = 0$. As a result, the total matrix, denoted as G_1, can be represented as follows:

$$G_1 = \begin{bmatrix} i(\omega - d) + \gamma & ia & 0 & ia \\ ia & i\omega + 2\gamma & ia & \gamma \\ 0 & ia & i(\omega + d) + \gamma & ia \\ ia & \gamma & ia & i\omega + 2\gamma \end{bmatrix} \tag{E.98}$$

Here,

$$\gamma = 9g_2, \tag{E.99}$$

where the quantities g_n were defined by Eq. (1.6) from paper [37]. The approximate value of g_2 is

$$g_2 \sim 2(\mu/T_i)^{1/2} N_i (\hbar/m_e)^{1/2}, \tag{E.100}$$

where μ is the reduced mass of the pair "radiator — perturbing ion". We note in passing that in the matrix (E.98), we corrected the sign of the elements $(G_1)_{24}$ and $(G_1)_{42}$ compared to paper [37].

After inverting the matrix G_1, we obtain the following normalized profiles for each component of the Ly_α line. For the π-component:

$$I(\omega) = (\gamma/\pi)/(\gamma^2 + \omega^2). \tag{E.101}$$

For the σ-component originating from state (001):

$$\begin{aligned} I(\omega) = (\gamma/\pi)\{&\omega^4 + 2d\omega^3 + (2a^2 + d^2 + 10\gamma^2)\omega^2 + 2d(4a^2 + 9\gamma^2)\omega \\ &+ 8a^4 + 6a^2d^2 + 18a^2\gamma^2 + 9d^2\gamma^2 + 9\gamma^4\}/\{\gamma^2(5\omega^2 - 4a^2 - 3d^2 \\ &- 3\gamma^2)^2 + \omega^2(\omega^2 - 4a^2 - d^2 - 7\gamma^2)^2\}. \end{aligned} \tag{E.102}$$

For the σ-component originating from the state (00–1), the normalized profile can be obtained from Eq. (E.102) by changing d to $-d$.

It is seen again that, while the profile of the π-component remains unaffected by the allowance for the spiraling trajectories of perturbing electrons, the profiles of the σ-components are significantly affected by the allowance for the spiraling trajectories of perturbing electrons — both qualitatively and quantitatively.

There are qualitatively different results depending on the interplay of the two dimensionless parameters d/γ and a/γ, as illustrated in Fig. E.26. The solid curve shows that when $d/\gamma > 1$ and $a/\gamma > 1$, each of the two σ-components looks like a triplet. The dashed curve demonstrates that when $d/\gamma > 1$, but $a/\gamma < 1$, each of the two σ-components looks like a shifted singlet. The dotted curve shows that when $d/\gamma \ll 1$ and $a/\gamma \ll 1$, each of the two σ-components looks like a practically unshifted singlet.

Figure E.27 shows a three-dimensional plot of various profiles of just one σ-component (originating from the state of the parabolic quantum numbers $n_1 = 0$, $n_2 = 0$, $m = 1$) of the Ly_α line for the case of $a = d$ and different values of d/γ. It is seen that as the ratio d/γ

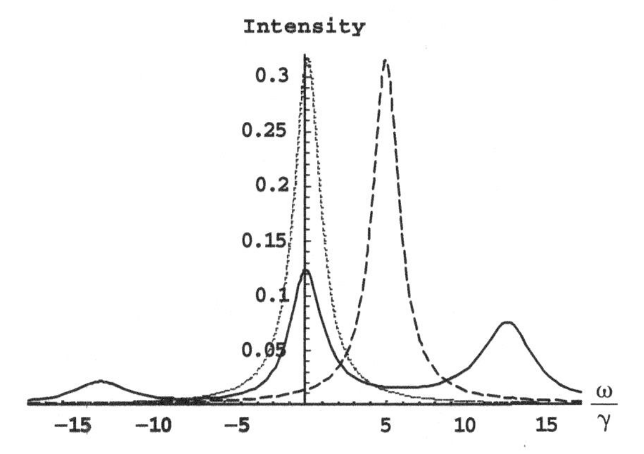

Fig. E.26. Normalized profile of just one σ-component (originating from the state of the parabolic quantum numbers $n_1 = 0$, $n_2 = 0$, $m = 1$) of the Ly_α line. The abscissa is in units ω/γ, where ω is the distance (in the frequency scale) from the unperturbed position of the spectral line. Here γ, defined in Eqs. (24), (25), is the ion dynamical "width". The profile of the σ-component originating from the state of $n_1 = 0$, $n_2 = 0$, $m = -1$ can be obtained by reflecting the above profile with respect to the ordinate axis. Solid curve: the case of $d/\gamma = 4$ and $a/\gamma = 6$, where d and a are defined in Eqs. (E.91) and (E.76). Dashed curve: $d/\gamma = 5$ and $a/\gamma = 0.25$. Dotted curve: $d/\gamma = 0.1$, $a/\gamma = 0.05$.

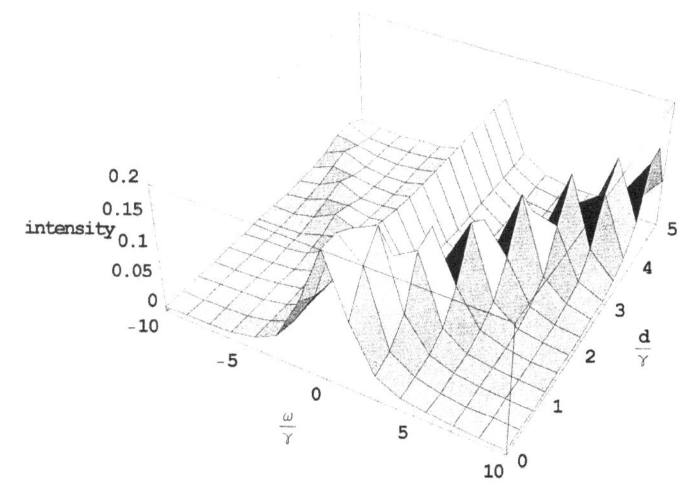

Fig. E.27. Three-dimensional plot of various profiles of just one σ-component (originating from the state of the parabolic quantum numbers $n_1 = 0$, $n_2 = 0$, $m = 1$) of the Ly_α line for the case of $a = d$ and different values of d/γ.

increases, there also appears an unshifted component. As the ratio d/γ further increases, there also appears a component shifted to the opposite (red) wing of the line, though it is less intense compared to the analogous situation depicted in Fig. E.23.

We remind once again that in the situation discussed here, where the primary contribution to the width of the Zeeman components is due to the ion impacts, those impacts do not contribute anything to the shift of the σ-components of the Ly_α line. The additional shift of the σ-components is still controlled by the ratio $2a/d$. Let us consider, as an example, parameters relevant to the edge plasmas of the future tokamak ITER (which is currently under construction in France): $B = 10$ Tesla, $N_e = 10^{15}\,\text{cm}^{-3}$, $T_e = 5\,\text{eV}$. In this case, the ratio is $2a/d = 0.4$, so that the allowance for the spiraling trajectories of the electrons can play a noticeable role in the ratio of the intensities and in the shift of the Zeeman components. We note again that at these parameters, the relatively strong magnetic field partially inhibits the ion dynamical broadening. As a result, the contribution of the latter to the width of the Zeeman components is only several times greater than the contribution from electrons, while at $B = 0$, the ratio of these contributions to the width would be $\sim (\mu/m_e)^{1/2} \gg 1$, where μ is the reduced mass of the pair "radiator — perturbing ion".

References

[1] A. Derevianko and E. Oks, *Phys. Rev. Lett.* **73** 2059 (1994).
[2] H.R. Griem, *Spectral Line Broadening by Plasmas*, Academic, New York (1974).
[3] V.S. Lisitsa, Sov. Phys. Uspekhi **122** 603 (1977).
[4] E. Oks, *Stark Broadening of Hydrogen and Hydrogenlike Spectral Lines in Plasmas: The Physical Insight*, Alpha Science International, Oxford, UK (2006).
[5] Ya. Ispolatov and E. Oks, *J. Quant. Spectrosc. Rad. Transfer* **51** 129 (1994).
[6] V.S. Lisitsa and G.V. Sholin, *Sov. Phys. JETP* **34** 484 (1972).
[7] A. Derevianko and E. Oks, in *Physics of Strongly Coupled Plasmas*, eds. W.D. Kraft and M. Schlanges (Singapore: World Scientific), pp. 286 and 291 (1996).

[8] J. Rosato, Y. Marandet, H. Capes, S. Ferri, C. Mosse, L. Godbert-Mouret, M. Koubiti and R. Stamm, *Phys. Rev. E* **79** 046408 (2009).
[9] G.V. Sholin, A.V. Demura and V.S. Lisitsa, *Sov. Phys. JETP* **37** 1057 (1973).
[10] E. Oks, *Intern. Rev. of Atom. and Mol. Phys.* **1** 169 (2010).
[11] E. Oks, in: *Atomic Processes in Basic and Applied Physics*, Springer Series on Atomic, Optical and Plasma Physics, vol. 68, V. Shevelko and H. Tawara (eds.), Springer, New York, p. 393 (2012).
[12] E. Oks, *J. Quant. Spectrosc. Rad. Transfer* **171** 15 (2016).
[13] J. Touma, E. Oks, S. Alexiou and A. Derevianko, *J. Quant. Spectrosc. Rad. Transfer* **65** 543 (2000).
[14] J. Rosato, H. Capes, L. Godbert-Mouret, M. Koubiti, Y. Marandet and R. Stamm, *J. Phys. B: At. Mol. Opt. Phys.* **45** 165701 (2012).
[15] E. Oks, A. Derevianko and Ya. Ispolatov, *J. Quant. Spectrosc. Rad. Transfer* **54** 307 (1995).
[16] N. Bleistein and R.A. Handelsman, *Asymptotic Expansions of Integrals*, (Dover, New York) (1986).
[17] Yu.I. Galushkin, *Sov. Astron.* **14** 301 (1970).
[18] R.C. Isler, *Phys. Rev. A* **14** 1015 (1976).
[19] C. Breton, C. De Michelis, M. Finkental and M. Mattioli, *J. Phys. B: Atom. Mol. Phys.* **13** 1703 (1980).
[20] Nguen Hoe, J. Grumberg, M. Caby, E. Leboucher and G. Couland, *Phys. Rev. A* **24** 438 (1981).
[21] S. Brilliant, G. Mathys and C. Stehle, *Astron. Astrophys.* **339** 286 (1998).
[22] C. Stehle, S. Brilliant and G. Mathys, *Eur. Phys. J. D* **11** 491 (2000).
[23] E. Oks, *Intern. Review of Atom. and Mol. Phys.* **4** 105 (2013).
[24] E. Oks, *J. Quant. Spectrosc. Rad. Transfer* **156** 24 (2015).
[25] J. Holtsmark, *Ann. Phys.* **58** 577 (1919).
[26] F.D. Rosenberg, U. Feldman and G.A. Doschek, *Astrophys. J.* **212** 905 (1977).
[27] U. Feldman and G.A. Doschek, *Astrophys. J.* **212** 913 (1977).
[28] B.L. Welch, H.R. Griem, J. Terry, C. Kurz, B. LaBombard, B. Lipschultz, E. Marmar and J. McCracken, *Phys. Plasmas* **2** 4246 (1995).
[29] U. Wagner, M. Tatarakis, A. Gopal *et al.*, *Phys. Rev. E* **70** 026401 (2004).
[30] S. Fujioka, Z. Zhang, K. Ishihara *et al.*, *Scientific Reports* **3** 1170 (2013).
[31] L.J. Silvers, *Phil. Trans. R. Soc. A* **366** 4453 (2008).
[32] A.A. Schekochihin and S.C. Cowley, in: *Magnetohydrodynamics — Historical Evolution and Trends*, S. Molokov, R. Moreau and H.K. Moffett (eds.) (Springer, Berlin), p. 85 (2007).

[33] E. Oks, *J. Quant. Spectrosc. Rad. Transfer* **171** 15 (2016).
[34] V.P. Gavrilenko, E. Oks and V.A. Rantsev-Kartinov, *JETP Lett.* **44** 404 (1986).
[35] E. Oks, *J. Phys. B: At. Mol. Opt. Phys., Fast Track Communication* **44** 101004 (2011).
[36] H. Pfennig, *Z. Naturforsch.* **26a** 1071 (1971).
[37] M.L. Strekalov and A. I. Burshtein, *Sov. Phys. JETP* **34** 53 (1972).

Appendix F

Stark Broadening by Low-frequency Electrostatic Turbulence

F.1 Neutral Radiators in Weakly-coupled Plasmas

Electrostatic turbulence frequently occurs in various kinds of laboratory and astrophysical plasmas [1, 2]. It can lead to anomalous transport phenomena in plasmas, the most important of which being usually the anomalous resistivity. Electrostatic turbulence is represented by Oscillatory Electric Fields (OEFs) sometimes also called collective electric fields: they correspond to collective degrees of freedom in plasmas — in distinction to the electron and ion microfields that correspond to individual degrees of freedom of charged particles.

At the absence of a magnetic field, there is only one type of a low-frequency electrostatic turbulence: ion acoustic waves — frequently called *ionic sound*. The corresponding OEF is a broadband field, whose frequency spectrum is below or of the order of the ion plasma frequency:

$$\omega_{pi} = (4\pi e^2 N_i Z^2/m_i)^{1/2} = 1.32 \times 10^3 Z(N_i m_p/m_i)^{1/2}, \qquad \text{(F.1)}$$

where N_i is the ion density, Z is the charge state; m_p and m_i are the proton and ion masses, respectively. In the "practical" parts of Eqs. (F.1–F.4), CGS units are used.

In magnetized plasmas, in addition to the ionic sound, propagating along the magnetic field **B**, two other types of low-frequency electrostatic turbulence are possible. One is *electrostatic ion cyclotron wave*, whose wave vector is nearly perpendicular to **B**. Its frequency is close to the ion cyclotron frequency:

$$\omega_{ci} = ZeB/(m_i c) = 9.58 \times 10^3 ZB(m_p/m_i)^{1/2}. \tag{F.2}$$

Another type is *lower hybrid oscillations* having the wave vector perpendicular to **B**. Its frequency is

$$\omega = 1/[(\omega_{ci}\omega_{ce})^{-1} + (\omega_{pi})^{-2}]^{1/2}, \tag{F.3}$$

where ω_{ce} is the electron cyclotron frequency

$$\omega_{ci} = eB/(m_e c) = 1.76 \times 10^7 B. \tag{F.4}$$

From Eq. (F.3), it is seen that $\omega < \omega_{pi}$ always. This means that frequencies of both ionic sound and lower hybrid oscillations are below or of the order of the ion plasma frequency ω_{pi}.

It is usually assumed that hydrogenic radiators perceive OEFs, associated with a low-frequency plasma turbulence as *quasistatic.* Let us discuss this assumption in more detail. The discussion is based on papers [3, 4], the results of which were recently summarized in paper [5].

The physics of the spectral line broadening in plasmas containing OEFs is very rich and complex due to the interplay of a large number of characteristic times and frequencies. There are seven characteristic frequencies, which can be considered as "elementary" parameters. For our specific discussion here, the following four frequencies are important.

1. $\Delta\omega$ — detuning from the unperturbed position of a given spectral line of the radiator. It affects the characteristic value of the argument τ of the correlation function.
2. ω — QEF frequency.
3. γ — homogeneous width of the power spectrum of OEF, which is also the inverse of the QEF *coherence time* τ_F.
4. $\delta_s(E_0)$ — instantaneous Stark shift at the amplitude value E_0 of OEF. For example, $\delta_s(E_0) = a_1 E_0$ in the case of the linear Stark effect or $\delta_s(E_0) = a_2 E_0^2$ in the case of the quadratic Stark effect; $a_1(k)$, $a_2(k)$ are Stark constants that depend on the set of

quantum numbers of the particular states of the radiator. *Here and below, the set of quantum numbers is denoted by k.*

On the basis of the above "elementary" frequencies, there occur two composite parameters that are characteristic times as follows.

1. $\tau_{\rm QS}(k, E_0, \omega)$ — characteristic time of the formation of Quasienergy States (QS):

$$\tau_{\rm QS}(k, E_0, \omega) \sim \min(1/(\omega^2 \delta_s)^{1/3}, 1/\omega). \tag{F.5}$$

Being subjected to OEF, the states of the radiator can oscillate with the OEF frequency ω. This effect is described as the emergence of Quasienergy States (QS), which were introduced in 1967 in papers [6] and [7] (independently of each other). The above formula for $\tau_{\rm QS}(k, E_0, \omega)$ was derived in paper [3]. So, for relatively weak OEF, the QS are formed at the timescale of the order of the period of the OEF $1/\omega$. However, for relatively strong OEF, the QS are formed at a much shorter time scale proportional to $1/E_0^{1/3}$ or to $1/E_0^{2/3}$ in the cases of the linear or quadratic Stark effect, respectively.

2. $\tau_{\rm life}(k, N_e, T_e, N_i, T_i, \gamma, \omega, E_0, \Delta\omega)$ — the lifetime of the exited state of the radiator:

$$\tau_{\rm life}(k, N_e, T_e, N_i, T_i, \gamma, \omega, E_0, \Delta\omega) \sim 1/\Gamma, \tag{F.6}$$

$$\Gamma = \gamma_e(k, N_e, T_e, \Delta\omega) + \gamma_i(k, N_i, T_i, N_e, \Delta\omega) + \gamma_F(k, \gamma, \omega, E_0). \tag{F.7}$$

Here, Γ is the sum of the *homogeneous* Stark widths due to electrons, dynamic part of ions, and OEF. The contribution $\gamma_F(k, \gamma, \omega, E_0)$ from OEF was calculated in paper [8].

A criterion for OEF to be considered as quasistatic is the following:

$$\tau_{\rm QS}(k, E_0, \omega) \gg \min[1/\delta_s(E_0), 1/\Delta\omega, \tau_{\rm life}(k, N_e, T_e, N_i, T_i, \gamma, \omega, E_0, \Delta\omega)], \tag{F.8}$$

For Balmer and Paschen lines emitted from magnetic fusion plasmas, the condition (F.8) is usually fulfilled for OEFs of low-frequency electrostatic turbulence. Therefore, from the spectroscopic point of view, one deals here with a hydrogenic atom/ion in crossed *static* electric ($\mathbf{F}$) and magnetic ($\mathbf{B}$) fields — the problem allowing an exact solution for each value of $\mathbf{F}$ and $\mathbf{B}$ [9].

If the emission is observed from a relatively small volume, within which the magnetic field can be considered homogeneous, then the spectral line profiles can be obtained by averaging the solution from [9] over the ensemble distribution $W(\mathbf{F})$ of the quasistatic field $\mathbf{F}$. In other words, the key part of the problem becomes the calculation of $W(\mathbf{F})$.

In magnetic fusion plasmas, typically Zeeman splitting in the field $\mathbf{B}$ is much greater than Stark splitting in the Lorentz field $\mathbf{v} \times \mathbf{B}/c$ because the following condition is fulfilled

$$v_{\mathrm{Ta}} \ll c/(205.5n), \tag{F.9}$$

where v_{Ta} is the thermal velocity of the radiating atoms; n is the principal quantum number of the upper level, from which the spectral line originates. Equation (F.9) can be re-written in terms of the temperature T_a of the radiating atoms as follows:

$$T_a \ll (10^4/n^2)\mathrm{eV}. \tag{F.10}$$

Therefore, the distribution $W(\mathbf{F})$ of the quasistatic fields is either the distribution $W_t(\mathbf{E}_t)$ of the turbulent fields or a convolution of $W_t(\mathbf{E}_t)$ with the distribution $W_i(\mathbf{F}_i)$ of the ion microfield, if the latter is also quasistatic.

The situation, where the turbulent field is quasistatic, but the ion microfield is not quasistatic, is possible even if their frequencies are of the same order of magnitude — as long as the average turbulent field E_0 is much greater than the characteristic ion microfield $\langle F_i \rangle$. Indeed, if the average amplitude E_0 of the OEF is sufficiently large, then the criterion (F.8) reduces to $1/[\omega^2 \delta_s(E_0)]^{1/3} \gg 1/\delta_s(E_0)$. For example, for the linear Stark effect (hydrogenic atoms/ions) this inequality is equivalent to

$$\omega \ll \delta_{\mathrm{s}}(E_0) = a_1(k)E_0, \tag{F.11}$$

which can be fulfilled even if the corresponding criterion for the ion microfield $\omega_i \ll a_1(k)\langle F_i \rangle$ would not be met. For example, in magnetic fusion plasmas, this kind of situation can occur for low-n hydrogen/deuterium spectral lines at densities $\ll 10^{15}\,\mathrm{cm}^{-3}$ if the average field of a low-frequency electrostatic turbulence would be of the order of or greater than $10\,\mathrm{kV/cm}$.

In magnetized plasmas, the distribution of the turbulent field should be axially symmetric, the axis of symmetry being along the magnetic field. It is Sholin–Oks' distribution [10] of the following form:

$$\begin{aligned} &W(E, \cos\theta) dE d(\cos\theta) \\ &\quad = [4/(\pi E_{\mathrm{par}}^2)]^{1/2} (E^2/E_{\mathrm{perp}}^2) \exp[-E^2/E_{\mathrm{perp}}^2 \\ &\qquad - \cos^2\theta (E^2/E_{\mathrm{par}}^2 - E^2/E_{\mathrm{perp}}^2)] dE d(\cos\theta), \end{aligned} \tag{F.12}$$

where E_{par} and E_{perp} are the root-mean-square values of the turbulent field components parallel and perpendicular to $\mathbf{B}$, respectively; θ is the angle between $\mathbf{E}$ and $\mathbf{B}$.

In any code designed for calculating spectral line profiles, an important task becomes the averaging over the ensemble distribution $W(\mathbf{E})$ of the *total* quasistatic field $\mathbf{E} = \mathbf{E}_t + \mathbf{F}_i$. In other words, the key part of the problem becomes the calculation of $W(\mathbf{E})$.

In the isotropic case, the distribution (F.12) can be represented in the following form [11]:

$$W_t(\alpha, x) dx = 3[6/\pi]^{1/2} \alpha^3 x^2 \exp(-3\alpha^2 x^2/2)] dx. \tag{F.13}$$

Here,

$$x = E_t/E_N \tag{F.14}$$

is the scaled turbulent field and

$$\alpha = E_N/E_R \tag{F.15}$$

is the ratio of the "standard" ion microfield E_N to the root-mean-square turbulent field E_R, where

$$\begin{aligned} E_N &= 2\pi(4/15)^{2/3} Z_p^{1/3} e N_e^{2/3} \\ &= 3.751 \times 10^{-7} Z_p^{1/3} [N_e(\mathrm{cm}^{-3})]^{2/3}\,\mathrm{V/cm}. \end{aligned} \tag{F.16}$$

The total quasistatic field $\mathbf{E}$ results from the vector summation of the two statistically independent contributions: $\mathbf{E} = \mathbf{E}_t + \mathbf{F}_i$. The justification of this has been given in papers [12, 13]. Therefore a *general* distribution $W_g(\mathbf{E}/E_N)$ of the total field is a convolution of the distribution $W_t(\mathbf{E}_t/E_N)$ of the turbulent field with the distribution $W_i(\mathbf{F}_i/E_N)$ of the ion microfield (subscript "g" in $W_g(\mathbf{E}/E_N)$ stands for "general"):

$$W_g(\boldsymbol{\beta})d\boldsymbol{\beta} = \left[\iint d\mathbf{x}d\mathbf{u}\, W_i(\mathbf{u})W_t(\mathbf{x})\delta(\beta - \beta_s)\right] d\boldsymbol{\beta},$$

$$\boldsymbol{\beta} = \mathbf{E}/E_N, \quad \mathbf{x} = \mathbf{E}_t/E_N, \quad \mathbf{u} = \mathbf{F}_i/E_N. \tag{F.17}$$

Here

$$\beta_s = |\mathbf{x} + \mathbf{u}| = (x^2 + u^2 - 2ux\cos\theta)^{1/2}, \tag{F.18}$$

where θ is the angle between vectors $\mathbf{u}$ and $\mathbf{x}$.

Equation (F.17) represents a general result valid for both isotropic and anisotropic distributions of the turbulent field. Here, we consider the case of the isotropic distribution of the turbulent field, so that it is represented by Eq. (F.11). In this case, the distribution of the total field $W(\beta)$ will be also isotropic (since both $W_i(\mathbf{u})$ and $W_t(\mathbf{x})$ are isotropic) and can be written as

$$W(\beta)d\beta = \left[\iint d\mathbf{x}d\mathbf{u}\, W_i(\mathbf{u})W_t(\mathbf{x})\delta(\beta - \beta_s)\right] d\beta. \tag{F.19}$$

In paper [11] the study was focused at so-called ideal (or "non-coupled") plasmas where the kinetic energy is by several orders of magnitude greater than the potential energy. It is manifested by a very large number of perturbing ions N_{iD} in a sphere of the electron Debye radius r_{De}:

$$N_{iD} = (4\pi/3)N_e r_{De}^3/Z_p \gg 1, \tag{F.20}$$

where

$$r_{De} = [T_e/(4\pi e^2 N_e)]^{1/2}. \tag{F.21}$$

A practical formula for the quantity N_{iD} is[1]:

$$N_{iD} = 1.7181 \times 10^9 [T_e(\mathrm{eV})]^{3/2} / \{Z_p [N_e(\mathrm{cm}^{-3})]^{1/2}\}. \tag{F.22}$$

The study in paper [11] was designed for application to experiments, where hydrogen line profiles were emitted from turbulent plasmas of electron densities $\sim 10^{14}$–$10^{15}\,\mathrm{cm}^{-3}$ and temperatures of several eV. For example, for $N_e = 3 \times 10^{14}\,\mathrm{cm}^{-3}$, $T_e = 4$–$5\,\mathrm{eV}$, and $Z_p = 1$, Eq. (F.22) yields $N_{iD} \sim 10^3$. For this kind of plasmas, where N_{iD} was by many orders of magnitude greater than unity, the ion microfield distribution was chosen in paper [11] as the Holtsmark distribution [14]:

$$W_H(u) = (2u/\pi) \int_0^\infty dx \sin ux \exp(-u^{3/2}). \tag{F.23}$$

The Holtsmark distribution describes a transition from the Gaussian distribution for weak fields $u \ll 1$ to the binary distribution (nearest-neighbor distribution) of strong fields $u \gg 1$, as noted in review [15]. Indeed, the weak-field part of the Holtsmark distribution $W_H \approx 4u^2/(3\pi)$ is due to a cumulative effect of large number of perturbers — therefore, like any sum of a large number of random quantities, it follows the Gaussian distribution (i.e., its starting part $\sim u^2$). In the opposite limit of $u \gg 1$, where $W_H = 15/[4(2\pi)^{1/2} u^{5/2}] = 1.496/u^{5/2}$, only the nearest-neighbor controls the distribution.

For the distribution of the total quasistatic field, defined by Eq. (F.19), the following result was obtained in paper [11] for ideal plasmas — by performing analytically several integrations in Eq. (F.19):

$$W(\alpha, \beta) = [3/(2\pi)^{1/2}]\alpha\beta \int_0^\infty du \{\exp[-3\alpha^2(\beta - u)^2] - \exp[-3\alpha^2(\beta + u)^2]\} W_H(u)/u. \tag{F.24}$$

[1] N_{iD} is related to another coupling parameter Γ used in plasma physics as follows: $N_{iD} = [Z_p/(3\Gamma)]^{3/2}$.

F.2 Charged Radiators in Moderately-coupled Plasmas

In this section, we focus at moderately-coupled plasmas, where the quantity N_{iD} is still greater than unity, but not dramatically greater: $N_{iD} \sim 10$. A particular application could be, e.g, the laser-produced plasma experiment described in papers [16, 17], characterized by $N_e \sim 3 \times 10^{20}\,\text{cm}^{-3}$, $T \sim 150\,\text{eV}$, the majority of radiating and perturbing ions having the charge $Z = 12$, so that $N_{iD} = 15$. In this situation, first of all, the ion microfield distribution differs significantly from the Holtsmark distribution. The second distinction is that the ion microfield distribution should be calculated at a point of the charge $Z \sim 10$, and in plasmas of $N_{iD} \sim 10$ this distribution would be noticeably different from the distribution at a neutral point because of ion–ion correlations.

Therefore, our starting point is the following expression for the distribution $W(\alpha, \beta)$ of the total quasistatic field, which is similar to Eq. (F.24), but without assuming any specific form of the ion microfield distribution $W_i(u)$:

$$W(\alpha, \beta) = [3/(2\pi)^{1/2}]\alpha\beta \int_0^\infty du\{\exp[-3\alpha^2(\beta - u)^2] - \exp[-3\alpha^2(\beta + u)^2]\}W_i(u)/u. \quad \text{(F.25)}$$

In other words, $W_i(u)$ could be any ion microfield distribution at a charged point, calculated by any appropriate code (e.g., by using the APEX method [18]).

Equation (F.25) can be rewritten in the form:

$$W(\alpha, \beta) = [3/(2\pi)^{1/2}]\alpha\beta \exp(-3\alpha^2\beta^2) \times \int_0^\infty du\{\exp[-3\alpha^2(u^2 - 2\beta u)/2] - \exp[-3\alpha^2(u^2 + 2\beta u)/2]\}W_i(u)/u. \quad \text{(F.26)}$$

In a frequently encountered situation, where the average turbulent field is much greater than the characteristic ion microfield ($\alpha = E_N/E_R \ll 1$), it is appropriate to expand both exponentials in the integrand in (F.26) in Taylor series. Keeping terms up to (including) these $\sim u^4$, we obtain:

$$W(\alpha, \beta) = W_t(\alpha, \beta)[1 - (3\alpha^2/2)(1 - \alpha^2\beta^2)M_{i2}], \tag{F.27}$$

where $W_t(\alpha, \beta)$ is the distribution given by Eq. (F.13) and M_{i2} is the second moment of the ion microfield distribution:

$$M_{i2} = \int_0^\infty du\, u^2 W_i(u). \tag{F.28}$$

Equation (F.27) shows that at $\alpha = E_N/E_R \ll 1$, in the first approximation the distribution of the total quasistatic field reduces to the distribution from Eq. (F.13), as should be expected. More important is that Eq. (F.27) also shows that the first non-vanishing correction to the Rayleigh distribution (the second term in brackets in (23)) is controlled by the second moment of the ion microfield distribution.

Two important comments should be made at this point. First, Eq. (F.27) has a *great computational advantage* compared to Eq. (F.25). Indeed, while employing Eq. (F.25), one would have to choose at least $\sim 10^2$ values of β and at least ~ 10 values of α. So, one would have to use $W_i(u)$, calculated numerically by some code, at least $\sim 10^3$ times. In distinction, while utilizing Eq. (F.27), one would have to use $W_i(u)$ *only once* — for calculating the second moment of $W_i(u)$. Thus, the employment of Eq. (F.27) instead of Eq. (F.21) makes *much more robust* any code for calculating spectral line profiles in turbulent plasmas.

The second comment, following from Eq. (F.27), is related to paper [19]. This paper was also devoted to calculations of the distribution of the total quasistatic field for the situation, where the average turbulent field is much greater than the characteristic ion microfield. The authors of [19] suggested an approximate method where the correction to the Rayleigh distribution was controlled by the behavior of the ion microfield distribution $W_i(u)$ *at small fields* ($u \ll 1$). However, from Eqs. (F.27), (F.28) it can be seen that this was a misconception. Indeed, the integral in (F.28) accumulates most of its value at $u \gg 1$: it converges only because at very large values of

u, the ion–ion correlations — the repulsion between the radiating and perturbing ions — "kill" the integral. Thus, in reality, the correction to the distribution from Eq. (F.13) is controlled by the behavior of the ion microfield distribution $W_i(u)$ *at large fields* rather than at small fields.

In view of the above second comment, we can use the well-known asymptotic of $W_i(u)$ at $u \gg 1$ at a charged point and obtain a universal analytical result for the second moment of $W_i(u)$ and thus for the correction to the Rayleigh distribution. Analytical results for the large-field asymptotic of the ion microfield distribution at a charged point were presented in papers [20, 21].[2] We consider here plasmas where the charge of radiating ions Z_r is the same as for perturbing ions: $Z_r = Z_p \equiv Z$. For this case, the corresponding formula, obtained in [20, 21] with the allowance for ion–ion correlations and for the screening of the ion field by plasma electrons, has the form:

$$\begin{aligned} W_{i,\mathrm{as}}(u) = {} & (q/Z^2)(u/Z + v^2/2)^{-5/2} \\ & \times \exp\{-[T_e/(2qT_i)]Z^2 v^2 (u/Z + v^2/2)^{1/2} \\ & \times \exp[-(1 + ZT_e/T_i)^{1/2}/(u/Z + v^2/2)^{1/2}]\}. \end{aligned} \tag{F.29}$$

Here,

$$\begin{aligned} q = 15/[4(2\pi)^{1/2}] = 1.496, \quad & v = r_0/r_{\mathrm{De}}, \\ & r_0 = [15/(4N_e)]^{1/3}/(2\pi)^{1/2}. \end{aligned} \tag{F.30}$$

(the quantity r_0 is defined such that it is close to the mean interionic distance).

The quantity v is yet another indicator of the proximity of a plasma to the non-ideality (i.e., the coupling indicator). It is related to the number of perturbing ions N_{iD} in a sphere of the electron Debye radius as follows:

$$N_{iD} = 0.9974/(Zv^3). \tag{F.31}$$

[2]The current status of the ion microfield distribution studies can be found in review [22].

It is seen that $v \ll 1/Z^{1/3}$ corresponds to ideal plasmas ($N_{iD} \gg 1$), while $v > 1/Z^{1/3}$ corresponds to strongly-coupled plasmas. A practical formula for the quantity v has the form:

$$v = 8.98 \times 10^{-2}[N_e(\mathrm{cm}^{-3})]^{1/6}/[T_e(K)]^{1/2}. \tag{F.32}$$

If the following condition is met:

$$u \gg Zv^2/2, \tag{F.33}$$

the asymptotic formula (F.29) simplifies to

$$W_{i,as}(u) = (qZ^{1/2}/u^{5/2})\exp(-ku^{1/2}), \tag{F.34}$$

where

$$k = T_e Z^{3/2} v^2/(2qT_i). \tag{F.35}$$

We emphasize that the condition (F.33) is usually much less restrictive than the validity condition ($u \gg 1$) of the formula (25). For example, for the laser-produced plasma experiment described in papers [16, 17], characterized by $N_e \sim 3 \times 10^{20}\,\mathrm{cm}^{-3}$, $T \sim 150\,\mathrm{eV}$, $Z = 12$, the inequality (F.33) yields $u \gg 0.2$. Thus, for a broad range of weakly coupled plasmas, the simplified asymptotic from Eq. (F.34) can be used with a very good accuracy instead of the asymptotic from Eq. (F.29).

Using this asymptotic, the second moment of the ion microfield distribution can be approximately represented in the form:

$$M_{i2} = I_1 + I_2, \quad I_1 = \int_0^C du\, u^2 W_i(u);$$

$$I_2 = \int_C^\infty du(qZ^{1/2}/u^{1/2})\exp(-ku^{1/2}), \tag{F.36}$$

where $C \sim 1$. The second integral I_2 can be calculated analytically yielding:

$$I_2 = (2qZ^{1/2}/k)\exp(-C^{1/2}k). \tag{F.37}$$

The central point of this section is that for a broad range of weakly coupled plasmas, the integral $I_2 \sim (2qZ^{1/2}/k) \gg 1$, while the first integral in Eq. (F.36) $I_1 \sim 1$. Therefore, the second moment of the ion microfield distribution, calculated by any appropriate code, can be well approximated by the analytical result (F.37) as long as $k < 1$ (so that I_2 from (F.37) is not sensitive to a particular choice of $C \sim 1$) and

$$2qZ^{1/2}/k = 4q^2T_i/(Zv^2T_e) \gg 1. \qquad \text{(F.38)}$$

Under this condition, which is satisfied for a broad range of weakly coupled plasmas, the second moment of the ion microfield distribution can be accurately calculated by the analytical formula (F.37) regardless of the particular behavior of this distribution at small fields. In this sense, formula (F.37) is *universal.*

As an illustration of the accuracy of the analytical result (F.37), we compare it with the exact calculation of the second moment for the nearest-neighbor (binary) distribution $W_{iB}(u)$ at a charged point.[3] The latter distribution in the normalized form is given by the following expression (under the condition (F.33)):

$$W_{iB}(u) = u^{-5/2}\exp(-u^{-3/2} - ku^{1/2}) \Big/ \int_0^\infty du \times [u^{-5/2}\exp(-u^{-3/2} - ku^{1/2})]. \qquad \text{(F.39)}$$

Figure F.1 shows two calculated dependences of the second moment of the above binary distribution versus parameter k. Solid line represents the result obtained using the large-field asymptotic with the choice of the lower limit $C = 1$, dashed line — the exact result. It is seen that the asymptotic method is very accurate as long as $k < 1$.

Figure F.2 shows dependences of the second moment of the above binary distribution, calculated by using the asymptotic formula (F.37), versus parameter k — for two different choices of the lower limit of the integration: $C = 0.5$ (solid line) and $C = 2$ (dashed line). It is seen that the results are practically insensitive to a particular

[3]This distribution has two important common features with ion microfield distributions calculated by any appropriate code for weakly coupled plasmas: the asymptotic at the large fields and a significant shift of the maximum toward small fields (compared to the Holtsmark distribution).

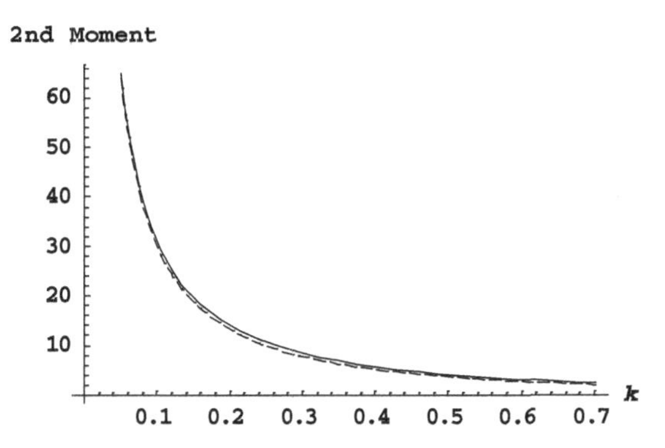

Fig. F.1. The second moment of the nearest-neighbor distribution at a charged point versus parameter $k = T_e Z^{3/2} v^2/(2qT_i)$, representing the degree of the non-ideality of plasmas: solid line — the result obtained using the large-field asymptotic of the nearest neighbor distribution with the choice of the lower limit $C = 1$; dashed line — the exact result. Here, Z is the charge of plasma ions, parameters v and q are defined in Eq. (F.30).

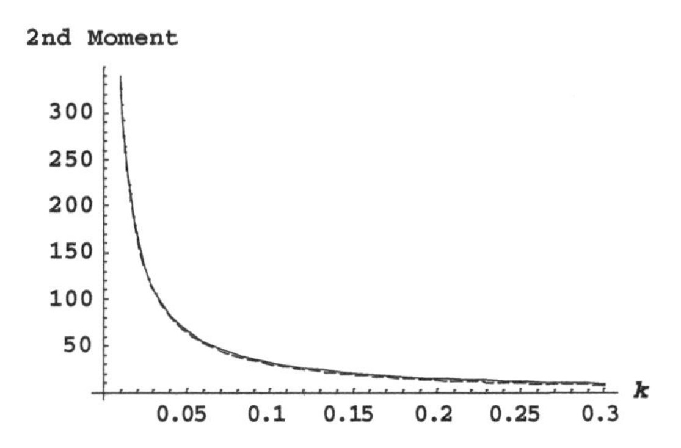

Fig. F.2. The second moment of the nearest-neighbor distribution at a charged point, calculated using the large-field asymptotic of this distribution, versus parameter $k = T_e Z^{3/2} v^2/(2qT_i)$, representing the degree of the non-ideality of plasmas: solid line — with the choice of the lower limit $C = 0.5$; dashed line — with the choice of the lower limit $C = 2$. Here, Z is the charge of plasma ions, parameters v and q are defined in Eq. (F.30).

choice of the lower limit of the integration C within the requirement $C \sim 1$ — as long as $k < 1$.

For completeness, we compare the approximate analytical result (F.37) with the second moment of more sophisticated, Hooper distributions [33]. Hooper distributions take into account correlations between perturbers and the Debye screening. Hooper distributions become noticeably different from the Holtsmark distribution when the number of perturbing ions N_{iD} in a sphere of the electron Debye radius falls to ~10 or less.

Different Hooper distributions correspond to different values of the ratio

$$a = r_{0i}/r_{De}, \tag{F.40}$$

where

$$r_{0i} = [3/(4\pi N_i)]^{1/3} \tag{F.41}$$

is the mean interionic distance and r_{De} is the Debye radius given by Eq. (F.21). In paper [33] Hooper presented distributions at a point of the charge $Z = 1$ for the following four values of the ratio a: 0.2, 0.4, 0.6 and 0.8. Here are the results of comparing the second moments.

For $a = 0.2$, the second moment of Hooper distribution is 222.7, while Eq. (F.37) yields 220.8, thus having an error of only 0.8%.

For $a = 0.4$, the second moment of Hooper distribution is 53.97, while Eq. (F.37) yields 53.04, thus having an error of only 1.7%.

For $a = 0.6$, the second moment of Hooper distribution is 22.65, while Eq. (F.37) yields 22.05, thus having an error of only 2.6%.

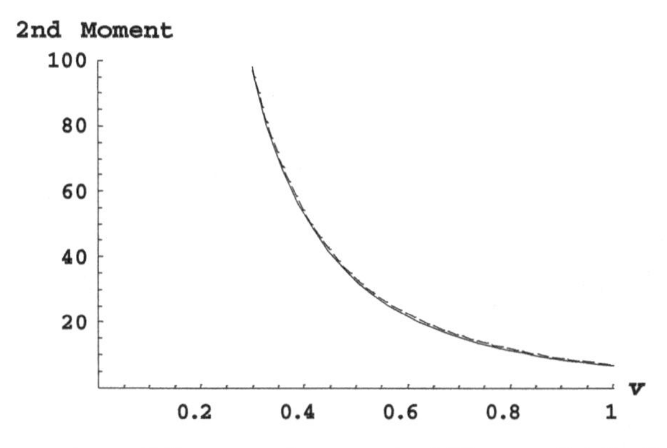

Fig. F.3. Comparison of the second moment of Hooper distributions (dashed line) with the corresponding analytical result from Eq. (F.37) at C = 1 (solid line) versus the parameter $v = r_0/r_{De}$.

For $a = 0.8$, the second moment of Hooper distribution is 11.81, while Eq. (F.37) yields 11.29, thus having an error of only 4.4%.

All of the above comparisons show that it is appropriate to calculate the second moment of the ion microfield distribution using the approximate analytical result (F.37) because the second moment enters only the correction term in Eq. (F.27) for the distribution of the total quasistatic field. Any correction to the analytical result (F.37) would enter Eq. (F.27) only as "a correction to the correction."

References

[1] V.N. Tsytovich, *Theory of Turbulent Plasmas*, Consultants Bureau, New York (1977).
[2] B.B. Kadomtsev, *Plasma Turbulence*, Academic, New York (1965).
[3] E. Oks, *Sov. Phys. Doklady* **29** 224 (1984).
[4] E. Oks, *Measurement Techniques* **29** 805 (1986).
[5] P. Sauvan, E. Dalimier, E. Oks, O. Renner, S. Weber and C. Riconda, *J. Phys. B: At. Mol. Opt. Phys.* **42** 195501 (2009).
[6] Ja.B. *Zel'dovich, Sov. Phys. JETP* **24** 1006 (1967).
[7] V.I. Ritus, *Sov. Phys. JETP* **24** 1041 (1967).
[8] E. Oks and G.V. Sholin, *Sov. Phys. JETP* **41** 482 (1975).
[9] Yu. Demkov, B. Monozon and V. Ostrovsky, *Sov. Phys. JETP* **30** 775 (1970).
[10] G.V. Sholin and E. Oks, *Sov. Phys. Doklady* **18** 254 (1973).
[11] E. Oks and G.V. Sholin, *Sov. Phys. Tech. Phys.* **21** 144 (1976).
[12] G.H. Ecker and K.G. Fisher, *Z. Naturforsch.* **26a** 1360 (1971).
[13] K.H. Spatschek, *Phys. Fluids* **17** 969 (1974).
[14] J. Holtsmark, *Ann. Phys.* **58** 577 (1919).
[15] V.S. Lisitsa, *Sov. Phys. Uspekhi* **122** 603 (1977).
[16] P. Sauvan, E. Dalimier, E. Oks, O. Renner, S. Weber and C. Riconda, *J. Phys. B* **42** 195001 (2009).
[17] P. Sauvan, E. Dalimier, E. Oks, O. Renner, S. Weber and C. Riconda, *Intern. Review Atom. and Mol. Phys.* **1** 123 (2010).
[18] C.A. Iglesias, J. Lebowitz and D. McGowan, *Phys. Rev. A* **28**, 1667 (1983).
[19] E. Sarid, Y. Maron, L. Troyansky, *Phys. Rev. E* **48** 1364 (1993).
[20] B. Held, *J. Physique* **45** 1731 (1984).
[21] B. Held, C. Deutsch and M.-M. Gombert, *Phys. Rev. A* **29** 880 (1984).
[22] A.V. Demura, *Intern. J. of Spectroscopy* **2010** 671073 (2010).
[23] C.F. Hooper, Jr., *Phys. Rev.* **165** 215 (1968).

Appendix G

Langmuir-Waves-Caused Dips in Hydrogenic Spectral Lines in Non/Weakly-Magnetized Plasmas

Following paper [1], we consider a radiative transition between levels of principal quantum numbers n and $n' < n$ of a hydrogenlike ion with a nuclear charge Z, in a quasistatic electric field $\mathbf{F}$ and in a single mode $\mathbf{E}_0 \cos \omega t$ of a quasimonochromatic field. The results can be easily extended to the case of multimode fields $\mathbf{E} = \sum_j \mathbf{E}_j \cos(\omega t + \varphi_j)$. It is sufficient to average the corresponding results for the single-mode case over a Rayleigh distribution $W_R(E_0, E_{av})$ of the amplitudes E_0, where

$$W_R(E_0, E_{av}) = (E_0/E_{av})^2 \exp[-E_0^2/(2E_{av})^2],$$
$$E_{av} = \left(\sum_j \mathbf{E}_j^2/2\right)^{1/2}. \tag{G.1}$$

In the dipole approximation under the action of the electric field F, the levels of the radiator with principal quantum numbers n and n' split into $(2n-1)$ and $(2n'-1)$ equidistant Stark sublevels separated from each other (in frequency units) by $\omega_F(n) = 3n\hbar F/(2Z_r m_e e)$ and $\omega_F(n')$, respectively. The ensemble of radiators yields some distribution $W(F)$ of the absolute values of F. If the quasistatic Stark splitting of the upper level $\omega_{F*}(n)$ satisfies the condition of resonance with the monochromatic field for some groups of radiators at the electric field F^*, we may write

$$3n\hbar F^*/2Z_r m_e e \approx k\omega, \qquad \text{(G.2)}$$

where $k = 1, 2, 3, \ldots$ is the number of quanta of the field involved in the resonance. Because of this resonance, local structures called L-dips appear in the quasistatic profile $I_{\alpha\beta}(\Delta\omega)$ of each lateral Stark component [2]. Here $\alpha = (n, q, m), \beta = (n', q', m'), q = n_1 - n_2$, and $q' = n_1' - n_2'$, where n_1, n_2, m and n_1', n_2', m' are the parabolic quantum numbers of the upper and lower Stark sublevels, respectively. The positions of these dips $\Delta\omega_\alpha^{\text{dip}}(k)$ are given by a simple relation [2]:

$$\Delta\omega_\alpha^{\text{dip}}(k) = 3(nq - n'q')\hbar F^*/(2Z_r m_e e) = n^{-1}(nq - n'q')k\omega. \qquad \text{(G.3)}$$

For some other group of radiators, the quasistatic Stark splitting of the lower level $\omega_{F*}(n')$ may also satisfy a condition analogous to Eq. (G.2), which leads to the appearance of a second set of dips in the profile $I_{\alpha\beta}(\Delta\omega)$ at the corresponding positions $\Delta\omega_\beta^{\text{dip}}(k)$ [2] .

It should be emphasized that for relatively weak oscillatory fields $E_0 \ll F^*$ (say, $E_0/F^* < 10^{-1}$) only the "first-order dips" arising from one-quantum resonances ($k = 1$) can be observed [2]. As a consequence, only two dips [at $\Delta\omega_\alpha^{\text{dip}}(1)$ and $\Delta\omega_\beta^{\text{dip}}(1)$] will appear in the profile of each lateral Stark component. Naturally, for Lyman lines, there is only one dip in the profile of each lateral Stark component [at $\Delta\omega_\alpha^{\text{dip}}(1)$] since the lower level ($n' = 1$) has no quasistatic Stark splitting.

We now consider this effect in a high-density plasma and assume that the low-frequency plasma turbulence is absent or its characteristic field strength F_{turb} is sufficiently low, i.e., $\text{F}_{\text{turb}} \ll F^*$. In this case, the quasistatic fields involved in the resonances are caused mainly by the ion microfield. For dense plasmas, one has to allow for a spatial non-uniformity of the ion microfield.

We include, therefore, into our consideration the quadrupole interaction of the radiator with the perturbing ions and treat the situation in the binary approximation in which only the nearest-neighbor perturber is taken into account. We point out that for lines from highly excited levels ($n \gg 1$), the influence of the ion microfield non-uniformity on dips might become essential already at lower densities. For this case, one can easily generalize the following results using a multiparticle description of the radiator–perturber quadrupole interaction in a way similar to Ref. [3], where such an approach was used for studying the asymmetry of spectral lines.

Under the action of a perturbing ion (having a charge Z_p) at a distance R, the separation $\delta\omega_\alpha^+$ between the neighboring Stark sublevels α and $\alpha+1$ (see Fig. G.1) can be expressed by (in frequency units) [4]:

$$\delta\omega_\alpha^+ = \delta a/r^2 + \delta b_\alpha^+/r^3, \tag{G.4}$$

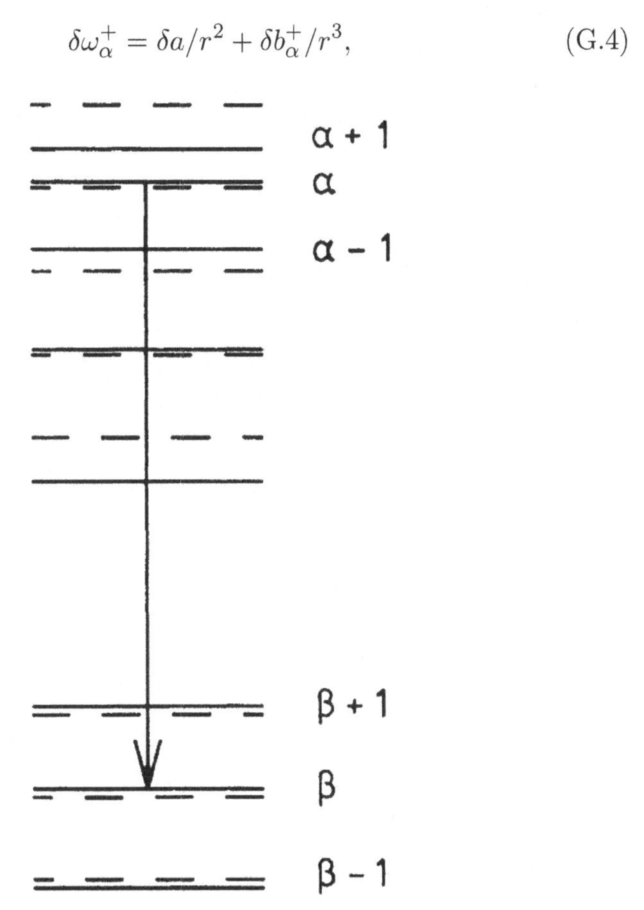

Fig. G.1. Scheme of the quasistatic splitting of upper and lower levels by the ion microfield without (dashed lines) and with (solid lines) the allowance for its nonuniformity. The radiative transition $\alpha \leftrightarrow \beta$ is indicated by the arrow.

where

$$r = R/R_0, \quad R_0 = [3Z_p/(4\pi N_e)]^{1/3},$$
$$\delta a = 3n\hbar Z_p/(2Z_r m_e R_0^2), \quad \delta b_\alpha^+ = -3n^2\hbar^3 Z_p(2q+1)/(Z_r^2 m_e^2 e^2 R_0^3), \tag{G.5}$$

and where it was taken into account that the sublevel $\alpha + 1$ has an electric quantum number $q + 1$.

In the case of a resonance of the oscillatory field with Stark sublevels α and $\alpha + 1$, we have (instead of (G.2))

$$\delta a/r^2 + \delta b_\alpha^+/r^3 = k\omega. \tag{G.6}$$

Then the position of dips $\delta\omega_\alpha^{\text{dip+}}(k)$ in the profile $I_{\alpha\beta}(\Delta\omega)$ of the same lateral Stark component is (instead of (G.3))

$$\delta\omega_\alpha^{\text{dip+}}(k) = n^{-1}(nq - n'q')k\omega + 2^{1/2}u_\alpha^+(k\omega)^{3/2}/(27n^3 Z_r Z_p \omega_a), \tag{G.7}$$

where

$$u_\alpha^+ = n^2(n^2 - 6q^2 - 1) + 6n(2q+1)(nq - n'q') - n'^2(n'^2 - 6q'^2 - 1),$$
$$\omega_a = me^4/\hbar^3 = 4.14 \times 10^{16} s^{-1} \tag{G.8}$$

(ω_a is the so-called "atomic frequency").

The separation $\delta\omega_\alpha^-$ between Stark sublevels α and $\alpha - 1$ (see Fig. G.1) differs from $\delta\omega_\alpha^+$ (in distinction to the pure dipole approximation):

$$\delta\omega_\alpha = \delta a/r^2 + \delta b_\alpha^-/r^3,$$
$$\delta b_\alpha^- = -3n^2\hbar^3 Z_p(2q-1)/(Z_r^2 m_e^2 e^2 R_0^3). \tag{G.9}$$

Therefore the oscillatory field resonates with sublevels α and $\alpha - 1$ at some other value of a reduced perturber-radiator distance r. This leads to the appearance of an additional set of dips in the profile $I_{\alpha\beta}(\Delta\omega)$ at the positions:

$$\delta\omega_\alpha^{\text{dip-}}(k) = n^{-1}(nq - n'q')k\omega + 2^{1/2}u_\alpha^-(k\omega)^{3/2}/(27n^3 Z_r Z_p \omega_a), \tag{G.10}$$

where

$$u_\alpha^- = n^2(n^2 - 6q^2 - 1) + 6n(2q - 1)(nq - n'q') - n'^2(n'^2 - 6q'^2 - 1). \tag{G.11}$$

The corresponding positions in the wavelength scale can be obtained by multiplying Eq. (G.10) by $\lambda_0/(2\pi)$, where λ_0 is the unperturbed wavelength of the spectral line.

Thus, for a fixed number k of field quanta involved in a resonance, a "doublet dip" arises in the vicinity of the point $\Delta\omega_\alpha^{\text{dip}}(k) = n^{-1}(nq - n'q')k\omega$ in the profile $I_{\alpha\beta}(\Delta\omega)$ of a lateral Stark component instead of a single dip. Both dips are significantly shifted (for most of the Stark components, the shift is to the blue wing).

It should be emphasized that a high spectral resolution is required for observing the doublet dip in experiments. Otherwise, the observation would show a singlet dip resulting from the merger of the components of the doublet dip.

Analogous qualitative changes occur in the vicinity of the points $\Delta\omega_\beta^{\text{dip}}(k) = n'^{-1}(nq - n'q')k\omega$ in the profile $I_{\alpha\beta}(\Delta\omega)$ of the same lateral Stark component where dips due to a resonance of the oscillatory field with a quasistatic Stark splitting of the lower multiplet would be located. Instead of this, now for each fixed number k, two dips may arise at the positions $\Delta\omega_\beta^{\text{dip}+}(k)$ and $\Delta\omega_\beta^{\text{dip}-}(k)$:

$$\Delta\omega_\beta^{\text{dip}\pm}(k) = n'^{-1}(nq - n'q')k\omega + 2^{1/2}u_\alpha^\pm(k\omega)^{3/2}/(27n'^3 Z_r Z_p \omega_a), \tag{G.12}$$

where

$$u_\alpha^\pm = n^2(n^2 - 6q^2 - 1) + 6n'(2q' \pm 1)(nq - n'q') - n'^2(n'^2 - 6q'^2 - 1). \tag{G.13}$$

Again, the corresponding positions in the wavelength scale can be obtained by multiplying Eq. (G.12) by $\lambda_0/(2\pi)$, where λ_0 is the unperturbed wavelength of the spectral line.

Now we consider consequences of the same resonances between a quasimonochromatic field and sublevels α, $\alpha \pm 1$ of the upper multiplet for the case in which the radiative transition $\alpha \leftrightarrow \beta$ corresponds to a central Stark component of a spectral line. If the quadrupole interaction between the perturber and the radiator were not taken into account, then the central component would be emitted at the position $\Delta\omega = 0$ for all values r of a reduced perturber–radiator separation including the resonance value $r^* = \delta a/(k\omega)^{1/2}$. In this dipole approximation, perturbing ions do not produce a so-called "inhomogeneous" SB in the vicinity of $\Delta\omega = 0$ — nor if they are quasistatic or impact — and without the inhomogeneous SB dips are not observable.

If we include the quadrupole perturber–radiator interaction, then at least in the wings of a central component, perturbing ions can produce quasistatic (and, consequently, the inhomogeneous) SB for dense plasmas, and resonances of the type (G6) now lead to the appearance of dips in the profile $I_{\alpha\beta}(\Delta\omega)$ of the central Stark component at the positions

$$\delta\omega_\alpha^{c\,\mathrm{dip}}(k) = (\delta b_\alpha - \delta b_\beta)/r^{*3}$$
$$\approx 2^{1/2}(27n^3 Z_r Z_p \omega_a)^{-1/2}(k\omega)^{3/2}u_c, \qquad \text{(G.14)}$$

where

$$u_c = n^2(n^2 - 6q^2 - 1) - n'^2(n'^2 - 6q'^2 - 1) \qquad \text{(G.15)}$$

Analogously, due to resonances between the quasimonochromatic electric field and sublevels β, $\beta \pm 1$ of the lower multiplet, another set of dips can appear in the profile $I_{\alpha\beta}(\Delta\omega)$ of the same central Stark component at the positions:

$$\delta\omega_\alpha^{c\,\mathrm{dip}}(k) = (27n'^3 Z_r Z_p \omega_a)^{-1/2} 2^{1/2}(k\omega)^{3/2}u_c. \qquad \text{(G.16)}$$

Thus, for each fixed number k of quanta involved in a resonance, a "doublet dip" can arise in the central Stark component profile $I_{\alpha\beta}(\Delta\omega,)$ as well as in the lateral Stark components. For central components, this pair of dips is always located in the blue wing.

The half-width $\delta\lambda_{1/2}$ of the L-dip (approximately equal to the separation between the primary minimum and the nearest bump) is controlled by the amplitude E_0 of the electric field of the Langmuir wave

$$\delta\lambda_{1/2} \approx (3/2)^{1/2}\lambda_0^2 n^2 E_0/(8\pi m_e e c Z_r), \qquad \text{(G.17)}$$

where λ_0 is the unperturbed wavelength of the spectral line. Thus, by measuring the experimental half-width of L-dips, one can determine the amplitude E_0 of the Langmuir wave.

It should be emphasized that the above theory did not take into account the fine structure $\Delta_{\rm fs}$ of hydrogenic spectral lines. This is legitimate for typical experimental situations where $\min[\omega, \delta\omega_s(E_0)] \gg \Delta_{fs}$ (here $\delta\omega_s(E_0)$ is the instantaneous Stark splitting in the static electric field of the strength E_0). The opposite case was considered by Gavrilenko *et al.* [5] using the example of the Lyman-alpha line of F IX. It resulted in a sequence of peaks (separated by troughs) whose positions depended on E_0 and ω in a complicated way.

References

[1] E. Oks, St. Böddeker and H.-J. Kunze, *Phys. Rev. A* **44** 8338 (1991).
[2] V.P. Gavrilenko and E. Oks, *Sov. Phys. J. Plasma Phys.* **13** 22 (1987).
[3] A.V. Demura and G.V. Sholin. *J. Quant. Spectrosc. Rad. Transfer* **15** 881 (1975).
[4] G.V. Sholin, *Optics Spectrosc.* **24** 275 (1969).
[5] V.P Gavrilenko, V.S. Belyaev, A.S. Kurilov *et al.*, *J. Phys. A: Math. Gen.* **39** 4353 (2006).

Appendix H

Effects of Langmuir Waves on Hydrogenic Spectral Lines under Strong Magnetic Fields

The highest-frequency electrostatic turbulence is Langmuir turbulence. Its connection to anomalous transport phenomena is known for many decades (see, e.g., paper [1]). The magnetic fusion research community is interested to find out whether Langmuir turbulence develops in magnetic fusion plasmas and, if it does, to determine its parameters. It is desirable to have spectroscopic diagnostics for this purpose, because they are "non-intrusive": they do not perturb parameters to be measured.

A number of spectroscopic methods for diagnosing Langmuir turbulence/oscillations in different kinds of plasmas have been developed and practically implemented by the author of this review and his collaborators, as presented in the book [2]. All these methods related to situations where the radiator (e.g., a hydrogen or deuterium atom) is subjected to a quasistatic electric field — in addition to the oscillatory electric field of Langmuir turbulence and to the broadband dynamic microfield due to plasma electrons. The quasistatic electric field was usually represented by the ion microfield (in the case where the latter was mostly quasistatic) and/or by a low-frequency electrostatic turbulence (e.g., by ionic sound).

In this situation, there occur the following two major effects of Langmuir turbulence on profiles of hydrogenic spectral lines. The first effect is an appearance of dips/depressions at distances from the unperturbed line position (in the frequency scale) that are proportional to the plasma electron frequency ω_p, the proportionality coefficients being rational numbers (expressed via the corresponding quantum numbers). Langmuir-wave-caused dips (hereafter, *L-dips*) in profiles of hydrogenic spectral lines were discovered experimentally in 1977 [3] and explained theoretically in papers [3–7] — see Appendix G. This effect was observed and used for diagnostics in a large number of experiments conducted by various experimental groups at different plasma sources (see, e.g., book [2]). The latest experimental results (obtained in a laser-produced plasma) can be found in papers [8, 9].

The second effect is an additional dynamical Stark Broadening (SB) [10, 11] (presented also in book [2]). In distinction to the first one, it was not widely used for diagnostics.

In all of the above experiments, magnetic fields did not play any substantial role. Therefore, for magnetic fusion plasmas, characterized by a strong magnetic field of several (up to 10) Tesla, possible effects of Langmuir turbulence on hydrogenic lines has been analyzed afresh in recent papers [12, 13]. Below, we follow these papers to present such analysis and methods for the spectroscopic diagnostics of Langmuir turbulence in magnetic fusion plasmas. The primary focus will be the additional dynamical broadening — for reasons explained below.

H.1 Broadening of Spectral Lines by Isotropic Langmuir Turbulence

In 1975 in paper [10], there were analytically derived additional contributions to the width and shift of hydrogenic spectral lines due to Langmuir turbulence for the case where the separation ω_F between sublevels of the principal quantum number n is caused by a quasistatic electric field F (hereafter, the "electric" case, for brevity). In this case, the separation between the Stark sublevels in the frequency scale is

$$\omega_F = 3n\hbar F/(2Z_r m_e e), \tag{H.1}$$

where Z_r is the nuclear charge of the radiator. The stochastic electric field of Langmuir turbulence was represented in [10] in the form:

$$\mathbf{E}_p(t) = \sum_{j=1}^{J} \mathbf{E}_j(t) \cos[\omega_j t + \varphi_j(t)], \tag{H.2}$$

where the phase $\varphi_j(t)$ and the amplitude $\mathbf{E}_j(t)$ change their values with the every change of the state of a Poisson process characterized by the average change frequency γ_p. Between the changes, the quantities $\varphi_j(t)$ and the components $\mathbf{E}_j^\sigma(t)$ are constant taking random values characterized by a certain distribution. In particular, the phase φ_j has a uniform distribution in the interval $(0, 2\pi)$ with the density $1/(2\pi)$. The stochastic function $\mathbf{E}_j(t)$ in (H.2) is the realization of a kangaroo-type uniform Markovian stationary stochastic process. A convenient characteristic is the root-mean-square average $E_0 = (\langle|\mathbf{E}_j(t)|^2\rangle)^{1/2}$, which is called for brevity the average amplitude.

The main frequencies ω_j are all approximately equal to the plasma electron frequency:

$$\omega_p = (4\pi e^2 N_e/m_e)^{1/2} = 5.641 \times 10^4[N_e(\text{cm}^{-3})]^{1/2}. \tag{H.3}$$

The frequency $\gamma_p < \omega_p$ is the sum of the characteristic frequencies of the following processes in plasmas: the electron-ion collision rate γ_{ei} (see, e.g., [14]), the average Landau damping rate γ_L (see, e,g, [15]), the characteristic frequency γ_{ind} of the nonlinear mechanism of the induced scattering of Langmuir plasmons on ions (see, e.g., [16]), the characteristic frequency γ_{dec} of the nonlinear decay of Langmuir plasmons into ion acoustic waves (see, e.g., [16]), and so on. The frequency γ_p is assumed to control the width of the power spectrum of Langmuir turbulence.

The additional contributions to the width and shift of hydrogenic spectral lines due to Langmuir turbulence, derived analytically in paper [10] for the case of a weak (or zero) magnetic field, depend on the separation between the Stark sublevels ω_F caused by a quasistatic electric field as follows:

$$\Gamma_{\alpha\beta} = \Gamma_\alpha + \Gamma_\beta - d_{\alpha\alpha} d_{\beta\beta} E_0^2 \gamma_p / [3\hbar^2(\gamma_p^2 + \omega_p^2)], \quad D_{\alpha\beta} = D_\alpha + D_\beta, \tag{H.4}$$

where,

$$\begin{aligned} \Gamma_\alpha &= [E_0^2\gamma_p/(12\hbar^2)]\{2\mathbf{d}_{\alpha\alpha}^2/(\gamma_p^2 + \omega_p^2) + (|\mathbf{d}_{\alpha,\alpha-1}|^2 + |\mathbf{d}_{\alpha,\alpha+1}|^2) \\ &\quad \times [1/(\gamma_p^2 + (\omega_F - \omega_p)^2) + 1/(\gamma_p^2 + (\omega_F + \omega_p)^2)]\}, & \text{(H.5)} \\ D_\alpha &= [E_0^2\gamma_p/(12\hbar^2)](|\mathbf{d}_{\alpha,\alpha-1}|^2 - |\mathbf{d}_{\alpha,\alpha+1}|^2)[(\omega_F - \omega_p)/(\gamma_p^2 + (\omega_F \\ &\quad - \omega_p)^2) + (\omega_F + \omega_p)/(\gamma_p^2 + (\omega_F + \omega_p)^2)]. & \text{(H.6)} \end{aligned}$$

Here, $\Gamma_{\alpha\beta} = -\mathrm{Re}\,\Phi_{\alpha\beta}$ and $D_{\alpha\beta} = -\mathrm{Im}\,\Phi_{\alpha\beta}$ the Langmuir-wave-caused contributions to the diagonal elements of the impact broadening operator Φ; α and β label sublevels of the upper (a) and lower (b) levels involved in the radiative transition. The matrix elements of the dipole moment operator in Eqs. (H.4)–(H.6) are

$$\begin{aligned} \mathbf{d}_{\alpha\alpha}^2 &= [3ea_0 n_\alpha q_\alpha/(2Z_r)]^2, |\mathbf{d}_{\alpha,\alpha-1}|^2 - |\mathbf{d}_{\alpha,\alpha+1}|^2 \\ &= \mathbf{d}_{\alpha\alpha}^2/q_\alpha, |\mathbf{d}_{\alpha,\alpha-1}|^2 + |\mathbf{d}_{\alpha,\alpha+1}|^2) \\ &= \mathbf{d}_{\alpha\alpha}^2(n^2 - q^2 - m^2 - 1)_\alpha/(2q_\alpha^2), & \text{(H.7)} \end{aligned}$$

where a_0 is the Bohr radius; $q = n_1 - n_2$; n_1, n_2 and m are the parabolic quantum numbers. In Eqs. (H.5), (H.6) in the subscripts, we used the notation $\alpha + 1$ and $\alpha - 1$ for the Stark sublevels of the energies $+\hbar\omega_B$ and $-\hbar\omega_B$, respectively (compared to the energy of the sublevel α). Formulas for Γ_β and D_β entering Eq. (H.4) can be obtained from Eqs. (H.5), (H.6) by substituting the superscript α by β. For brevity we call $\Gamma_{\alpha\beta}$ and $D_{\alpha\beta}$ the width and the shift, respectively.

However, for the conditions typical for magnetic fusion plasmas — in particular, in the tokamak divertor region — the ion microfield is not quasistatic for the most intense (low-n) hydrogen/deuterium spectral lines (HDSL). Therefore, at the absence of a strong low-frequency electrostatic turbulence, the separation between sublevels of the principal quantum number n is caused by a relatively strong magnetic field B (so that in this case, these are Zeeman sublevels rather than the Stark sublevels):

$$\omega_B = eB/(2m_e c). \tag{H.8}$$

Thus, in this "magnetic" case, the Langmuir-wave-caused contributions to the diagonal elements $\Gamma_{\alpha\beta} = -Re\,\Phi_{\alpha\beta}$ and $D_{\alpha\beta} = -Im\,\Phi_{\alpha\beta}$ of the impact broadening operator Φ can be obtained from the corresponding results from Eqs. (H.4)–(H.6) by substituting ω_F by ω_B:

$$\Gamma_{\alpha\beta} = \Gamma_\alpha + \Gamma_\beta - d_{\alpha\alpha} d_{\beta\beta} E_0^2 \gamma_p / [3\hbar^2(\gamma_p^2 + \omega_p^2)], \quad D_{\alpha\beta} = D_\alpha + D_\beta, \tag{H.9}$$

where

$$\begin{aligned} \Gamma_\alpha = {} & [E_0^2 \gamma_p/(12\hbar^2)]\{2\mathbf{d}_{\alpha\alpha}^2/(\gamma_p^2 + \omega_p^2) + (|\mathbf{d}_{\alpha,\alpha-1}|^2 + |\mathbf{d}_{\alpha,\alpha+1}|^2) \\ & \times [1/(\gamma_p^2 + (\omega_B - \omega_p)^2) + 1/(\gamma_p^2 + (\omega_B + \omega_p)^2)]\}, \end{aligned} \tag{H.10}$$

$$\begin{aligned} D_\alpha = {} & [E_0^2 \gamma_p/(12\hbar^2)](|\mathbf{d}_{\alpha,\alpha-1}|^2 - |\mathbf{d}_{\alpha,\alpha+1}|^2)[(\omega_B - \omega_p)/(\gamma_p^2 + (\omega_B \\ & - \omega_p)^2) + (\omega_B + \omega_p)/(\gamma_p^2 + (\omega_B + \omega_p)^2)]. \end{aligned} \tag{H.11}$$

Formulas for Γ_β and D_β entering Eq. (H.9) can be obtained from Eqs. (H.10), (H.11) by substituting the superscript α by β.

Let us analyze the above results for the width — because it is often practically more important than the shift. The expressions for the width demonstrate the following two characteristic features.

For relatively large magnetic fields, such that

$$\omega_B \gg \omega_p, \tag{H.12}$$

the term containing the diagonal matrix elements $\mathbf{d}_{\alpha\alpha}^2 + \mathbf{d}_{\beta\beta}^2$ predominates, so that the other term can be neglected. The dominating term is the *adiabatic* contribution: it does not couple (by virtual transitions) different Zeeman sublevels — in distinction to the neglected term. Under the same condition (H.12), the non-diagonal matrix elements of the impact broadening operator become much

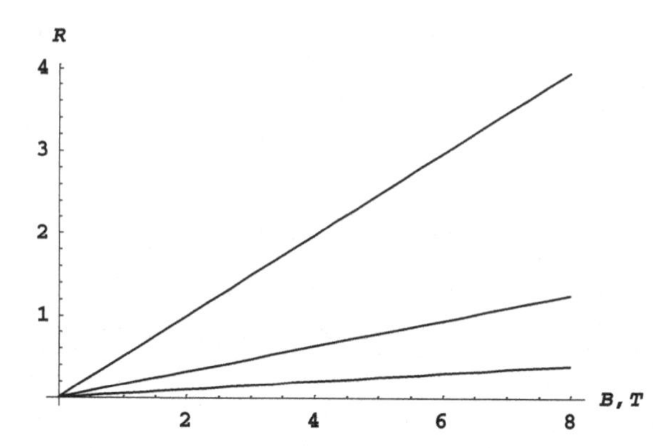

Fig. H.1. The ratio of the frequencies $R = \omega_B/\omega_p$ versus the magnetic field B in tesla for three different electron densities N_e : $10^{13}\,\mathrm{cm}^{-3}$ (the upper line), $10^{14}\,\mathrm{cm}^{-3}$ (the middle line), $10^{15}\,\mathrm{cm}^{-3}$ (the lower line).

smaller than the diagonal elements, so that the quantity $\Gamma_{\alpha\beta}$ from Eq. (H.9) becomes a "true width."

Figure H.1 shows the ratio $R = \omega_B/\omega_p$ versus the magnetic field B for three different electron densities N_e. It is seen that even for $N_e = 10^{13}\,\mathrm{cm}^{-3}$, which is usually considered as the lowest electron density relevant to magnetic fusion plasmas, the fulfillment of the condition (H.12) requires magnetic fields greater than 10 tesla.[1]

[1] In principle, there might also exist another adiabatic effect of the stochastic electric field of Langmuir turbulence if $\gamma_p \ll \omega_p$: the formation of satellites separated by $\pm k\omega_p$ (in the frequency scale) from each component of the Zeeman triplet ($k = 1, 2, 3, \ldots$). For the case, where Langmuir turbulence develops anisotropically in such a way, that its electric field is linearly-polarized, the satellite intensities were calculated analytically in paper [99] (see also book [62]). However, the satellite intensities are relatively small. Even for the most intense satellite ($k = 1$), the ratio of its intensity I_s to the intensity of the corresponding component of the Zeeman triplet I_0 is

$$I_s/I_0 \sim (n^2 T_e/U_{Hi})[E_0^2/(8\pi N_e T_e)].$$

Here, $U_{Hi} = 13.6\,\mathrm{eV}$ is the ionization potential hydrogen/deuterium atoms, T_e is the electron temperature; the quantity $E_0^2/(8\pi N_e T_e)$, which is called the degree of turbulence, is the ratio of the energy density of the Langmuir turbulence to the thermal energy density of the plasma. The latter ratio is always much smaller than unity: usually, it is in the range 10^{-2}–10^{-4}. Given that for spectroscopic experiments related to tokamak divertors, where the most intense hydrogenic

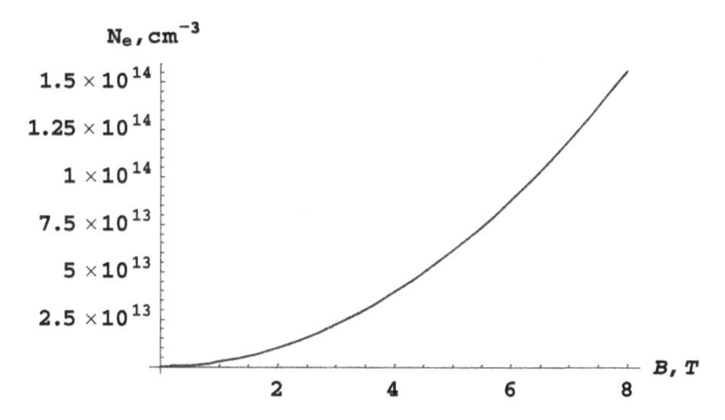

Fig. H.2. The line (the geometric set of points) in the plane (B, N_e) corresponding to the resonance: $\omega_B = \omega_p$. Here, B is the magnetic field in tesla, N_e is the electron density in cm^{-3}.

The most interesting is another scenario, where

$$\omega_B = \omega_p. \tag{H.13}$$

This resonance can occur exactly or approximately for a number of pairs (B, N_e) typical for the conditions of tokamak divertors. Indeed, from Fig. H.2, which shows the line (the geometric set of points) in the plane (B, N_e) corresponding to the resonance (H.13), it follows that the resonance takes place, e.g., for $B = 2$ T and $N_e = 10^{13}\,\mathrm{cm}^{-3}$, or for $B = 5$ T and $N_e = 6 \times 10^{13}\,\mathrm{cm}^{-3}$, or for $B = 8$ T and $N_e = 1.6 \times 10^{14}\,\mathrm{cm}^{-3}$.

In the conditions close to the resonance (H.13), the Langmuir-wave-caused Stark width dramatically increases. Neglecting the non-resonance terms in Eqs. (H.10), (H.11), it can be represented in the form:

$$\Gamma_{\alpha\beta} = (|\mathbf{d}_{\alpha,\alpha-1}|^2 + |\mathbf{d}_{\alpha,\alpha+1}|^2 + |\mathbf{d}_{\beta,\beta-1}|^2 + |\mathbf{d}_{\beta,\beta+1}|^2) E_0^2/(12\hbar^2\gamma_p). \tag{H.14}$$

lines are used $(L_\alpha, L_\beta, H_\alpha, H_\beta)$ one has $n^2 T_e/U_{Hi} \sim 1$, it is seen that indeed $I_s/I_0 \sim (10^{-2} - 10^{-4}) \ll 1$. Thus, these satellites do not seem to be useful for diagnostics of magnetic fusion plasmas unless highly-excited hydrogenic lines $(n \gg 1)$ are employed.

We note that all terms in Eq. (H.14) correspond to the *non-adiabatic* contribution: they couple by virtual transitions different Zeeman sublevels.

Now let us compare $\Gamma_{\alpha\beta}$ from Eq. (H.14) to the width due to the competing Stark Broadening (SB) mechanism. For the typical parameters of tokamak divertors, the magnetic field causes a significant decrease of the non-adiabatic contribution to the dynamical Stark width due to ions — as discussed in detail above in Appendix E. Thus, the total contribution to the dynamical Stark width due to ions can be well represented by the adiabatic contribution given by Eq. (E.2) of the generalized theory.

The ratio of $\Gamma_{\alpha\beta}$ from Eq. (H.14) to the corresponding contribution due to the dynamical broadening by ions from Eq. (E.2) can be represented as the product of five dimensionless factors as follows:

$$\begin{aligned}\Gamma_{\alpha\beta}/\gamma_{\alpha\beta} = \{&\pi^{1/2}(|\mathbf{d}_{\alpha,\alpha-1}|^2 + |\mathbf{d}_{\alpha,\alpha+1}|^2 + |\mathbf{d}_{\beta,\beta-1}|^2 \\ &+ |\mathbf{d}_{\beta,\beta+1}|^2)/[272^{1/2}(ea_0X_{\alpha\beta})^2 I(R_i)]\} \\ &\times (m_e/M)^{1/2}[T_e\rho_{\mathrm{D}}/e^2](\omega_p/\gamma_p)[E_0^2/(8\pi N_e T_e)]. \quad \text{(H.15)}\end{aligned}$$

The quantity $E_0^2/(8\pi N_e T_e)$ is the degree of turbulence: the ratio of the energy density of Langmuir turbulence to the thermal energy density of the plasma. We note that the right side of Eq. (H.15) can be simplified to a more explicit scaling: $\Gamma_{\alpha\beta}/\gamma_{\alpha\beta}$ is proportional to $E_0^2 T_e^{1/2}/(N_e\gamma_p M^{1/2})$, if $T_i = T_e$. However, the representation of $\Gamma_{\alpha\beta}/\gamma_{\alpha\beta}$ as the product of the four dimensionless factors in (H.15) provides a better physical understanding and is more convenient for estimates. A practical formula for the product of the second and third factors in the right side of Eq. (H.15) is

$$\begin{aligned}&(m_e/M)^{1/2}[T_e\rho_{\mathrm{D}}/e^2] \\ &\quad = 1.204 \times 10^8[T_e(\mathrm{eV})]^{3/2}[N_e(\mathrm{cm}^{-3})]^{-1/2}(M_p/M)^{1/2}. \quad \text{(H.16)}\end{aligned}$$

Let us estimate the ratio $\Gamma_{\alpha\beta}/\gamma_{\alpha\beta}$ for a hydrogen plasma (so that $M = M_p/2$) of the electron density $N_e = 6 \times 10^{13}\,\mathrm{cm}^{-3}$ and of the temperature $T_e = 5\,\mathrm{eV}$. From Eq. (H.16), we get: $(m_e/M)^{1/2}[T_e\rho_D/e^2] = 246 \gg 1$, so that $\Gamma_{\alpha\beta}/\gamma_{\alpha\beta} \sim 2\times 10^2(\omega_p/\gamma_p)[E_0^2/(8\pi N_e T_e)]$. The ratio ω_p/γ_p is a large quantity — typically in the range of (10^2–10^4), while the degree of turbulence $E_0^2/(8\pi N_e T_e)$ is a small quantity — typically in the range of (10^{-4}–10^{-2}). Since, the first factor in (H.15) is typically ~0.5 for magnetic fusion plasmas, we obtain the following range: $\Gamma_{\alpha\beta}/\gamma_{\alpha\beta} \sim (1 - 10^4)$.

This example shows that for magnetic fusion plasmas, the contribution to the dynamical Stark width due to Langmuir turbulence can dominate over the competing dynamical SB by ions, so that the half-width-at-half-maximum of a hydrogenic line will be $\delta\lambda_{1/2} = [\lambda_0^2/(2\pi c)]\Gamma_{\alpha\beta}$, where λ_0 is the unperturbed wavelength and $\Gamma_{\alpha\beta}$ is given by Eq. (H.14). Thus, it can be used for diagnostics of Langmuir turbulence. Specifically, from the experimentally measured Stark width of hydrogenic spectral lines in the conditions close to the resonance, it is possible to determine the quantity E_0^2/γ_p — as it is seen from Eq. (H.14).

H.2 Polarization Analysis of the Spectral Line Broadening by Anisotropic Langmuir Turbulence

The above results are obtained for the case where Langmuir turbulence developed isotropically. Now we address the situation where it develops *anisotropically*. In this case, the spectroscopic diagnostic can be significantly enhanced by the polarization analysis, allowing to obtain an additional information — such as the *degree of anisotropy* of the distribution of Langmuir turbulence.

In 1973, in paper [17] a theory of optical polarization measurements of low-frequency electrostatic turbulence in plasmas was presented. It was successfully implemented in a later experiment [18]. In 1977, in paper [11], was developed a theory of optical polarization measurements of Langmuir turbulence for the "electric" case, i.e., for the situation where the separation ω_F between sublevels of the principal quantum number n is caused by a quasistatic electric field F. However, formulas presented below are actually valid also for the "magnetic" case, i.e., for the situation where the separation ω_B between sublevels of the principal quantum number n is caused by a relatively strong magnetic field B.

Following paper [17], we consider a typical situation where the directionality diagram of Langmuir turbulence has the axial symmetry. Profiles of a chosen spectral line are observed at two directions of a linear polarizer: 1 — parallel to the axis of symmetry, 2 — perpendicular to the axis of symmetry. Hydrogenic spectral lines consist of π- and σ-components, having different polarization properties. Therefore, in accordance to paper [17], there are the following four functions controlling the angular dependence of the characteristics of the π- and σ-components of the spectral line observed at the polarizer directions 1 and 2

$$f_{1\pi} = \cos^2\theta, \quad f_{2\pi} = f_{1\sigma} = 3(1 - \cos^2\theta)/2, \quad f_{2\sigma} = 3(1 + \cos^2\theta)/4, \tag{H.17}$$

where θ is the polar angle with respect to the axis of symmetry, which is the direction of the magnetic field $\mathbf{B}$. The axially-symmetric Sholin–Oks' distribution $W(E, \cos\theta)$ of the amplitude of the electric field of Langmuir turbulence from Eq. (F.12) can be re-written in the form:

$$W(E, \cos\theta) = [2E^2/(\pi^{1/2}\eta E^3_{\text{perp}})] \exp\{-(E/E_{\text{perp}})^2 \times [1 - \cos^2\theta(1 - 1/\eta^2)]\}. \tag{H.18}$$

Here, η is the degree of anisotropy defined as:

$$\eta = 2^{1/2} E_{\text{par}}/E_{\text{perp}}, \tag{H.19}$$

where E_{par} and E_{perp} are the root-mean-squared values of the amplitude in the directions parallel and perpendicular to the axis of symmetry, respectively.

Using Eqs. (H.17)–(H.19), we obtain the following results for the relative difference of the Langmuir-wave-caused contributions to the dynamical Stark width of π- and σ-components of the spectral line in two perpendicular polarizations:

$$P_\pi(\eta) = (\Gamma_{1\pi} - \Gamma_{2\pi})/(\Gamma_{1\pi} + \Gamma_{2\pi}) = (\eta^2 - 1)/(\eta^2 + 1), \tag{H.20}$$

$$P_\sigma(\eta) = (\Gamma_{1\sigma} - \Gamma_{2\sigma})/(\Gamma_{1\sigma} + \Gamma_{2\sigma}) = (1 - \eta^2)/(3 + \eta^2). \tag{H.21}$$

Thus, from the experimental polarization difference of the Langmuir-wave-caused dynamical Stark width, it is possible to deduce the degree of anisotropy η — in addition to the quantity E_0^2/γ_p.

H.3 Langmuir-wave-caused Dips in the "Magnetic" Case

The central point of the L-dip phenomenon was a resonant coupling between a quasistatic electric field **F** and an oscillatory electric field of the Langmuir wave — as presented in detail in Appendix G. For example, in the profile of the Stark component of the Lyman line originating from the sublevel q, the resonance could manifest, generally, as two dips (L^+-dip and L^--dip) located at the following distances $\Delta\lambda_{\pm}^{\text{dip}}$ from the unperturbed wavelength λ_0 of this Lyman line:

$$\Delta\lambda_{\pm}^{\text{dip}} = -[\lambda_0^2/(2\pi c)]\{q\omega_p + [2\omega_p^3/(27n^3 Z_r Z_i \omega_{\text{at}})]^{1/2} \times [n^2(n^2 - 6q^2 - 1) + 12n^2q^2 \pm 6n^2q]\}. \quad \text{(H.22)}$$

Here, $\omega_{\text{at}} = m_e e^4/\hbar^3 \cong 4.14 \times 10^{16}\,\text{s}^{-1}$ is the atomic unit of frequency, n is the principal quantum numbers. The first, primary term in braces reflects the dipole interaction with the ion microfield. The second term in braces takes into account — via the quadrupole interaction — a spatial nonuniformity of the ion microfield. This second term is, generally speaking, a correction to the first term — except for the case of the central Stark component ($q = 0$), for which the first term vanishes. We note that in the profile of the central Stark component, there could be only one L-dip (hereafter, "central L-dip") since the term $\pm 6n^2q$ vanishes. A formula for the L-dip positions in profiles of hydrogenic lines from other spectral series (Balmer, etc.) can be found in Appendix G.

It is important to emphasize the following. For a given electron density N_e, the value of the plasma electron frequency ω_p is fixed — in accordance to Eq. (H.3). The resonance occurs when the separation between the Stark sublevels of the principal quantum number n

caused by the field F

$$\omega_F = 3n\hbar F/(2Z_r m_e e) \tag{H.23}$$

is equal to ω_p:

$$\omega_F = \omega_p. \tag{H.24}$$

The quasistatic electric field in plasmas has a broad distribution over the ensemble of radiators — regardless of whether this field represents the ion microfield or the low-frequency electrostatic turbulence. Therefore, if the ion microfield is mostly quasistatic or a low-frequency electrostatic turbulence has been developed in the plasma, then there would always be a fraction of radiators, for which the resonance condition (H.24) is satisfied.

However, for the conditions typical for magnetic fusion plasmas — in particular, in the tokamaks divertor region — the ion microfield is not quasistatic. Therefore, at the absence of a low-frequency electrostatic turbulence, the separation $\omega_B = eB/(2m_e c)$ between sublevels of the principal quantum number n is caused by a relatively strong magnetic field B (so that in this case these are Zeeman sublevels rather than the Stark sublevels). Then the resonance condition is given by

$$\omega_B = \omega_p. \tag{H.25}$$

instead of Eq. (H.24).

In this situation, the following two conditions are necessary for observing L-dips. First, the magnetic field should have a noticeable nonuniformity ΔB across the region, from which a particular hydrogenic line is emitted:

$$\Delta B/B > [\lambda_0/(2\pi c)]n^2\hbar E_0/(m_e e Z_r), \tag{H.26}$$

Second, the Langmuir electric field should not be too strong:

$$n^2\hbar E_0/(m_e e Z_r) < \gamma_p. \tag{H.27}$$

Under conditions (H.26), (H.27), it could be possible to observe an L-dip in the profile of each component of the Zeeman triplet.

The halfwidth of the L-dip $\delta\lambda_{1/2}$ would be controlled only by one parameter of Langmuir turbulence — by the averaged amplitude E_0,

$$\delta\lambda_{1/2} \cong (3/2)^{1/2}\lambda_0^2 n^2 \hbar E_0/(8\pi m_e ecZ_r), \tag{H.28}$$

so that the other parameter, namely γ_p, would not enter formula (H.28).

Therefore, the following diagnostic method can be proposed. If L-dips are observed in the profiles of the components of the Zeeman triplet, one can first deduce the averaged amplitude E_0 of the Langmuir electric field from the experimental halfwidth of the L-dip using Eq. (H.28). Then from the experimental halfwidth of the components of the Zeeman triplet, one can infer the quantity E_0^2/γ_p via Eq. (H.14) and thus (since E_0 would be already determined) the characteristic frequency γ_p of the nonlinear process controlling the width of the power spectrum of Langmuir turbulence. Finally, the polarization analysis can allow deducing the degree of anisotropy of Langmuir turbulence $\eta = 2^{1/2}E_{\text{par}}/E_{\text{perp}}$.

Thus, the combination of these diagnostics can allow the experimental determination of the following parameters of Langmuir turbulence: the average amplitude E_0, the characteristic frequency γ_p of the nonlinear process controlling the width of the power spectrum of Langmuir turbulence, and the degree of anisotropy $\eta = 2^{1/2}E_{\text{par}}/E_{\text{perp}}$.

We emphasize that one of the experimental manifestations of the dynamical SB by Langmuir turbulence would be unusually long Lorentzian-shape wings of hydrogen or deuterium spectral lines.

References

[1] V.N. Tsytovich and L. Stenflo, *Phys. Scripta* **12** 323 (1975).
[2] E. Oks, *Plasma Spectroscopy: The Influence of Microwave and Laser Fields*, Springer Series on Atoms and Plasmas, vol. 9, Springer, New York (1995).
[3] A.I. Zhuzhunashvili and E. Oks, *Sov. Phys. JETP* **46** 2142 (1977).
[4] E. Oks and V.A. Rantsev-Kartinov, *Sov. Phys. JETP* **52** 50 (1980).
[5] V.P. Gavrilenko and E. Oks, *Sov. Phys. JETP* **53** 1122 (1981).
[6] V.P. Gavrilenko and E. Oks, *Sov. Phys. Plasma Phys.* **13** 22 (1987).

[7] E. Oks, St. Böddeker and H.-J. Kunze, *Phys. Rev. A* **44** 8338 (1991).
[8] O.Renner, E. Dalimier, E. Oks, F. Krasniqi, E. Dufour, R. Schott and E. Förster, *J. Quant. Spectr. Rad. Transfer* **99** 439 (2006).
[9] E.Oks, E. Dalimier, A.Ya. Faenov *et al.*, Fast track communications, *J. Phys. B: At. Mol. Opt. Phys.* **47** 221001 (2014).
[10] E. Oks and G.V. Sholin, *Sov. Phys. JETP* **41** 482 (1975).
[11] E. Oks and G.V. Sholin, *Opt. Spectrosc.* **42** 434 (1977).
[12] E. Oks, *J. Phys. B: Atom. Mol. Opt. Phys.*, Fast Track Communication **44** 101004 (2011).
[13] E. Oks, *Intern. Review of Atom. and Mol. Phys.* **1** 169 (2010).
[14] J.D. Huba, *NRL Plasma Formulary*, Naval Research Laboratory: Washington, DC (2013).
[15] P.M. Bellan, 2006 *Fundamentals of Plasma Physics*, Cambridge University Press, Cambridge (2006).
[16] B.B. Kadomtsev, *Collective Phenomena in Plasma*, Nauka, Moscow (1988).
[17] G.V. Sholin and E. Oks, *Sov. Phys. Doklady* **18** 254 (1973).
[18] M.V. Babykin, A.I. Zhuzhunashvili, E. Oks, V.V. Shapkin and G.V. Sholin, *Sov. Phys. JETP* **38** 86 (1974).

Appendix I

Stark Broadening of Hydrogen Lines at Super-High Densities: Effects of Plasma Turbulence at the Thermal Level

Benchmark experiments, i.e., experiments where plasma parameters were measured independently of the Stark Broadening (SB), play a very important role in spectroscopic diagnostics of plasmas. As a new benchmark experiment was performed at some novel plasma source at the range of the electron densities N_e higher than for the previous benchmark experiment performed at a different plasma source, almost always discrepancies were found with existing theories. In this way, benchmark experiments stimulated developing more advanced theories — the theories allowing for various high-density effects. There is a huge amount of literature on this subject. By limiting ourselves by books and reviews published in the last 10 years, we point out books [1,2], review [3], and references in these publications.

The most recent (year 2014) benchmark experiment by Kielkopf and Allard (hereafter, KA) [4], where the SB of Hydrogen Spectral Lines (SBHSL) was tested using the H_α line, was performed at a laser-produced pure-hydrogen plasma reaching $N_e = 1.4 \times 10^{20}\,\mathrm{cm}^{-3}$. This exceeded by two orders of magnitude the highest values of $N_e(3-4) \times 10^{18}\,\mathrm{cm}^{-3}$ reached by the corresponding previous benchmark experiments: by Kunze group (Büscher *et al.* [5]) at the gas-liner pinch[1] and by Vitel group (Flih *et al.* [6]) at the flash tube plasma.

At the electron densities reached in KA experiment [4], no theoretical calculations of the Full Width at Half Maximum (FWHM) of the H_α line existed. Indeed, the highest value of N_e in the tables of FWHM of the H_α line by Gigosos and Cardenoso [8], produced by fully-numerical simulations, was 4.64×10^{18} cm^{-3}(their simulations are considered by the research community as the most advanced). In frames of the so-called conventional analytical theory, Kepple and Griem [9] calculated the FWHM of the H_α line up to $N_e = 10^{19}$ cm^{-3}, because at higher values of N_e the conventional theory becomes invalid. (The primary distinction between the conventional analytical theory [9] and Gigosos-Cardenoso simulations [8] is that the latter allowed for the ion dynamics in distinction to the former; however, the role of the ion dynamics diminishes as N_e increases and becomes practically insignificant at values of $N_e \sim 10^{19}$ cm^{-3} and higher.) All other simulations and analytical methods (except the one discussed in the next paragraph), reviews of which can be found, e.g., in Appendix A, as well as in book [1] and paper [3], listed the FWHM of the H_α line either up to $N_e \sim 4 \times 10^{18}$ cm^{-3} or lower.

As for the theory by Kielkopf and Allard in their paper [4] extended above $N_e > 10^{20}$ cm^{-3}, it is inconsistent because it completely neglects the contribution of plasma electrons to the SB. If Kielkopf and Allard would have attempted including the broadening by electrons, their theoretical widths would have increased by about (50–60)% and thus would have overestimated the experimental widths at $N_e > 10^{19}$ cm^{-3} by about (50–60)%. By the way, at the

[1]In the earlier experiment at the gas-liner pinch (Böddeker *et al.* [7]), the densities up to $N_e \sim 10^{19}$ cm^{-3} had been reached. However, the experiment by Böddeker *et al.* [7] had deficiencies, which were addressed and eliminated in the experiment by Büscher *et al.* [5]. In distinction to the former experiment, in the latter one: (a) the spectroscopic measurements were performed simultaneously with the diagnostics; (b) highly reproducible discharge condition was used where the H_α line was measured spatially resolved along the discharge axis indicating that no inhomogeneities along the axis existed; (c) high care has been taken to prevent the optical thickness.

lowest density of Kielkopf–Allard experiment $N_e = 8.65 \times 10^{17}$ cm^{-3}, the experimental width is 50% greater than their theoretical width, so that the inclusion of the broadening by electrons would bring their theoretical width in agreement with the experimental width for $N_e < 10^{19}$ cm^{-3}, and thus in agreement with other theories for $N_e < 10^{19}$ cm^{-3}; this fact further underscores the inconsistent nature of Kielkopf–Allard theory. The contributions of ions and electrons to the broadening are of the same order of magnitude. For the above reason, the "agreement" of their theory with their experimental widths at $N_e > 10^{19}$ cm^{-3} is fortuitous.

Therefore, here we present a consistent analytical theory that is relevant to the range of the electron densities reached in KA experiment [4]. At this range of N_e, a new factor becomes significant for the SBHSL — the factor never taken into account in any previous simulations or analytical theories of the SBHSL. This new factor is a rising contribution of the Electrostatic Plasma Turbulence (EPT) at the *thermal* level of its energy density.

The EPT at any level of its energy density is represented by oscillatory electric fields F_t arising when the waves of the separation of charges propagate through plasmas: they correspond to *collective degrees of freedom* in plasmas — in distinction to the electron and ion microfields that correspond to individual degrees of freedom of charged particles. In relatively low density plasmas, various kinds of the EPT at the *supra-thermal* levels of its energy density, specifically at the levels several orders of magnitude higher than the thermal level, were discovered experimentally via the enhanced ("anomalous") SBHSL in numerous experiments performed by different groups at various plasma sources [10–25], some of these experiments being summarized in book [26].

As for the EPT at the thermal level of its energy density (hereafter, "thermal EPT"), their contribution to the SBHSL in relatively low density plasmas is by several orders of magnitude smaller than the contribution of the electron and ion microfields, so that their effect was negligibly small and therefore never detected spectroscopically. However, at the range of the electron densities reached in KA experiment [4], the contribution to the SBHSL from the thermal EPT becomes comparable to the contribution of the electron and ion microfields.

Here, we take into account the contribution to the SBHSL from the thermal EPT. As a result, the theoretical FWHM of the H_α line becomes in a very good agreement with the experimental FWHM of the H_α line by KA [4] in the entire range of their electron densities, including the highest electron density $N_e = 1.4 \times 10^{20}$ cm^{-3}, as shown below.

According to Bohm and Pines [27], the number of collective degrees of freedom in a unit volume of a plasma is $N_{\text{coll}} = 1/(6\pi^2 r_D^3)$, where r_D is the Debye radius. Therefore the energy density of the oscillatory electric fields at the thermal level is $F_t^2/(8\pi) = N_{\text{coll}}T/2$, so that

$$F_t^2 = 16\pi^{1/2} e^3 N_e^{3/2}/(3T^{1/2}), \tag{I.1}$$

where T is the temperature and e is the electron charge.

At the absence of a magnetic field, there are only two types of the EPT: Langmuir waves/turbulence and ion acoustic waves/turbulence (a.k.a. ionic sound). Langmuir waves are the high-frequency branch of the EPT. Its frequency is approximately the plasma electron frequency

$$\omega_{pe} = (4\pi e^2 N_e/m_e)^{1/2} = 5.64 \times 10^4 N_e^{1/2}. \tag{I.2}$$

Ion acoustic waves are the low-frequency branch of the EPT. They are represented by a broadband oscillatory electric field, whose frequency spectrum is below or of the order of the ion plasma frequency

$$\omega_{pi} = (4\pi e^2 N_i Z^2/m_i)^{1/2} = 1.32 \times 10^3 Z(N_i m_p/m_i)^{1/2}, \tag{I.3}$$

where N_i is the ion density, Z is the charge state; m_e, m_p and m_i are the electron, proton and ion masses, respectively. In the "practical" parts of Eqs. (I.2) and (I.3), CGS units are used. Below we set $Z = 1$, so that $N_i = N_e$.

The thermal energy density of the collective degrees of freedom $E_{\text{tot}}^2/(8\pi) = N_{\text{coll}}T/2$ is distributed in equal parts between the high- and low-frequency branches:

$$F_0^2 = E_0^2 = E_{\text{tot}}^2/2, \tag{I.4}$$

where F_0, E_0 and E_{tot} are the root-mean-square (rms) thermal electric fields of the ion-acoustic turbulence, the Langmuir turbulence, and the total turbulence, respectively.

Let us first discuss the contribution of the thermal ion-acoustic turbulence to the SBHSL. It is useful to begin by estimating a ratio of the rms thermal electric field F_0 of the ion-acoustic turbulence to the standard characteristic value F_N of the ion microfield, where

$$\begin{aligned} F_N &= 2\pi(4/15)^{2/3} e N_e^{2/3} = 2.603 e N_e^{2/3} \\ &= 3.751 \times 10^{-7} [N_e(\text{cm}^{-3})]^{2/3} \text{V/cm} \end{aligned} \tag{I.5}$$

Using Eqs. (I.1), (I.4), (I.5), we obtain:

$$F_0/F_N = 0.1689 N_e^{1/12}/[T(K)]^{1/4}, \tag{I.6}$$

where the temperature T is in Kelvin. At the highest density point of KA experiment [4] ($N_e = 1.39 \times 10^{20}$ cm^{-3}, $T = 34486$ K), Eq. (I.6) yields $F_0/F_N = 0.59$. This shows that in the conditions of KA experiment [4], the rms thermal electric field of the ion-acoustic turbulence becomes comparable to the standard ion microfield.

In a broad range of plasma parameters, especially at the range of densities of KA experiment [4], radiating hydrogen atoms perceive oscillatory electric fields of the ion-acoustic turbulence as quasistatic. In any code designed for calculating shapes of spectral lines from plasmas, an important task becomes the averaging over the ensemble distribution $W(\mathbf{F})$ of the total quasistatic field $\mathbf{F} = \mathbf{F}_t + \mathbf{F}_i$, where $\mathbf{F}_i$ is the quasistatic part of the ion microfiled (for the range of densities of KA [4] almost the entire ion microfiled is quasistatic). In other words, the key part of the problem becomes the calculation of $W(\mathbf{F})$.

The distribution of a low-frequency turbulent field was derived in paper [28]. In the isotropic case, it can be represented in the following form:

$$W_t(a, x)dx = 3[6/\pi]^{1/2}a^3x^2\exp(-3a^2x^2/2)]dx, \tag{I.7}$$

where

$$x = F_t/F_N, \quad a = F_0/F_N. \tag{I.8}$$

The total quasistatic field **F** results from the vector summation of the two statistically independent contributions: $\mathbf{F} = \mathbf{F}_t + \mathbf{F}_i$. The justification of this has been given in papers by Ecker and Fisher [29] and Spatscheck [30]. Therefore the distribution $W(F/F_N)$ of the total field is a convolution of the distribution $W_t(F_t/F_N)$ of the turbulent field with the distribution $W_i(F_i/F_N)$ of the ion microfield:

$$W(\beta)d\beta = \left[\int\int \mathbf{dxdu}\ W_i(\mathbf{u})W_t(\mathbf{x})\delta(\beta - \beta_s)\right] d\beta,$$
$$\beta = F/F_N, \quad \mathbf{x} = \mathbf{F}_t/F_N, \quad \mathbf{u} = \mathbf{F}_i/F_N. \tag{I.9}$$

Here,

$$\beta_s = |\mathbf{x} + \mathbf{u}| = (x^2 + u^2 - 2ux \cos\theta)^{1/2}, \tag{I.10}$$

where θ is the angle between vectors **u** and **x**. Using the properties of the δ-function, Eq. (I.9) has been simplified in paper [28] as follows:

$$W(a, \beta) = [3/(2\pi)^{1/2}]a\beta \int_o^\infty du\{\exp[-3a^2(\beta - u)^2]$$
$$-\exp[-3a^2(\beta + u)^2]\}W_i(u)/u. \tag{I.11}$$

For the distribution of the quasistatic part of the ion microfield $W_i(u)$ in Eq. (I.11) we use here the APEX distribution [31].

Now let us discuss the contribution of the thermal Langmuir turbulence to the SBHSL. The Langmuir turbulence, being the high-frequency one, causes a *dynamical* SBHSL — similar to the dynamical SBHSL by the electron microfield. In paper [32], there was derived analytically the Langmuir-turbulence-caused contribution (additional to the electron microfield contribution) to the real part

$\Gamma = -\mathrm{Re}\Phi$ of the dynamical broadening operator Φ. In particular, diagonal elements of Γ have the form

$$\Gamma_{\alpha\beta} = \Gamma_\alpha + \Gamma_\beta - d_{\alpha\alpha} d_{\beta\beta} E_0^2 \gamma_p / [3\hbar^2(\gamma_p^2 + \omega_{pe}^2)], \tag{I.12}$$

where

$$\Gamma_\alpha = [E_0^2\gamma_p/(12\hbar^2)]\{2\mathbf{d}_{\alpha\alpha}^2/(\gamma_p^2 + \omega_{pe}^2) + (|\mathbf{d}_{\alpha,\alpha-1}|^2 + |\mathbf{d}_{\alpha,\alpha+1}|^2)$$
$$\times[1/(\gamma_p^2 + (\omega_F - \omega_{\mathrm{pe}})^2) + 1/(\gamma_p^2 + (\omega_F + \omega_{pe})^2)]\}, \tag{I.13}$$

The formula for Γ_β entering Eq. (I.12) can be obtained from Eqs. (I.13) by substituting the subscript α by β. Here, α and β label Stark sublevels of the upper (a) and lower (b) levels involved in the radiative transition, respectively; $\omega_F = 3n_\alpha\hbar F/(2m_e e)$ is the separation between the Stark sublevels caused by the total quasistatic electric field F; the matrix elements of the dipole moment operator are

$$\begin{aligned}\mathbf{d}_{\alpha\alpha}^2 &= [3ea_0 n_\alpha q_\alpha/(2)]^2, |\mathbf{d}_{\alpha,\alpha-1}^2| - |\mathbf{d}_{\alpha,\alpha+1}|^2 \\ &= \mathbf{d}_{\alpha\alpha}^2/q_\alpha, |\mathbf{d}_{\alpha,\alpha-1}|^2 + |\mathbf{d}_{\alpha,\alpha+1}|^2 \\ &= \mathbf{d}_{\alpha\alpha}^2(n^2 - q^2 - m^2 - 1)_\alpha/(2q_\alpha^2),\end{aligned} \tag{I.14}$$

where a_0 is the Bohr radius, n is the principal quantum number, $q = n_1 - n_2$ is the electric quantum number; n_1, n_2 and m are the parabolic quantum numbers. In Eq. (I.14) in the subscripts, we used the notation $\alpha + 1$ and $\alpha - 1$ for the Stark sublevels of the energies $+\hbar\omega_F$ and $-\hbar\omega_F$, respectively (compared to the energy of the sublevel α).

The quantity γ_p in Eqs. (I.12), (I.13) is the sum of the characteristic frequencies of the following processes in plasmas: the electron-ion collision rate γ_{ei} (see, e.g., [33]), the average Landau damping rate γ_L (see, e.g., [34]), and the characteristic frequency γ_{ind} of the nonlinear mechanism of the induced scattering of Langmuir plasmons on ions (see, e.g., [35]):

$$\gamma_p = \gamma_{\mathrm{ei}} + \gamma_{\mathrm{L}} + \gamma_{\mathrm{ind}}. \tag{I.15}$$

The frequency γ_p controls the width of the power spectrum of the Langmuir turbulence.

At the highest density point of KA experiment [4] ($N_e = 1.39 \times 10^{20}$ cm^{-3}, $T = 34486$ K), the ratio of the thermal-Langmuir-turbulence-caused contribution to the dynamical Stark width of the H_α line to the corresponding contribution by the electron microfield reaches the value ~0.1.

We calculated Stark profiles of the H_α line with the allowance for the above two effects of the thermal EPT. We used the formalism of the core generalized theory of the SBHSL [1,36] modified according to [37] to allow for incomplete collisions — see Appendix C[2].

Here, we modified the formalism of the generalized theory from [37], to allow for the thermal EPT. Namely, we used the distribution of the total quasistatic microfield given by Eq. (I.11), thus allowing for the low-frequency thermal EPT, and also added the contribution of the high-frequency thermal EPT to the dynamical broadening operator Φ.

Figure I.1 shows the comparison of the experimental FWHM of the H_α line from KA experiment [4] (dots) with the corresponding FWHM yielded by our present analytical theory (solid line). It is seen that the agreement is very good. Even at the highest density point of KA experiment [4] ($N_e = 1.39 \times 10^{20}$ cm^{-3}, $T = 34486$ K), our theoretical FWHM differs by just 4.5% from the most probable experimental value and is well within the experimental error margin.

The present theory can be also used for calculating Stark profiles and the FWHM of other hydrogen spectral lines. It should be kept in mind though that in the range of $N_e \sim (10^{19}$–$2 \times 10^{20})$ cm^{-3}, only three hydrogen spectral lines "survive": H_α, Ly_α and Ly_β. All higher spectral lines of hydrogen merge into a quasicontinuum because of the large SB at these range of densities. Similarly, in the range of $N_e \sim (2 \times 10^{20}$–$10^{22}$ cm^{-3}) cm^{-3}, the only one "surviving" spectral line of hydrogen would be Ly_α.

[2]While some later additions to the generalized theory, such as, e.g., the effect of the acceleration of perturbing electrons by the ion field, caused a difference of opinions in the literature, in the rigorous analytical results of the core generalized theory there was never found any flaw.

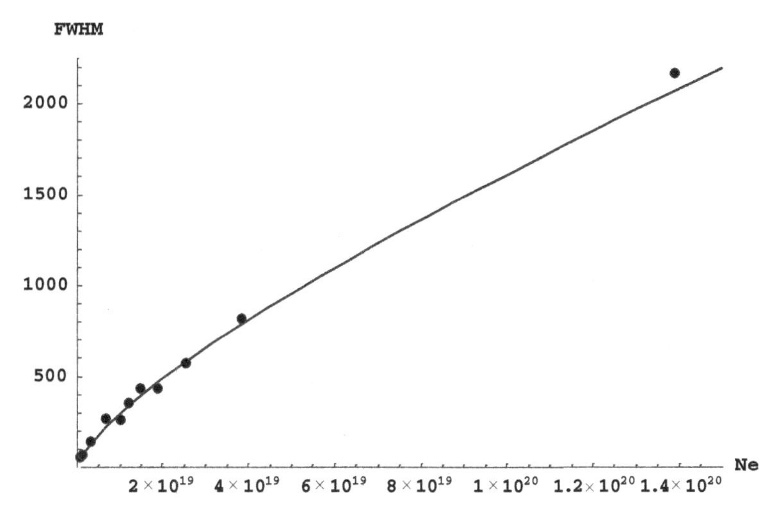

Fig. I.1. Comparison of the experimental FWHM of the H_α line from Kielkopf–Allard experiment [4] (dots) with the corresponding FWHM yielded by our present analytical theory (solid line). The FWHM is measured in Angstrom, while N_e–in cm^{-3}.

References

[1] E. Oks, *Stark Broadening of Hydrogen and Hydrogenlike Spectral Lines in Plasmas: The Physical Insight*, Alpha Science International, Oxford (2006).

[2] H.-J. Kunze, *Introduction to Plasma Spectroscopy*, Springer: Berlin (2009).

[3] M.A. Gigosos, *J. Phys. D: Appl. Phys.* **47** 343001 (2014).

[4] J.F. Kielkopf and N.F. Allard, *J. Phys. B: At. Mol. Opt. Phys.* **47** 155701 (2014).

[5] S. Büscher, T. Wrubel, S. Ferri and H.-J. Kunze, *J. Phys. B: At. Mol. Opt. Phys.* **35** 2889 (2002).

[6] S.A. Flih, E. Oks and Y. Vitel, *J. Phys. B: At. Mol. Opt. Phys.* **36** 283 (2003).

[7] St. Böddeker, S. Günter, A. Könies, L. Hitzschke and H.-J. Kunze, *Phys. Rev.E* **47** 2785 (1993).

[8] M.A. Gigosos and V. Cardenoso, *J. Phys. B: At. Mol. Opt. Phys.* **29** 4795 (1996).

[9] P. Kepple and H.R. Griem, *Phys. Rev.* **173** 317 (1968).

[10] A.S. Antonov, O.A. Zinov'ev, V.D. Rusanov and A.V. Titov, *Zh. Exper. Teor. Fiz. (Sov. Phys. JETP)* **58** 1567 (1970).

[11] S.P. Zagorodnikov, G.E. Smolkin, E.A. Striganova and G.V. Sholin, *Pis'ma v Zh. Exper. Teor. Fiz. (Sov. Phys. JETP Lett.)* **11** 475 (1970).
[12] S.P. Zagorodnikov, G.E. Smolkin, E.A. Striganova and G.V. Sholin, *Doklady Akad. Nauk SSSR (Sov. Phys. Doklady)* **195** 1065 (1970).
[13] E.K. Zavojskij, J.G. Kalinin, V.A. Skorjupin, V.V. Shapkin and G.V. Sholin, *Doklady Akad. Nauk SSSR (Sov. Phys. Doklady)* **194** 55 (1970).
[14] E.K. Zavojskij, Ju.G. Kalinin, V.A. Skorjupin, V.V. Shapkin and G.V. Sholin, *Pis'ma v Zh. Exper. Teor. Fiz. (Sov. Phys. JETP Lett.)* **13** 19 (1970).
[15] M.A. Levine and C.C. Gallagher, *Phys. Lett.* **A32** 14 (1970).
[16] N. Ben-Yosef and A.G. Rubin, *Phys. Lett.* **A33** 222 (1970).
[17] L.P. Zakatov, A.G. Plakhov, V.V. Shapkin and G.V. Sholin, *Doklady Akad. Nauk SSSR (Sov. Phys. Doklady)* **198** 1306 (1971).
[18] A.B. Berezin, A.V. Dubovoj and B.V. Ljublin, *Zh. Tekhnich. Fiz. (Sov. Phys. Tech. Phys.)* **41** 2323 (1971).
[19] M.V. Babykin, A.I. Zhuzhunashvili, E. Oks, V.V. Shapkin and G.V. Sholin, *Zh. Expcr. Teor. Fiz. (Sov. Phys. JETP)* **65** 175 (1973).
[20] J.F. Volkov, V.G. Djatlov and A.I. Mitina, *Zh. Tekhnich. Fiz. (Sov. Phys. Tech. Phys.)* **44** 1448 (1974).
[21] A.I. Zhuzhunashvili and E. Oks, *Sov. Phys. JETP* **46** 2142 (1977).
[22] E. Oks and V.A. Rantsev-Kartinov, *Sov. Phys. JETP* **52** 50 (1980).
[23] A.B. Berezin, B.V. Ljublin and D.G. Jakovlev, *Zh. Tekhnich. Fiz. (Sov. Phys. Tech. Phys.)* **53** 642 (1983).
[24] A.N. Koval and E. Oks, Bull. *Crimean Astrophys. Observ.* **67** 78 (1983).
[25] E. Oks, St. Böddeker and H.-J. *Kunze, Phys. Rev. A* **44** 8338 (1991).
[26] E. Oks, *Plasma Spectroscopy: The Influence of Microwave and Laser Fields*, Springer Series on Atoms and Plasmas, vol. 9, Springer: Berlin (1995).
[27] D. Bohm and D. Pines, *Phys. Rev.* **92** 609 (1953).
[28] E. Oks and G.V. Sholin, *Sov. Phys. Tech. Phys.* **21** 144 (1976).
[29] G.H. Ecker and K.G. Fisher, *Z. Naturforsch.* **26a** 1360 (1971).
[30] K.H. Spatschek, *Phys. Fluids* **17** 969 (1974).
[31] C.A. Iglesias, H.E. DeWitt, J.L. Lebowitz, D. MacGowan and W.B. Hubbard, *Phys. Rev. A* **31** 1698 (1985).
[32] E. Oks and G.V. Sholin, *Sov. Phys. JETP* **41** 482 (1975).
[33] J.D. Huba, *NRL Plasma Formulary*, Naval Research Laboratory, Washington, DC (2013).
[34] P.M. Bellan, *Fundamentals of Plasma Physics*, Cambridge, Cambridge University Press, (2006).

[35] B.B. Kadomtsev, *Collective Phenomena in Plasma*, Nauka, Moscow (1988).

[36] Ya. Ispolatov and E. Oks, *J. Quant. Spectrosc. Rad. Transfer* **51** 129 (1994).

[37] E. Oks, 18 th *Internation. Conference. "Spectral Line Shapes"* (Auburn, AL), Invited Paper, *AIP Conference. Proceedings.* **874** 19 (2006).

Appendix J

Satellites of Dipole-Forbidden Spectral Lines of Helium, Lithium and of the Corresponding Ions, Caused by Quasimonochromatic Electric Fields in Plasmas

J.1 Weak Field Approximation

Baranger and Mozer [1] considered a helium or lithium atom (or a He-like or Li-like ion), in whose energy spectrum, one could select a system of three levels 0, 1, 2 having the following properties. Levels 1 and 2 are relatively close to each other and are coupled by a dipole matrix element; the radiative transition from level 1 to a distant level 0 is dipole-forbidden while the radiative transition from level 2 to level 0 is dipole-allowed (Fig. J.1). Baranger and Mozer showed that under a Quasimonochromatic Electric Field (QEF) of the frequency ω, two satellites can appear at the frequencies $\omega_{\text{sat}\pm} = \omega_{10} \pm \omega$, where ω_{10} is the frequency where the dipole-forbidden spectral line would appear if the electric field would be static.

Baranger and Mozer [1] calculated the intensities of the satellites $S_{\pm}$ relative to the intensity of the allowed line I_a for the case of an isotropic multimode QEF using the standard time-dependent

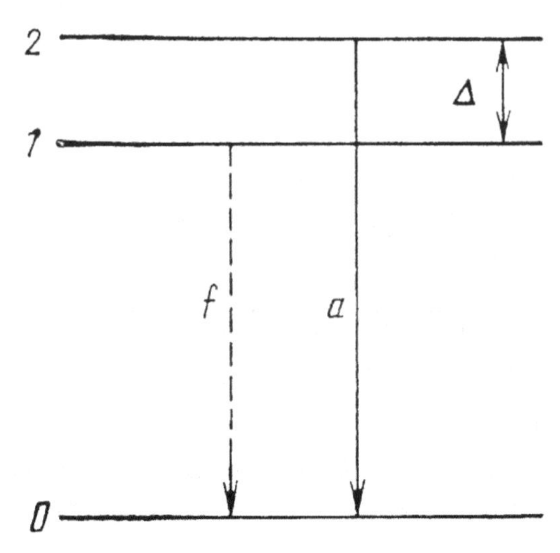

Fig. J.1. Dipole-forbidden (f) and dipole- allowed radiative transitions in a three-level system selected from the energy spectrum of a helium or lithium atom (or a He-like or Li-like ion).

perturbation theory:

$$S_{\pm}/I_a = [6(\Delta \pm \omega)^2(2l_2+1)]^{-1}\langle E^2\rangle \times \max(l_1, l_2)\left[\int_0^{\infty} R_{l1}(r)R_{l2}(r)r^3 dr\right]^2. \quad \text{(J.1)}$$

Here, $\langle E\rangle^2$ is the root-mean-square amplitude of the QEF, Δ is the separation between levels 1 and 2, l_1 and l_2 are their orbital quantum numbers, $R_{l1}(r)$ and $R_{l2}(r)$ are the radial parts of the spherical wave functions. Here and below, the atomic units are used: $\hbar = m_e = e = 1$.

Cooper and Ringler [2] calculated satellites intensities in the same three-level model, but for the case of a linearly polarized single-mode QEF (such as a laser or maser field) of the form $\mathbf{E}(t) = \mathbf{E}_0 \cos\omega t$. They also used the standard time-dependent perturbation theory. Because of the axial (rather than spherical) symmetry of this setup, the result depends on the direction of the observation. If the z-axis was chosen along $\mathbf{E}_0$, then for the observation perpendicular to $\mathbf{E}_0$, the result was as follows:

$$S_\pm/I_a = \{E_0^2/[4(\Delta \pm \omega)^2]\}\left[\sum_{m0,m1,m2} |z_{12}|^2(|y_{20}|^2 + |z_{20}|^2)\right] \Big/$$

$$\left[\sum_{m0,m2} (|y_{20}|^2 + |z_{20}|^2)\right]. \tag{J.2}$$

The summations in Eq. (J.2) reflect the fact that to each of the levels 0, 1 and 2 may belong several wave functions that differ by magnetic quantum numbers m_0, m_1 and m_2.

Two conditions restrict the validity of Eqs. (J.1) and (J.2). First, there should be no resonances where $q\omega$ is approximately equal to Δ (q is an integer). Second, the QEF should be relatively weak: $\max(\omega, \Delta) \gg |r_{12}|\langle E^2\rangle^{1/2}, |r_{12}|E_0$.

The practical application, suggested by Baranger and Mozer [J.1], was to measure a QEF amplitude $\langle E^2\rangle^{1/2}$ via the comparison of the experimental ratio $S_\pm/I_a$ with the corresponding theoretical ratio. This idea has been used in early experiments. However, later experiments dealt with relatively strong QEFs, for which the standard time-dependent perturbation theory breaks down. The controlling parameter is

$$\alpha = 2E_0 z_{12}/\Delta. \tag{J.3}$$

In those later experiments, the parameter α was greater or of the order of unity.

The majority of the later experiments were performed in the low-frequency situation, where $\omega \ll \Delta$. In this case, the problem can be solved analytically even for $\alpha > 1$, as long as $\alpha\omega/\Delta \ll 1$. The corresponding analytical solution based on the adiabatic perturbation theory from the paper [3] is presented below.

J.2 Stronger Fields

In the adiabatic perturbation theory, instead of the unperturbed eigenfunctions ψ_1 and ψ_2 of levels 1 and 2 corresponding to the unperturbed Hamiltonian H_0, we use the well-known [4] *instantaneous* eigenfunctions of the two-level system for the perturbed Hamiltonian $H(t) = H_0 + zE_0 \cos \omega t$:

$$\begin{aligned} \chi_1(t) &= \psi_1 \cos(\beta/2) - \psi_2 \sin(\beta/2), \\ \chi_2(t) &= \psi_1 \sin(\beta/2) + \psi_2 \cos(\beta/2), \\ \beta &= \arctan(\alpha \cos \omega t). \end{aligned} \tag{J.4}$$

The corresponding *instantaneous* eigenvalues are

$$\begin{aligned} \omega_1 &= [\omega_1^{(0)} + \omega_2^{(0)} - \Delta(1 + \alpha^2 \cos^2 \omega t)^{1/2}]/2, \\ \omega_2 &= [\omega_1^{(0)} + \omega_2^{(0)} + \Delta(1 + \alpha^2 \cos^2 \omega t)^{1/2}]/2, \end{aligned} \tag{J.5}$$

where $\omega_1^{(0)}$ and $\omega_2^{(0)}$ are the unperturbed energies of the levels 1 and 2 (we remind that we use the atomic units, so that $\hbar = 1$).

We seek the solution of the Schrödinger equations:

$$i\partial\psi/\partial t = (H_0 + zE_0 cos\omega t)\psi, \tag{J.6}$$

in the form

$$\psi(t) = \sum_{j=1}^{2} C_j(t)\chi_j(t) \exp\left[-i \int_0^T d\tau \omega_j(\tau)\right]. \tag{J.7}$$

By substituting (J.7) into (J.6) and solving the resulting system of equations with respect to $C_{1,2}(t)$ with the initial conditions $C_1(0) = 1$ and $C_2(0) = 0$, we get

$$C_2(t) = (1/2) \int_0^t d\tau [d\beta(\tau)/d\tau] \exp\left[i \int_0^T dt_1 \omega_{21}(t_1)\right]. \tag{J.8}$$

Finally, for the spectra of the two satellites we obtained:

$$S_{\pm}(\Delta\omega) = \sum_{m0,m1,m2} \sigma_{\pm} |\mathbf{r}_{20}\mathbf{e}|^2 \delta\left[\Delta\omega - (\omega_1^{(0)} + \omega_2^{(0)} - \Delta_1)/2 \pm \omega\right]. \tag{J.9}$$

Here, $\mathbf{r}_{20} = \langle\psi_2|\mathbf{r}|\psi_0\rangle$, $\mathbf{e}$ is the unit vector of the photon polarization, and

$$\Delta_1 = (2/\pi)\Delta(1 + \alpha^2)^{1/2}\mathbf{E}_{ell}(k), \quad k = \alpha/(1 + \alpha^2)^{1/2}, \tag{J.10}$$

where $\mathbf{E}_{\text{ell}}(k)$ is the complete elliptic integral. As for the functions $\sigma_\pm$ in Eq. (J.9), they are defined as follows

$$\begin{aligned}
\sigma_- &= \{(J_0 - J_1)B_1/2 + [(2-a_2)J_0(\varepsilon_2/(2\omega)) \\
&\quad + 2J_1(\varepsilon_2/(2\omega))]J_0 k A_0 \omega/[8(\Delta_1 - \omega)] \\
&\quad - (2A_0J_1 + A_2J_0)J_0(\varepsilon_2/(2\omega))k\omega/[8(\Delta_1+\omega)]^2, \\
\sigma_+ &= \{(J_0 + J_1)B_1/2 - [(2-a_2)J_0(\varepsilon_2/(2\omega)) \\
&\quad - 2J_1(\varepsilon_2/(2\omega))]J_0 k A_0 \omega/[8(\Delta_1 + \omega)] \\
&\quad - (2A_0J_1 - A_2J_0)J_0(\varepsilon_2/(2\omega))k\omega/[8(\Delta_1-\omega)]^2, \qquad \text{(J.11)}
\end{aligned}$$

where the arguments of the Bessel functions J_0 and J_1, whenever omitted in Eq. (J.11) for brevity, is $\varepsilon_2/(4\omega)$. The quantities a_2 and ε_2 in Eq. (J.11) are

$$\begin{aligned}
a_2 &= (2/k^2)[(1-k^2)^{1/2} - 1]^2, \\
\varepsilon_2 &= [4/(3\pi)]\Delta(1+\alpha^2)^{1/2}[\mathbf{E}_{\text{ell}}(k) - 2(1-k^2)\mathbf{D}_{\text{ell}}(k)],
\end{aligned} \qquad \text{(J.12)}$$

where $\mathbf{D}_{\text{ell}}(k)$ is another complete elliptic integral. The quantity B_1 is the coefficient of the Fourier expansion of $\sin[\beta(t)/2]$ at the first harmonic of ω, while the quantities A_0 and A_2 are the coefficients of the Fourier expansion of $\cos[\beta(t)/2]$ at the zeroth and second harmonics of ω, respectively. Figure J.2 shows the dependence of B_1, A_0 and A_2 on k^2. For $k \ll 1$, the Taylor expansions of these coefficients are

$$A_0 = 1 - k^2/16, \quad B_1 = k/2 + 7k^3/64, \quad A_2 = -k^2/16. \qquad \text{(J.13)}$$

Below, we provide examples of applying the above analytical results for the spectral lines He I 492.2 nm and He I 447.1 nm that are frequently used in plasma diagnostics. Figure J.3 shows the analytically calculated intensity ratios S_+/I_a and S_-/I_a versus the root-mean-square field $E_{\text{rms}} = E_0/2^{1/2}$ for the transitions $(4^1D, 4^1F) \rightarrow 2^1P$ in a helium atom at the following two frequencies: $\omega/(2\pi c) = 0.5\,\text{cm}^{-1}$ and $\omega/(2\pi c) = 2\,\text{cm}^{-1}$ (the results are summed up over the two polarizations). These analytical results are compared in Fig. J.3 with simulations from paper [5]. It should be noted

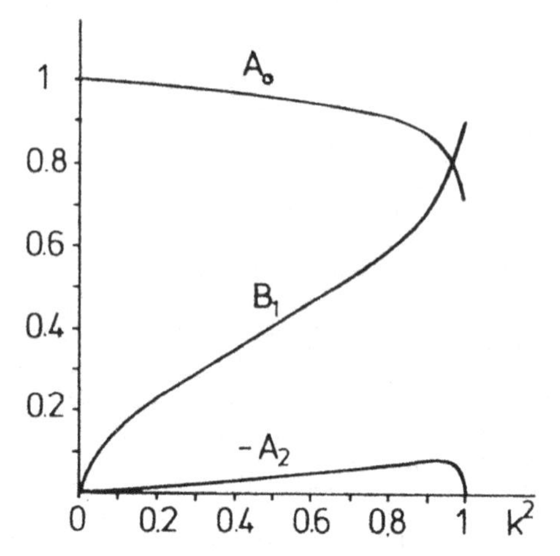

Fig. J.2. The coefficients B_1, A_0 and A_2 of the Fourier expansions of $\sin[\beta(t)/2]$ and $\cos[\beta(t)/2]$, which enter the adiabatic wave functions from Eq. (J.4) versus the parameter k^2 defined in Eq. (J.10).

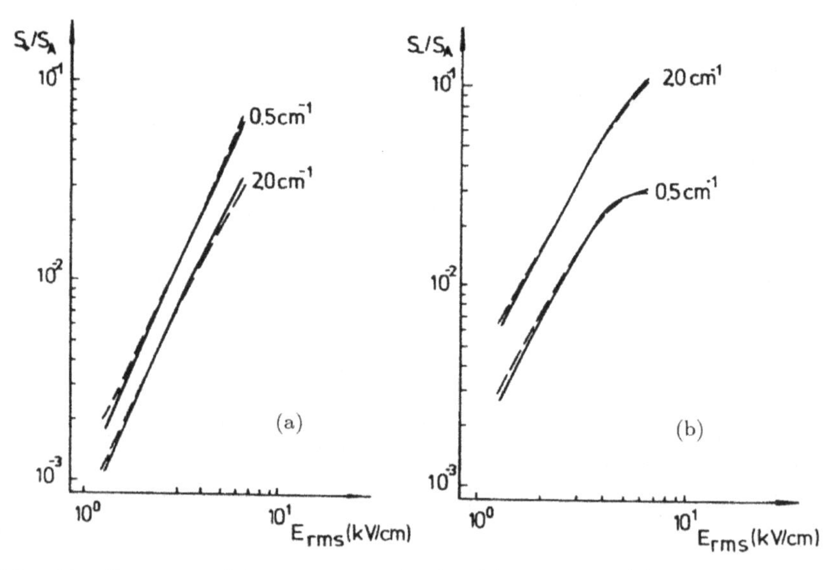

Fig. J.3. The ratio of intensities of the far (a) and near (b) satellites to the intensity of the allowed line He I 492.2 nm versus the root-mean-square field $E_{\rm rms} = E_0/2^{1/2}$ for two frequencies: $\omega/(2\pi c) = 0.5\,\text{cm}^{-1}$ and $\omega/(2\pi c) = 2\,\text{cm}^{-1}$. Solid lines are our analytical results based on the adiabatic perturbation theory, dashed lines are simulations from the paper [5].

that in our analytical calculations, we used the more accurate value $\Delta/(2\pi c) = 5.43\,\text{cm}^{-1}$ from the reference data in [6] instead of $5.63\,\text{cm}^{-1}$ from [5].

Typical situations in many experiments involve an unknown spatial form factor: the satellites are emitted from a relatively small volume while the allowed line is emitted from a significantly larger volume. In this situations, it would be useless trying to deduce E_0 from the experimental ratio S_+/I_a or S_-/I_a. In paper [3], we suggested using the experimental ratio of intensities of the two satellites to each other S_-/S_+. Figure J.4 shows this ratio S_-/S_+ versus E_0 for the lines He I 492.2 nm and He I 447.1 nm, calculated by the adiabatic perturbation theory for the frequency $\omega/(2\pi c) = 1.28\,\text{cm}^{-1}$. It should be emphasized that according to Eqs. (J.1) or (J.2), obtained by the standard time-dependent perturbation theory with the allowance for the terms $\sim E_0^2$, the ratio S_-/S_+ is constant, so that those results are useless for the typical situations involving

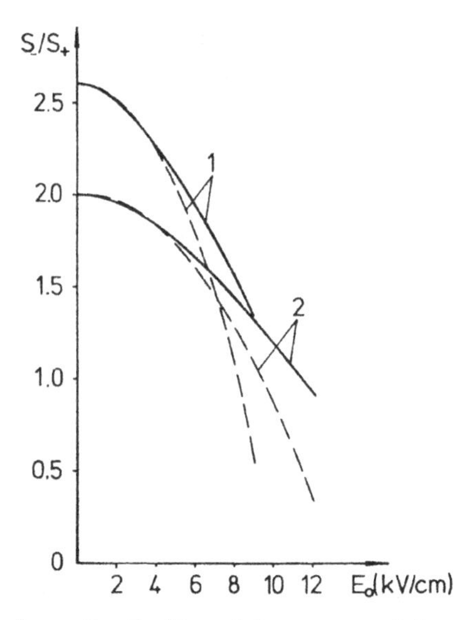

Fig. J.4. The intensity ratio S_-/S_+ of the near and far satellites versus the amplitude E_0 of the QEF of the frequency $\omega/(2\pi c) = 1.28\,\text{cm}^{-1}$ for the lines He I 492.2 nm and He I 447.1 nm. Solid lines — results by the adiabatic perturbation theory, dashed lines — results by the standard time-dependent perturbation theory, which are valid only at relatively small E_0.

a spatial form factor. For completeness, we also calculated the ratio S_-/S_+ by the standard time-dependent perturbation theory taking into account term up to the order of E_0^4 and presented them in Fig. J.4. It is seen that even with the allowance for the terms $\sim E_0^4$ the standard time-dependent perturbation theory is valid only for relatively weak fields.

The bottom line is that the analytical results obtained by using the adiabatic perturbation theory allows to extend this diagnostic to relatively strong fields where the standard time-dependent perturbation theory fails.

J.3 Very Strong Fields

For measuring very strong QEFs, such as ~50–300 kV/cm in the microwave range using neutral helium or neutral lithium lines, one could employ radiative transitions (3^1P, 3^1D) $\rightarrow$ 2^1S or (3^1P, 3^1D) $\rightarrow$ 2^1P. Very strong QEFs produce new features compared to weak QEFs. First, there will be multi-satellite structure both around the forbidden line and around the allowed line — in distinction to weak QEFs producing only two satellites around the forbidden line and no satellites around the allowed line. Second, the multi-satellite structure around the forbidden line will show a significant asymmetry of intensities. The corresponding analytical results can be found in book [7]. In case, where the spectral resolution would be insufficient for resolving individual satellites, one would observe the envelope of the multi-satellite structures.

One of the ways to measure very strong QEFs is based on the experimental determination of the ratio of the total intensity J_F of all satellites of the forbidden line to the total intensity of all satellites of the allowed line J_A. As an example, Fig. J.5 shows the ratio J_F/J_A versus the QEF amplitude E_0 for the case of the forbidden line He I 504.2 nm (2^1S–3^1D) and the allowed line He I 501.6 nm (2^1S–3^1P).

In the situations, involving the unknown spatial form factor (i.e., when the satellites are emitted from a relatively small volume while the allowed line is emitted from a significantly larger volume), the QEF amplitude can be deduced from the shape of the asymmetric

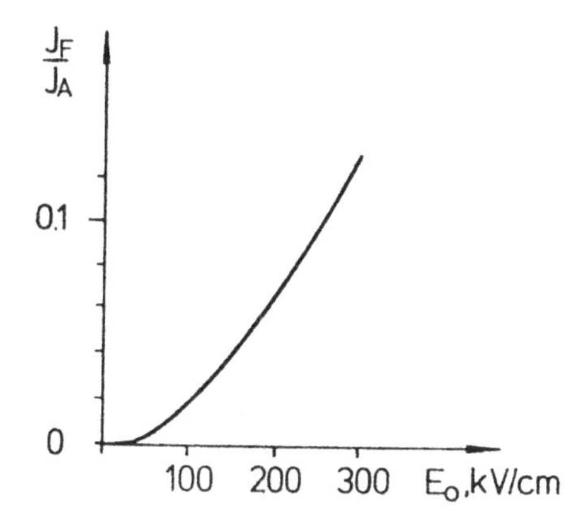

Fig. J.5. The ratio of the total intensity J_F of all satellites of the forbidden line to the total intensity of all satellites of the allowed line J_A versus the QEF amplitude E_0 for the case of the forbidden line He I 504.2 nm (2^1S–3^1D) and the allowed line He I 501.6 nm (2^1S–3^1P). The ratio was calculated for the polarization $\mathbf{e}||\mathbf{E}_0$. The result is valid for the QEF frequency range $\omega/(2\pi c) < 600$ GHz.

envelope of the multi-satellite structure around the forbidden line using the corresponding analytical results presented in the book [7]. As an example, Fig. J.6 illustrates the significant asymmetry of the multi-satellite structure around the forbidden line He I 504.2 nm in distinction to the symmetric multi-satellite structure in the vicinity of the allowed line He I 501.5 nm — for the QEF amplitude $E_0 =$ 150 kV/cm and the frequency $\omega/(2\pi) = 9.4$ GHz, the polarization being $\mathbf{e}||\mathbf{E}_0$.

J.4 Polarization Analysis of Satellites

By using polarizers (that select a particular linear polarization of the light) one can obtain more information about the QEF. Specifically, it allows measuring the degree of anisotropy of the QEF distribution or the angle between $\mathbf{E}_0$ and the polarizer transmission axis [3,7]. Let us consider the following geometry of the experiment. Two orthogonal orientation of the polarizer transmission axis are denoted as z- and x-axes, while the observation is along the y-axis. The vector $\mathbf{E}_0$ is in the xz-plane and constitutes an angle γ (to be determined) with the z-axis. From Eq. (J.9) for the satellite intensities $S_{\pm}^{(z)}$ (polarization

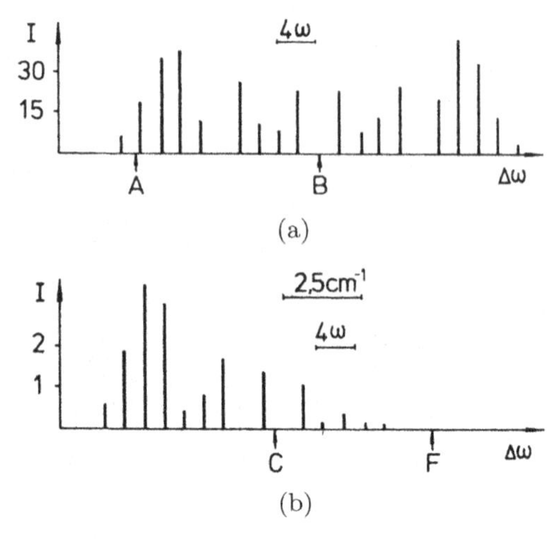

Fig. J.6. Spectra of the allowed line He I 501.5 nm (a) and the forbidden line He I 504.2 nm (b) under the QEF of the amplitude $E_0 = 150$ kV/cm and the frequency $\omega/(2\pi) = 9.4$ GHz, the polarization being $\mathbf{e}||\mathbf{E}_0$. Letter A marks the unperturbed position of the allowed line, letter B — the position of the center of symmetry of the multi-satellite structure around the allowed line, letter F — the unperturbed position of the forbidden line, letter C — the position such that the segment CF is equal to the segment AB.

along z) and $S_\pm^{(x)}$ (polarization along x), we obtain

$$\begin{aligned} S_\pm^{(z)} &= \sum_{m0,m1,m2} \sigma_\pm(|x_{20}|^2 \sin^2\gamma + |z_{20}|^2 \cos^2\gamma), \\ S_\pm^{(x)} &= \sum_{m0,m1,m2} \sigma_\pm(|x_{20}|^2 \cos^2\gamma + |z_{20}|^2 \sin^2\gamma). \end{aligned} \tag{J.14}$$

It is practically useful measuring the ratio of intensities of the same satellite at the two orientations of the polarizer:

$$\begin{aligned} &S_\pm^{(z)}/S_\pm^{(x)} = [1 + f_\pm(E_0)\cot^2\gamma]/[\cot^2\gamma + f_\pm(E_0)], \\ &f_\pm(E_0) = \left(\sum_{m0,m1,m2} \sigma_\pm(|z_{20}|^2)\right) \Bigg/ \left(\sum_{m0,m1,m2} \sigma_\pm(|x_{20}|^2)\right). \end{aligned} \tag{J.15}$$

Let us consider, as an example, radiative transitions (4^1F, 4^1D) $\to 2^1P$ in a helium or a lithium atom, or in a He-like or a Li-like ion. Since for weak fields one gets,

$$\sigma_\pm = \{E_0 z_{12}/[2(\Delta \pm \omega)]\}^2, \tag{J.16}$$

then Eq. (J.15) yields $f_-(E_0) = f_+(E_0) = 4/3$ — in agreement with the result from the paper [2]. This means that for weak fields, the outcome of the polarization measurements does not depend on field amplitude E_0, so that E_0 can be determined experimentally in this way. (This means also that E_0 cannot be determined from Baranger–Mozer' theory [1] or Cooper–Ringler's theory [2] because these theories are valid only for weak fields where $f_-(E_0)$ and $f_+(E_0)$ do not depend on E_0).

However, in strong fields, the ratio $S_\pm^{(z)}/S_\pm^{(x)}$ depends both on the angle γ and the field amplitude E_0, so that both γ and E_0 can be determined experimentally by the polarization analysis. Figure J.7 shows the functions $f_+(E_0)$ and $f_-(E_0)$ calculated according to Eqs. (J.15) and (J.11).

Thus, after determining experimentally E_0 by methods described in Sec. J.2, it is possible to determine experimentally also the angle

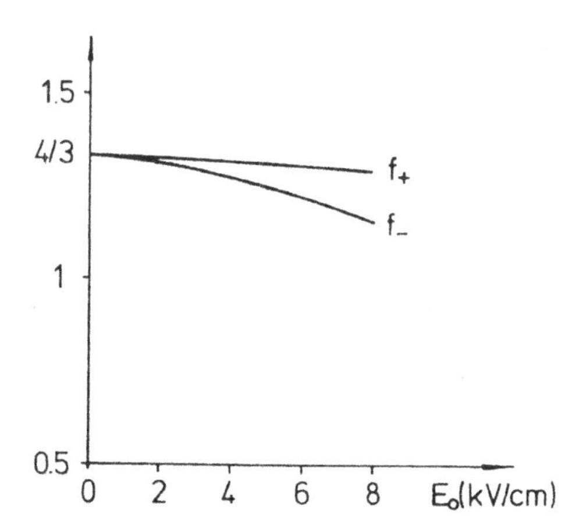

Fig. J.7. Functions $f_+(E_0)$ and $f_-(E_0)$ — from Eq. (J.15) — that determine the intensity ratio of the same satellite in two mutually perpendicular orientations of the polarizer.

γ by the polarization analysis. It should be noted that in very strong fields, both $f_+(E_0)$ and $f_-(E_0)$ are close to unity, so that the accuracy of the experimental determination of the angle γ would diminish.

J.5 Allowance for the Mixing of Singlet and Triplet Terms

Magnetic interactions, such as spin-orbit coupling and spin–spin coupling, can significantly affect intensities of satellites of He-like and Li-like ions of a relatively large nuclear charge Z in high temperature plasmas. Here is an example: the system of levels $0 \leftrightarrow 1^1S_0$, $1 \leftrightarrow 2^1S_0$, $2 \leftrightarrow 2^1P_1$. One should now take into account also level $3 \leftrightarrow 2^3P_1$: the intensity of the forbidden transition $1^1S_0 \leftrightarrow 2^1S_0$, will be affected not only by level 2 (as for low Z radiators), but also by level 3 due to magnetic interactions.

Indeed, the wave functions ψ_1, ψ_2, ψ_3 of these levels are linear combinations of the "unperturbed" wave functions φ_1, φ_2, φ_3 of the same levels, where "unperturbed" means disregarding magnetic interactions [8]:

$$\begin{gathered}\psi_1 = \varphi_1, \quad \psi_2 = (\varphi_2 + \beta\varphi_3)/(1+\beta^2)^{1/2}, \\ \psi_3 = (\varphi_3 - \beta\varphi_2)/(1+\beta^2)^{1/2},\end{gathered} \tag{J.17}$$

where β is the mixing coefficient. The mixing reflected by a non-zero value of β translates into a coupling (by dipole matrix elements) of level 1 with both level 2 and level 3:

$$\begin{aligned}\langle\psi_1|z|\psi_2\rangle &= z_{12}/(1+\beta^2)^{1/2}, \\ \langle\psi_1|z|\psi_3\rangle &= -z_{12}\beta/(1+\beta^2)^{1/2}.\end{aligned} \tag{J.18}$$

Equation (J.18) shows that the intensity of the allowed line decreases due to the factor $1/(1+\beta^2)$ and the intensities of the satellites of the forbidden line should differ from the case of $\beta = 0$. Namely, the intensities of the satellites of the forbidden line are now determined by the values $|C_+(\beta)|^2$, $|C_-(\beta)|^2$, where [7]

$$C_\pm(\beta) = -z_{12}E_0[1/(\omega_{21}^{(0)} \pm \omega) + \beta^2/(\omega_{31}^{(0)} \pm \omega)]/[2(1+\beta^2)]. \tag{J.19}$$

Interestingly enough, the satellites intensities can significantly differ from the unperturbed (no magnetic interaction) case even if $\beta^2 \ll 1$. This is because, for quite a range of the nuclear charge Z, level 3 is much closer to level 1 than level 2:

	C V	O VIII	Al XII	S XV	Ca XIX	Ti XXI	Fe XXV
$10^3 \beta^2$	0.0034	0.17	2.7	8.7	28	45	93
$10^3 \omega_{31}^{(0)}/\omega_{21}^{(0)}$	−3.5	−52	−89	−79	−49	−35	−11

Here is an example: for Ca XIX at $\omega/\omega_{31}^{(0)} = 5$, even the small $\beta^2 = 0.0028$ changes the ratio of satellite intensities to each other by 60% compared to the $\beta = 0$ case.

Further, we discuss the case of a special interest where: (1) the contribution of level 2 to the satellites intensities is much smaller than that of level 3, and (2) the QEF frequency ω is in the range $|\omega_{31}| \ll \omega \ll |\omega_{21}|$. In this case, the perturbed wave function of level 1 can be approximated as follows:

$$\psi_1 = \left\{ \varphi_1 \cos(\varepsilon \sin \omega t - \varphi_2 \beta (1+\beta^2)^{-1/2} \right.$$
$$\left. \sum_{p=-\infty}^{\infty} J_{2p-1}(\varepsilon) \exp[i(2p-1)\omega t] \right\} \exp(-i\omega t), \qquad \text{(J.20)}$$
$$\varepsilon = z_{12} E_0 \beta / [(1+\beta^2)^{1/2} \omega].$$

Equation (J.20) shows that the intensities of the satellites of the forbidden line are proportional to the squares of the Bessel functions $J_{2p-1}^2(\varepsilon)$ and that intensities of the satellites of the allowed line $2 \to 0$ are proportional to $J_{2p}^2(\varepsilon)$. Therefore, in this case the QEF amplitude E_0 can be determined from the experimental ratio of intensities $J_k^2(\varepsilon)/J_q^2(\varepsilon)$ for satellites of different numbers k and q. This method applies obviously also to the situations involving an unknown spatial form factor, i.e., when the satellites are emitted from a relatively small volume while the allowed line is emitted from a significantly larger volume.

References

[1] M. Baranger and B. Mozer, *Phys. Rev.* **123** 25 (1961).
[2] W.S. Cooper and H. Ringler, *Phys. Rev.* **179** 226 (1969).
[3] E. Oks and V.P. Gavrilenko, *Sov. Phys. Tech. Lett.* **9** 111 (1983).
[4] A.S. Davydov, *Quantum Mechanics*, Pergamon: Oxford, Sect. 49 (1965).
[5] W.W. Hicks, R.A. Hess and W.S. Cooper, *Phys. Rev. A* **5** 490 (1972).
[6] W.C. Martin, *J. Phys. Chem. Ref. Data* **23** 257 (1973).
[7] E. Oks, *Plasma Spectroscopy: The Influence of Microwave and Laser Fields*, Springer Series on Atoms and Plasmas, vol. 9, Springer: Berlin (1995).
[8] I.I. Sobelman, *Atomic Spectra and Radiative Transitions*, Springer Series Atoms and Plasmas, v. 12, Springer: Berlin (1992).

Appendix K

Floquet–Liouville Formalism

The Liouville space, usually employed to deal with the calculation of Stark profiles in dense plasmas, and the Floquet theory, developed to solve time periodic problems, have been joined together to solve the time-dependent Liouville equation in a so-called Floquet–Liouville formalism [1]. In the following, we summarize the main steps of this theory and briefly discuss the control of the accuracy for the convergence of the simulated profiles.

The starting point of the line shape calculation is the determination of the dipole correlation function

$$C(t) = \int Q(\boldsymbol{F})\mathrm{T}r_{ae}[\boldsymbol{d}U(\boldsymbol{F},t)(\rho^{(a)}\rho^{(e)}\boldsymbol{d})]\, d^3\boldsymbol{F}. \tag{K.1}$$

In Eq. (K.1) $\rho^{(a)}$ and $\rho^{(e)}$ stand for the emitter and the free-electrons density matrix, respectively. The trace Tr_{ae} involves all emitters and free electron states; d is the electric dipole moment operator. The ions are considered as quasistatic during the time of interest, and the ionic micro-field **F** is then taken constant during the radiative transitions. Profiles of the same spectral line emitted by different radiators depend on the field **F** as a parameter. The final profile of the spectral line is obtained by averaging over the ionic micro-field distribution Q(**F**).

The evolution operator $U(\mathbf{F},t)$ depends parametrically on F. In the Liouville formalism, it is expressed as follows:

$$U(\boldsymbol{F},t) = \exp\left(-i\int_o^t L(\boldsymbol{F},t')\,dt'\right). \tag{K.2}$$

Here, $L(\boldsymbol{F},t)$ is the Liouville operator, or more precisely, a sum of two Liouville operators: one operating on the emitter states $L_r(\boldsymbol{F},t)$ and the second operating only on free-electron states $L_e(t)$:

$$\begin{aligned} L(\boldsymbol{F},t) &= L_r(\boldsymbol{F},t) + L_e(t), \\ L_r(\boldsymbol{F},t) &\equiv L_r(t) = L_a + L_i(\boldsymbol{F}) + L_{of}(t). \end{aligned} \tag{K.3}$$

Here, L_a, L_i, L_{of}, L_e stand for the Liouville operators for the isolated emitters, the interaction of the emitter with the ion micro-field, the interaction of the emitter with the QEF, and the interaction of the emitter with the free electrons, respectively.

Following the assumption that the effect of QEF on the free-electron field is neglected, the trace over electron states of the commutator $[L_r, L_e]$ vanishes. Thus the trace in the correlation function (K.1) acts on a product of exponentials, so that (K.2) can be rewritten as

$$\begin{aligned} U(\boldsymbol{F},t) &= \exp\left(-i\int_o^t L_r(\boldsymbol{F},t')dt' - \phi_e t\right) \\ &= \exp\left(-i\int_o^t (L_r(\boldsymbol{F},t') - i\phi_e)dt'\right) \end{aligned} \tag{K.4}$$

Equation (K.4) involves the time-independent EBO φ_e(under the impact approximation) and the averaging over all initial times.

The Liouville operator $L(\boldsymbol{F},t)$, depending on the oscillatory field, can be transformed into a time-independent Floquet–Liouville L_F. This new operator satisfies, as demonstrated in [38], the eigenvalue equation involving *a matrix of infinite dimension* (as it was shown in [2]):

$$\begin{aligned} &\sum_{\sigma\tau}\sum_k \langle\alpha\beta;n|L_F|\sigma\tau;k\rangle\langle\sigma\tau;k|\Omega_{\mu v,m}\rangle \\ &\qquad = \Omega_{\mu\nu,m}\,\langle\alpha\beta;n\mid\Omega_{\mu\nu,m}\rangle\,. \end{aligned} \tag{K.5}$$

In (K.5), $|\Omega_{\mu v}, w\rangle$ represents the eigenvector corresponding to the eigenvalue $\Omega_{\mu\nu,m}$, and the states $|\alpha\beta; n\rangle = |\alpha\beta\rangle \otimes |n\rangle$ are expressed on the generalized tetradic-Fourier basis. The time-independent operator L_F matrix elements can be written more explicitly as:

$$\langle\alpha\beta; n|L_F|\mu v; m\rangle = L^{(m-n)}_{\alpha\beta,\mu\nu} + m\omega\delta_{\alpha\mu}\delta_{\beta\nu}\delta_{nm}, \tag{K.6}$$

where

$$L^{(k)}_{\alpha\beta,\mu\nu} = H^{(k)}_{\alpha\mu}\delta_{\beta\nu} - H^{(k)}_{\beta\nu}\delta_{\alpha\mu} - i\phi^{el}_{\alpha\beta,\mu\nu}\delta_{k0}, \tag{K.7}$$

$$H^{(k)}_{\alpha\beta} = H_{\alpha\beta}\delta_{k,0} + V_{\alpha\beta}\left(\delta_{k,1} + \delta_{k,-1}\right), \tag{K.8}$$

$$H_{\alpha\beta} = E_\alpha\delta_{\alpha\beta} - \langle\alpha|\, \boldsymbol{d}.\boldsymbol{F}\,|\beta\rangle, \tag{K.9}$$

$$V_{\alpha\beta} = -\frac{1}{2}\langle\alpha|\, \boldsymbol{d}.\boldsymbol{E}\,|\beta\rangle. \tag{K.10}$$

In these expressions, $H_{\alpha\beta}$ is the ionic Stark Hamiltonian, and $V_{\alpha\beta}$ the interaction with the QEF; **d** is the dipole moment of the emitter, and **F** and **E** are the ionic micro-field and the QEF, respectively.

The matrix (K.15) has periodic properties [2], i.e.:

$$\Omega_{\mu\nu,m+k} = \Omega_{\mu\nu,m} + k\omega, \tag{K.11}$$

$$\langle\alpha\beta; n + k \mid \Omega_{\mu\nu,m+k}\rangle = \langle\alpha\beta; n \mid \Omega_{\mu\nu,m}\rangle. \tag{K.12}$$

Due to the non-Hermitian nature of L_F, there exists a bi-orthogonal relationship between left and right eigenvectors:

$$\left\langle\Omega^{-1}_{\mu\nu,m} \;\middle|\; \Omega_{\alpha\beta,n}\right\rangle = \delta_{\alpha\mu}\delta_{\beta\nu}\delta_{nm}. \tag{K.13}$$

The following key-points of the Floquet's formalism should be emphasized. The evolution operator related to the density operator ρ is expressed via two matrix, $\phi(t)$, a matrix of eigenstates of L_F, and a diagonal matrix Q, the elements of which are the quasienergies $\Omega_{\alpha\beta,n}$ [3], [4], i.e.,

$$U(t, t_0) = \rho(t)\rho^{-1}(t_0), \tag{K.14}$$

$$\rho(t) = \phi(t)e^{-iQt}. \tag{K.15}$$

Taking into account the periodic property (K.12) and the bi-orthogonal relationship (K.13), the evolution operator can be expanded compactly as:

$$U(t,t_0) = \sum_{\substack{\alpha\beta \\ n}} \sum_{\mu\nu} |\alpha\beta;n\rangle\langle\alpha\beta;n| \exp\left(-iL_F(t-t_0)\right) |\mu v;0\rangle \langle\mu\nu;0|. \tag{K.16}$$

Consequently, its matrix elements are

$$U_{ab,a'b'}(t,t_0) = \sum_n \langle ab;n| \exp\left(-iL_F(t-t_0)\right) |a'b';0\rangle e^{in\omega_L t}, \tag{K.17}$$

where,

$$\langle ab|\alpha\beta;n\rangle = \langle ab|\alpha\beta\rangle \otimes |n\rangle \equiv \langle ab|\alpha\beta\rangle e^{in\omega_L t}. \tag{K.18}$$

In Eq. (K.17), the periodic property of the eigenvector allowed the removing of the sum over $|m\rangle$ states, and setting the state $|m\rangle = |0\rangle$, arbitrarily.

From Eq. (K.17), it is seen that the evolution operator is defined on the finite tetradic basis $|\alpha\beta\rangle$, and the Floquet–Liouville operator is defined on an infinite generalized tetradic Fourier basis $|\alpha\beta;n\rangle$. The reduction of the tetradic Fourier basis $|\alpha\beta;n\rangle$ on the tetradic basis $|\alpha\beta\rangle$ is performed by summing over all projections of the Floquet–Liouville operator on the subspace with a fixed mode n.

The evolution operator given by Eq. (K.4) involves an average over electron states and a time-independent EBO. Therefore, it has to be averaged over all initial times t_0, keeping constant the elapsed time $s = t - t_0$. For this purpose, the time evolution operation is expressed in terms of t_0 and s as

$$U_{\alpha\beta,\mu\nu}(s,t_0) = \sum_n \langle\alpha\beta\,;n| \exp\left(-iL_F s\right) \, |\mu v;0\rangle \, e^{in\omega_L s} e^{in\omega_L t_0}. \tag{K.19}$$

Since, the only dependence on t_0 is in the exponential, the non-zero contribution for the time-average evolution operator $\overline{U}_{\alpha\beta,\mu\nu}(s)$ will be only for $n = 0$, i.e.,

$$\overline{U}_{\alpha\beta,\mu\nu}(s) = \langle\alpha\beta\,;0| \exp\left(-iL_F s\right) \, |\mu v;0\rangle \,. \tag{K.20}$$

Finally, the average over initial times has removed the sum over Fourier modes, and only the mode $n = 0$ remains. Thus, all necessary operators for calculating spectral line profiles are defined.

The presence of the QEF introduces a polarization of the space, while the ionic microfield **F** may have any orientation with respect to the linearly-polarized QEF **E**. We introduce the complex vector basis $\boldsymbol{e}_0 = \boldsymbol{e}_z$, $\boldsymbol{e}_1 = -\frac{1}{\sqrt{2}}(\boldsymbol{e}_x + i\boldsymbol{e}_y)$, $\boldsymbol{e}_{-1} = \frac{1}{\sqrt{2}}(\boldsymbol{e}_x - i\boldsymbol{e}_y)$, by choosing the z-axis along **E**. The electric dipole moment operator d entering Eq. (K.3), the ionic micro-field **F** and the QEF **E** can be decomposed on the vector basis so that the scalar products in (K.9) and (K.10) will be expressed as $\boldsymbol{d} \cdot \boldsymbol{E} = d_0 E$ and $\boldsymbol{d} \cdot \boldsymbol{F} = d_0 F_0 - d_1 F_{-1} - d_{-1} F_1$. For the comparison with experimental results, it is useful to introduce parallel and perpendicular components of **F** with respect to **E**:

$$F_0 = F_{\|}, F_1 = -\frac{1}{\sqrt{2}} F_{\perp} e^{i\phi} \quad \text{and} \quad F_{-1} = \frac{1}{\sqrt{2}} F_{\perp} e^{-i\phi}. \tag{K.21}$$

Due to rotational invariance with respect to the z-axis, one can choose $\varphi = 0$. The correlation function $C_q(t)$ can be related to each polarization state distinguished by the suffix q.

Based on the above results, the intensity for each polarization q can be expressed as follows:

$$I_q\omega = \frac{1}{\pi}\mathrm{Re}\int_0^{\infty} e^{i\omega t} \int Q(F)\langle d_q|U(F,t)|\rho^{(a)} d_q\rangle d^3\boldsymbol{F} dt. \tag{K.22}$$

In accordance to [1], this can be re-written as:

$$I_q(\omega) = \int Q(\boldsymbol{F}) \\ \times \left[\frac{1}{\pi}\mathrm{Re}\int_0^{\infty} e^{i\omega t}\left\langle d_q^K \middle| P_0 \exp(-iL_F t) P_0 \middle| \rho^{(a)} d_q \right\rangle dt\right] d^3\boldsymbol{F}, \tag{K.23}$$

where P_0 is the projector on the Floquet subspace $n = 0$

Following the procedure from [5] for the numerical treatment of the double summation involved in (K.23), one diagonalizes the Floquet–Liouville operator $L_F(\mathbf{F})$ for every ionic field **F**. The Fourier transform can be performed easily, leading to an infinite diagonal matrix [1] in which all Floquet modes n are considered. In the numerical calculation, the infinite dimension of this matrix has to be reduced to a finite number of modes n. It has been shown in [1] that the number of modes to be included in the calculation can be estimated for any desired accuracy. For the integration over the ionic micro-field, a discretization of both the micro-field intensity and the direction had to be done, involving Gauss–Legendre quadrature [1].

In summary, the Floquet–Liouville operator, introduced to solve the time-dependent problem (Stark effect in a time-dependent QEF), leads to a time-independent treatment carried out via an operator of the infinite dimension. Nevertheless, the periodic property of this operator allows one to consider only the projection of the Floquet–Liouville operator on the Floquet subspace of size n. Depending on the coupling values between the Floquet subspaces for the Floquet–Liouville operator, for any desired accuracy of the line shape calculation, there will exist a value of n, at which the Floquet subspace can be truncated.

References

[1] P. Sauvan and E. Dalimier, *Phys. Rev. E* **79** 036405 (2009).
[2] T.S. Ho, K. Wang and Shih-I Chu, *Phys. Rev. A* **33** 1798 (1986).
[3] Ja.B. Zel'dovich, *Sov. Phys. JETP* **24** 1006 (1967).
[4] V.I. Ritus, *Sov. Phys. JETP* **24** 1041 (1967).
[5] A. Calisti, F.Khelfaoui, R. Stamm, B. Talin and R.W. Lee, *Phys. Rev. A* **42**, 5433 (1990).

Appendix L

Satellites of Hydrogenic Spectral Lines

In 1933, Blochinzew considered the splitting of a model hydrogen line, consisting of just one Stark component, under a linearly-polarized electric field $\mathbf{E}_0 \cos \omega t$ and showed that it splits in satellites separated by $p\omega(p = \pm 1, \pm 2, \pm 3, \ldots)$ from the unperturbed frequency ω_0 of the spectral line [1]. In paper [2], Blochinzew's result was generalized to profiles of real, multicomponent hydrogenic spectral lines in the "reduced frequency" scale as follows (presented later also in book [3], Sec. 3.1):

$$S(\Delta\omega/\omega) = \sum_{p=-\infty}^{+\infty} I(p, \varepsilon)\delta(\Delta\omega/\omega) - p),$$

$$I(p, \varepsilon) = \left[f_0\, \delta_{p0} + 2 \sum_{k=1}^{k\,\max} f_k J_p^2(X_k \varepsilon) \right] \Big/ (f_0 + 2 \textstyle\sum f_k),$$

$$X_k = nq - n_0 q_0. \tag{L.1}$$

Here, f_0 is the total intensity of all central Stark components, f_k is the intensity of the lateral Stark component with the number $k = 1, 2, \ldots, k\max$; n, q and n_0, q_0 are the principal and electric quantum numbers of the upper and lower energy levels, respectively, involved in the radiative transition; $J_p(z)$ are the Bessel functions; ε is the scaled amplitude of the field:

$$\varepsilon = 3E_0/(2Z_r m_e e\omega). \tag{L.2}$$

Physically, the larger is the product $X_k\varepsilon$, the greater is the phase modulation of the atomic oscillator.

In the practically important case of the strong modulation ($X_k\varepsilon \gg 1$), Eq. (L.1) predicts numerous satellites. Frequently the individual satellites merge together by broadening mechanisms, so that only the envelope of these satellites can be observed. The most intense part of the satellites envelope has the shape of the Airy function, as shown in paper [2] and reproduced in book {3], Sec. 3.1. Based on this analytical results, the following practical formulas have been derived and presented in [2] and [3] (Sec. 3.1) for the position $p_{\max}$ of the satellite having the maximum intensity (and thus corresponding to the experimental peak) and for the half width at the half maximum $p_{1/2} = \Delta\omega_{1/2}/\omega$ for many hydrogenic spectral lines for the observation perpendicular to the vector $\mathbf{E}_0$, as follows.

$$p_{\max}(H_\beta) = 4\varepsilon - 1.28\varepsilon^{1/3}, \quad p_{1/2}(H_\beta) = 8\varepsilon, \tag{L.3}$$

$$p_{\max}(H_\delta) = 6\varepsilon - 1.47\varepsilon^{1/3}, \quad p_{1/2}(H_\delta) = 6\varepsilon, \tag{L.4}$$

$$p_{\max}(H_\zeta) = 8\varepsilon - 1.62\varepsilon^{1/3}, \quad p_{1/2}(H_\zeta) = 8\varepsilon, \tag{L.5}$$

$$p_{\max}(Ly_\beta) = 3\varepsilon - 1.17\varepsilon^{1/3}, \quad p_{1/2}(Ly_\beta) = 6\varepsilon, \tag{L.6}$$

$$p_{\max}(Ly_\delta) = 5\varepsilon - 1.28\varepsilon^{1/3}, \quad p_{1/2}(Ly_\delta) = 20\varepsilon. \tag{L.7}$$

The results given by Eqs. (L.1)–(L.7) were derived for a single-mode monochromatic linearly-polarized electric field. The case of a multi-mode monochromatic linearly-polarized electric field (i.e., superposition of a large number of modes $\mathbf{E}_j \cos(\omega t + \varphi_j)$ with randomly distributed phases) was analyzed by Lifshitz for a model hydrogen line, consisting of just one Stark component [4]. For the multi-satellite case, corresponding to the strong modulation of the atomic oscillator ($X_k\varepsilon \gg 1$), the envelope of the satellites has the Gaussian shape.

All of the above results were obtained by neglecting the fine structure Δ_{fs} of hydrogenic spectral lines. This is legitimate for typical experimental situations, where $\min(\omega, X_k \varepsilon \omega) \gg \Delta_{fs}$. The opposite case was considered by Gavrilenko *et al.* [5] using the example of the Lyman-alpha line of F IX.

References

[1] D.I. Blochinzew, *Phys. Z. Sov. Union* **4** 501 (1933).

[2] E. Oks, *Sov. Phys. Doklady* **29** 224 (1984).

[3] E. Oks, *Plasma Spectroscopy: The Influence of Microwave and Laser Fields*, Springer Series on Atoms and Plasmas, vol. 9, Springer: Berlin (1995).

[4] E.V. Lifshitz, *Sov. Phys. JETP* **26** 570 (1958).

[5] V.P Gavrilenko, V.S. Belyaev, A.S. Kurilov *et al.*, *J. Phys. A: Math. Gen.* **39** 4353 (2006).

Appendix M

Advanced Methods of Using Laser-induced Fluorescence for Mapping the Distribution of Quasimonochromatic Electric Fields (Microwaves, Langmuir Waves, Bernstein Modes, Infra-red Laser Fields) in Plasmas: The Theoretical Basis

It is very important to emphasize upfront that theoretical basis of the Laser-Induced Fluorescence (LIF) method presented here opens up various experimental opportunities. In low-temperature plasmas, by the laser stimulation of spectral lines in the optical range, it can be used not only for mapping the distribution of a microwave field in the plasma, but also of Langmuir waves or Bernstein modes. In high-temperature plasmas, it makes it possible to map the distribution of a powerful infra-red laser field in the plasma by using a near-UV laser tuned to the corresponding spectral lines of hydrogenlike. or helium-like, or lithium-like ions.

So in general, it is about using a QES of a higher frequency, tuned to the frequency of a spectral line (and thus strongly coupled to the radiating atom/ion), to map the distribution (in plasmas) of a strong QEF of a lower frequency. Below, just for definiteness, the strong QEF of a lower frequency is called microwave.

These advanced methods were developed in the papers [1,2], some of the results being presented also in book [3], Sec. 3.4. One of the methods was practically implemented in the experiment [4] presented in Chapter 8. For mapping, the microwave distribution in plasmas, sometimes the LIF from a hydrogen or deuterium beam might be used, but generally, the methods do not require the employment of beams.

From the theoretical point of view, the method is based on our analytical solution for an atom interacting with two strong electromagnetic waves characterized by significantly different frequencies. Therefore, the solution obtained is valid in a much broader range of field strengths than an analogous solution that could have been found by the usual time-dependent perturbation theory.

The laser frequency ω_L should be tuned to a *multi-quantum resonance*

$$\omega_L \cong \varepsilon_{21} + k\omega_M - \mu, \quad k = 0, \pm 1, \pm 2, \ldots, \tag{M.1}$$

where ε_{21} is the atomic transition frequency between the lower level 1 and the upper level 2, ω_L and ω_M are the laser and microwave frequencies, respectively, and μ is a detuning ($|\mu| \ll \omega_M$). Equation (M.1) means that an atom can get excited from the state 1 to the state 2 by simultaneously absorbing one quantum of the laser field and absorbing (at $k > 0$) or emitting (at $k < 0$) k quanta of the microwave field.

The resonance condition (M.1) allows using the well-known rotating wave approximation. It was shown in paper [2] that in this approximation, one deals with the following *effective* quantum system of two levels a and b: the levels are separated by $\varepsilon_{21} + k\omega_M$ and they are coupled by the *effective* dipole matrix elements:

$$\zeta_{\mathrm{ab}} = J_k(\Delta\beta_{\mathrm{ba}})z_{\mathrm{ab}}. \tag{M.2}$$

Here, $J_k(x)$ are the Bessel functions and

$$\Delta\beta_{\mathrm{ba}} = (z_{\mathrm{aa}} - z_{\mathrm{bb}})E_{0M}/\omega_M, \quad z_{pq} = \langle\varphi_p|z|\varphi_q\rangle, \tag{M.3}$$

where φ_p and φ_q are the wave functions of the levels p and q in the parabolic coordinates with the quantization axis $Oz \| \mathbf{E}_{0L} \| \mathbf{E}_{0M}$, $\mathbf{E}_{0L}$

and $\mathbf{E}_{0M}$ being the amplitude vectors of the laser and microwave field, respectively.

As a result, the increase of the population of the level 2 due to the laser stimulation can be expressed as follows [2]:

$$\Delta N_2 = N_2 - N_{20}$$
$$= (1/2)G_k(N_1/s_1 - N_2/s_2)/[1 + (\varepsilon_{21} + k\omega_M - \omega_L)^2\tau_{12}^2 + G_k], \quad \text{(M.4)}$$

where,

$$G_k = E_{0L}^2 \sum_{a,b} z_{\text{ab}}^2 J_k^2(\Delta\beta_{\text{ba}})\tau_{12}/\Gamma. \quad \text{(M.5)}$$

Here, s_1 and s_2 are degrees of degeneracy of the levels 1 and 2, τ_{12} and $1/\Gamma$ are transverse and longitudinal relaxation times, respectively.

Equations (M.4) and (M.5) can be used for measuring the microwave amplitude E_{0M} (or generally, for measuring the amplitude of the QEF of the lower frequency) in plasmas. For this purpose, one should record the wavelength-integrated intensity of the LIF I_f = const. ΔN_2 at the transition from the upper level 2 to one of the lower levels, versus the laser intensity I_L = const E_{0L}^2:

$$I_f^{-1} = \text{const}(1 + I_{k,\text{satur}}I_L^{-1}), \quad I_{k,\text{satur}}I_L^{-1} = G_k^{-1}(E_{0M}). \quad \text{(M.6)}$$

Here, k is the index of the resonance from Eq. (M.1). the detuning is assumed to be zero.

The amplitude E_{0M} can be measured in two ways. In the first way, one could tune the laser frequency into the exact resonance (M.1), measure the ratio of slopes of experimental linear dependences $I_f^{-1}(I_L^{-1})$ for some known value E_{0M}^{known} and for the value E_{0M} to be determined, and then use the dependence $G_k(E_{0M})$ from Eq. (M.5). For $k = 0$, the known value E_{0M} could be, e.g., zero, thus corresponding to the absence of the microwave field. It is this way that was experimentally implemented in Ref. [M.4]. For hydrogen lines H-alpha and H-beta, the scaled dependence $G_k(E_{0M})$ versus the scaled amplitude of the microwave field $\varepsilon = 3E_{0M}/(2\omega_M)$ for

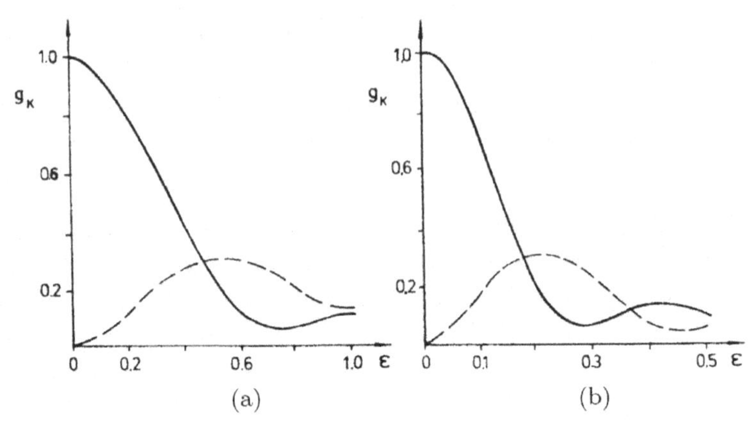

Fig. M.1. Dependence of the scaled saturation parameter $g_k = G_k(E_{0M})/G_0(0)$ on the scaled amplitude of the microwave field $\varepsilon = 3E_{0M}/(2\omega_M)$ for $k = 0$ (solid curve) and $k = \pm 1$ (dashed curve), the expression for ε being in atomic units. Here, k is the number of quanta of the microwave field involved in the resonance (M.1). The function $Gk(E_{0M})$ is defined by Eq. (M.5). (a) for the H-alpha line; (b) for the H-beta line.

$k = 0$ and $k = \pm 1$ is shown in Fig. M.1 (the expression for ε is in atomic units).

The above version *required a calibration* at the zero microwave field. However, the regime "microwaves on" and the regime "microwaves off" could be characterized by significantly different plasma parameters. In distinction to that, another method [2] described below, *does not require a calibration* at the zero microwave field and also is *more accurate*.

In the second way, one could measure the ratio of slopes of experimental dependences $I_f^{-1}(I_L^{-1})$ at the same value of E_{0M} to be determined, but at two different values of $k(k = k_1$ and $k = k_2)$, and use the dependence of the ratio G_{k1}/G_{k2} on E_{0M}. Again, the curves given in Fig. M.1 could be employed in this way by choosing $k_1 = 0$ and $k_2 = \pm 1$.

One more alternative way was developed in paper [2], where there was obtained analytically a *dynamic Stark shift* of the multi-quantum resonance (M.1). Namely, by measuring the ratio of the *experimental Stark shifts of the resonance* (M.1) at two different values of k one can also determine experimentally the microwave amplitude E_{0M}

(or generally, the amplitude of the QEF of the lower frequency) inside the plasma. Further details of this alternative way, can be found in paper [2].

References

[1] V.P. Gavrilenko and E. Oks, *Sov. Phys. Tech. Phys. Lett.* **10** 609 (1984).
[2] V.P. Gavrilenko and E. Oks, *Rev. Sci. Instr.* **70** 363 (1999).
[3] E. Oks, *Plasma Spectroscopy: The Influence of Microwave and Laser Fields*, Springer Series on Atoms and Plasmas, vol. 9, Springer: Berlin (1995).
[4] I.N. Polushkin, M.Yu. Ryabikin, Yu.M. Shagiev and V.V. Yazenkov, *Sov. Phys. JETP* **62** 953 (1985).

Subject Index

Printed in the United States
By Bookmasters